BIOINFORMATICS

BIOINFORMATICS

Kenneth Baclawski
Tianhua Niu

JAICO PUBLISHING HOUSE

Mumbai • Delhi • Bangalore • Kolkata
Hyderabad • Chennai • Ahmedabad • Bhopal

Published by Jaico Publishing House
121 Mahatma Gandhi Road
Mumbai - 400 001
jaicopub@vsnl.com
www.jaicobooks.com

Published in arrangement with
The MIT Press
Cambridge, MA, USA

Not for sale or shipment outside India, Bangladesh,
Burma, Nepal, Sri Lanka, Bhutan, and Pakistan.

BIOINFORMATICS
ISBN 81-7992-642-7

First Jaico Impression: 2007

Printed by
Anubha Printers
B-48, Sector-7, Noida - 201301

Contents

Preface

With recent advances in biotechnology spurred by the Human Genome Project, tremendous amounts of sequence, gene, protein, and pathway data have been accumulating at an exponential rate. Ontologies are emerging as an increasingly critical framework for coping with the onslaught of information encountered in genomics, transcriptomics, and proteomics. This onslaught involves not only an increase in sheer volume but also increases in both complexity and diversity. An ontology is a precise formulation of the concepts that form the basis for communication within a specific field. Because of this it is expected that the use of ontology and ontology languages will rise substantially in the postgenomic era. This book introduces the basic concepts and applications of ontologies and ontology languages in bioinformatics.

Distilling biological knowledge is primarily focused on unveiling the fundamental hidden structure as well as the grammatical and semantic rules behind the inherently related genomic, transcriptomic, and proteomic data within the boundary of a biological organism. Sharing vocabulary constitutes only the first step toward information retrieval and knowledge discovery. Once data have been represented in terms of an ontology, it is often necessary to transform the data into other representations which can serve very different purposes. Such transformations are crucial for conducting logical and critical analyses of existing facts and models, as well as deriving biologically sensible and testable hypotheses. This is especially important for bioinformatics because of the high degree of heterogeneity of both the format and the data models of the myriads of existing genomic and transcriptomic databases. This book presents not only how ontologies can be constructed but also how they can be used in reasoning, querying, and combining information. This includes transforming data to serve diverse purposes as well as combining information from diverse sources.

Our purpose in writing this book is to provide an introductory, yet in-depth analysis of ontologies and ontology languages to bioinformaticists, computer scientists, and other biomedical researchers who have intensive interests in exploring the meaning of the gigantic amounts of data generated by high-throughput technologies. Thus, this book serves as a guidebook for how one could approach questions like ontology development, inference, and reasoning in bioinformatics using contemporary information technologies and tools.

One of the most common ways that people cope with complexity is to classify into categories and then organize the categories hierarchically. This is a powerful technique, and modern ontologies make considerable use of it. Accordingly, classification into hierarchies is the starting point of the book.

The main division of the book is in three parts. We think of the parts as answering three questions: What ontologies are, How ontologies are used, and What ontologies could be. The actual titles are less colorful, but more informative. Since the audience of the book consists of scientists, the last part focuses on how ontologies could be used to represent techniques for reasoning with uncertainty.

The first part introduces the notion of an ontology, starting from hierarchically organized ontologies to the more general network organizations. It ends with a survey of the best-known ontologies in biology and medicine.

The second part shows how to use and construct ontologies. Ontologies have many uses. One might build an ontology just to have a better understanding of the concepts in a field. However, most uses are related in some way to the problem of coping with the large amount of information being generated by modern bioinformatics technologies. Such uses can be classified into three main categories: querying, viewing, and transforming. The first of these can be done using either imprecise natural language queries or precise queries using a formal query language. The second is actually a special case of the third, and this is explained in the first chapter in the subpart devoted to transformations. The other two chapters on transformations show two different approaches to transformations. The last part covers how to create an ontology.

The first two parts of the book consider only one style of reasoning: deductive or Boolean logic. The third part of the book considers the process of thinking in which a conclusion is made based on observation, also known as inductive reasoning. The goal of this part is to achieve a synthesis that supports both inductive and deductive reasoning. It begins by contrasting inductive and deductive reasoning. Then it covers Bayesian networks, a popular formalism that shows great promise as a means of expressing uncertainty. One important activity of science is the process of combining multiple independent observations of phenomena. The third chapter in this part gives a brief introduction to this very large subject. The final chapter of the part and the book is the most speculative. It proposes that the World Wide Web can be extended to support reasoning with uncertainty, as expressed using Bayesian networks. The result is an inductive reasoning web which we have named the Bayesian Web.

The authors would like to thank the many friends and colleagues who contributed their time and expertise. We especially appreciate John Bottoms who read the manuscript more than once and contributed many insightful suggestions. We wish to thank JoAnn Manson, Simin Liu, and the Division of Preventive Medicine, Brigham and Women's Hospital, for their help and encouragement. We also appreciate the contributions by our many colleagues at Northeastern University, Versatile Information Systems, and Composable Logic, including Mitch Kokar, Jerzy Letkowski, and Jeff Smith. We thank Xiaobin Wang at Children's Memorial Hospital in Chicago for sharing with us the microarray data on preterm delivery. KB would like to acknowledge his debt to his mentors, the late Gian-Carlo Rota and Mark Kac. Robert Prior and Katherine Almeida deserve special praise for their patience in what turned out to be a rather larger project than we originally anticipated. Finally, we wish to thank our families for their love, support and encouragement to complete this work.

Throughout the book there are many references to web resources. These references are Uniform Resource Identifiers (URIs). A Uniform Resource Locator (URL) is a special case of a URI that specifies the location of a web resource. A URL is used by a web browser to find and download a resource, such as a webpage. A URI is a unique identifier of a web resource and need not correspond to a downloadable resource, although they often do. Some web resources have a URL that is not the same as its URI. This is becoming an increasingly common practice for ontologies and schemas. The "typewriter" font was used in this book for URIs. Most URLs begin with `http://`. This initial part of the URL specifies the protocol for obtaining the resource. When the protocol is omitted, one obtains the Uniform Resource Name (URN). Most web browsers are capable of finding a resource even when the protocol has not been specified. In this book we will usually use the URN rather than the URL to save space. For typographical purposes, some URIs (and other constructs) in this book have been split so as to fit in the space available.

The URI for this book is `ontobio.org`, and this URI is also the URL of the website for the book. Because URIs are constantly changing, the website for the book has updated information about the URIs that appear in the book as well as new ones that may be of interest to readers. The book website also has additional exercises and solutions to them.

PART I

Introduction to Ontologies

Recent technological advances have resulted in an onslaught of biological information that is accessible online. In the postgenomic era, a major bottleneck is the coherent integration of all these public, online resources. Online bioinformatics databases are especially difficult to integrate because they are complex, highly heterogeneous, dispersed, and incessantly evolving. Scientific discovery increasingly involves accessing multiple heterogeneous data sources, integrating the results of complex queries, and applying further analysis and visualization applications in order to acquire new knowledge. However, online data are often described only in human-readable formats (most commonly free text) that are difficult for computers to analyze due to the lack of standardized structures. An ontology is a computer-readable system for encoding knowledge which specifies not only the concepts in a given field but also the relationships among those concepts. Ontologies provide insight into the nature of information produced by that field and are an essential ingredient in any attempts to arrive at a shared understanding of concepts in a field. Thus the development of ontologies for biological information and the sharing of those ontologies within the bioinformatics community are pivotal for biologists who rely heavily on online data.

The first part of the book introduces the basic concepts and semantics of ontologies. It begins with the fundamental notions of hierarchies and relationships. The web is becoming the primary mechanism for the exchange of information and data. Accordingly, the emphasis in this book is on the eXtensible Markup Language (XML) and XML-based languages. The next three chapters explain the semantics of XML, rules and the newly proposed Semantic Web, a layer above the World Wide Web that adds meaning to hypertext links. Finally, chapter 5 provides a survey of ontologies and databases in Bioinformatics. This chapter covers the major bio-ontologies used in computational biology, including the Gene Ontology and open biological ontologies.

1 *Hierarchies and Relationships*

1.1 Traditional Record Structures

One of the most common ways to represent information with computers is to use "records." Records are stored in a file, with one record per line. Such a file is called a *flat file*. A record consists of a series of data items, called "fields" or "columns." Here are some records from a health study (NHS 2004):

```
011500   18.66  0  0 62   46.271020111   25.220010
011500   26.93  0  1 63   68.951521001   32.651010
020100   33.95  1  0 65   92.532041101   18.930110
020100   17.38  0  0 67   50.351111100   42.160001
```

The actual records are considerably longer. It should be apparent that one cannot have any understanding of the meaning of the records without some explanation such as the following:

```
NAME     LENGTH  RANGE          FORMAT  MEAN OR CODES
instudy  6                      MMDDYY
bmi      8       13.25-60.07    Num     26.03
obesity  3       0-1            Num     0=No 1=Yes
ovrwt    8       0-1            Num     0=No 1=Yes
Height   3       49-79          Num     64.62
Wtkgs    8       38.1-175.1     Num     70.2
Weight   3       84-386         Num     154.75
```

```
NAME     LABEL
instudy  Date of randomization into study
bmi      Body Mass Index.  Weight(kgs)/height(m)**2
```

```
obesity  Obesity (30.0 <= BMI)
ovrwt    Overweight (25 <= BMI < 30)
Height   Height (inches)
Wtkgs    Weight (kilograms)
Weight   Weight (pounds)
```

The explanation of what the fields mean is called *metadata*. In general, metadata are any "data about data," such as the names of the fields, the kind of values that are allowed, the range of values, and explanations of what the fields mean.

In this case each field has a fixed number of characters, and each record has a fixed total number of characters. This is called the *fixed-width format* or *fixed-column format*. This format simplifies the processing of the file, but it limits what can be said within each field. If the text that should be in a field does not fit, then it must be abbreviated or truncated. There are other file formats that eliminate these limitations. One commonly used format is to use commas or tabs to delimit the fields. This allows the fields to have varying size. However, it complicates processing when the delimiting character (i.e., the comma or tab) must be used within a field.

The information in the record is often highly redundant. For example, the *obesity* and *ovrwt* fields are unnecessary because they can be computed from the *bmi* field. Similarly, the *bmi* field can be computed from the *Height* and *Weight* fields. Another common feature of flat files is that the field formats are often inappropriate. For example, the *obesity* field can only have the values "yes" or "no," but it is represented using numbers.

Each field of a flat file is defined by features such as its name, format, description, and so on. A *database* is a collection of flat files (called *tables*) with auxiliary structures (e.g., indexes) that improve performance for certain commonly used operations. The description of the fields of one or more flat files is called the *schema*.

A database schema is an example of an *ontology*. In general, whenever data are structured, the description of their structure is the ontology for the data. A glance at the example record makes it clear that the raw data record is completely useless without the ontology. The ontology is what gives the raw data their meaning. The same is true for any kind of data, whether they be electronic data used by a computer or audiovisual data sensed by a person. Ontologies are the means by which a person or some other agent understands its world, as well as the means by which a person or agent communicates with others.

Summary

- A flat file is a collection of records.
- A record consists of fields.
- Each record in a flat file has the same number and kinds of fields as any other record in the same file.
- The schema of a flat file describes the structure (i.e., the kinds of fields) of each record.
- A schema is an example of an ontology.

1.2 The eXtensible Markup Language

Flat files are simple and easy to process. A typical program using and producing flat files simply performs the same operation on each record. However, flat files are limited to relatively simple forms of data. They are not well suited to the complex information of genomics, proteomics, and so on. Accordingly, a new approach is necessary.

The eXtensible Markup Language (XML) is a powerful and flexible mechanism that can be used to represent bioinformatic data and facilitates communication. Unlike flat files, an XML document is *self-describing*: the name of each attribute is specified in addition to the value of the attribute. The health study record shown above could be written like this in XML:

```
<Interview RandomizationDate="2000-01-15" BMI="18.66" Height="62".../>
<Interview RandomizationDate="2000-01-15" BMI="26.93" Height="63".../>
<Interview RandomizationDate="2000-02-01" BMI="33.95" Height="65".../>
<Interview RandomizationDate="2000-02-01" BMI="17.38" Height="67".../>
```

The basic unit of an XML document is called an *element*. It is analogous to a record in a flat file, except that a single XML document can have many kinds of element. One would need a large collection of flat files (or a database with many tables) to represent the elements of a *single* XML document, and even that would not capture all of it, because the kinds of element in an XML document can be intermixed. Each kind of element is labeled by a name called its *tag*. The example given above is an `Interview` element.

The fields of an XML element are called its *attributes*. Flat files generally distinguish fields from one another by their positions in the record. XML attributes can appear in any order, and an attribute that is not needed by an element is not written at all.

An attribute in general is a property or characteristic of an entity. Linguistically, attributes are adjectives that describe entities. For example, a person may be overweight or obese, and the `BMI` attribute makes the description quantitative rather than qualitative. The notion of attribute represents two somewhat different concepts: the attribute in general and the attribute of a specific entity. `BMI` is an example of an attribute, but one would also speak of a `BMI` equal to 18.66 for a specific person as being an attribute. To avoid confusion we will refer to the former as the *attribute name*, while the latter is an *attribute value*.

```
<!ATTLIST molecule
          title       CDATA    #IMPLIED
          id          CDATA    #IMPLIED
          convention  CDATA    "CML"
          dictRef     CDATA    #IMPLIED
          count       CDATA    #REQUIRED
>
```

Figure 1.1 Part of the Chemical Markup Language DTD. This defines the attribute names that are allowed in a `molecule` element.

Just as a database is described by its schema, an XML document is described by its Document Type Definition (DTD). The DTD specifies the attribute names that are allowed for each kind of element. For example, in the Chemical Markup Language (CML) (CML 2003), a `molecule` can have a title, identifier, convention, dictionary reference, and count. Figure 1.1 shows how this is specified in the CML DTD. A `#REQUIRED` attribute is one that must be specified in every element of this kind; an `#IMPLIED` attribute is optional. If a value is specified in the DTD, then it is the default value of the attribute. For example, if no `convention` is specified, then it has the value "`CML`." CDATA means "character data" which means that the attribute value can use any kind of text except for elements.

One enters or updates data for an XML element in the same manner that one enters or updates data for a database table. An example of such a data entry screen is given in figure 1.2.

XML reserves two characters for indicating the presence of markup. The left angle bracket (<) is used by XML to mark the beginning of each element.

It is also used to show where an element ends. To include the left angle bracket in ordinary text, write it as "<". Writing a special character like the left angle bracket as "<" is called *escaping*. The ampersand character (&) is also reserved by XML, and it must be written as "&".

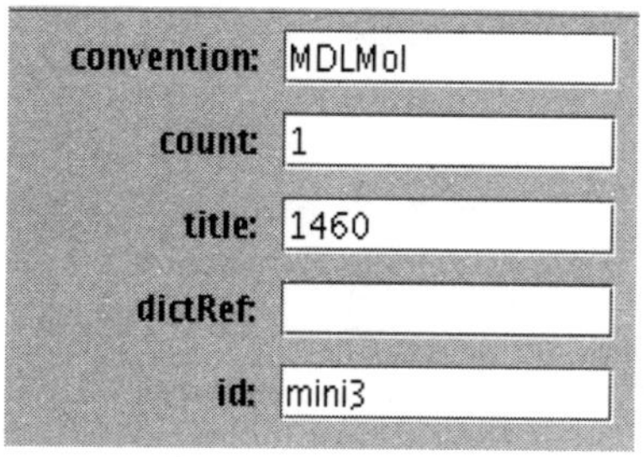

Figure 1.2 Data entry screen for the `molecule` element of the Chemical Markup Language.

Summary

- XML is a format for representing data.
- An XML element is analogous to a record in a flat file.
- An XML attribute is analogous to a field of a record.
- An XML DTD is a schema that describes the structure of the elements of an XML file.

1.3 Hierarchical Organization

Modern biology and medicine, like much of society, is currently faced with overwhelming amounts of raw data being produced by new information-gathering techniques. In a relatively short period of time information has gone from being relatively scarce and expensive to being plentiful and inexpensive. As a consequence, the traditional methods for dealing with information are overwhelmed by the sheer volume of information available. The traditional methods were developed when information was scarce, and they cannot handle the enormous scale of information.

The first and most natural reaction by people to this situation is to attempt to categorize and classify. People are especially good at this task. We are constantly categorizing objects, experiences, and people. We do it effortlessly and unconsciously. The very words we use to express ourselves represent categories. It is only when a categorization is problematic that we notice that we have been categorizing at all. Biology was the first discipline to engage in systematic, large-scale classification because of the enormous complexity of its domain.

```
<bioml>
 <organism name="Homo sapiens (human)">
  <chromosome name="Chromosome 11" number="11">
   <locus name="HUMINS locus">
    <reference name="Sequence databases">
     <db_entry name="Genbank sequence" entry="v00565" format="GENBANK"/>
     <db_entry name="EMBL sequence" format="EMBL" entry="V00565"/>
    </reference>
    <gene name="Insulin gene">
     <dna name="Complete HUMINS sequence" start="1" end="4992">
     1 ctcgaggggc ctagacattg ccctccagag agagcaccca acaccctcca ggcttgaccg
     ...
     </dna>
     <ddomain name="flanking domain" start="1" end="2185"/>
     <ddomain name="polymorphic domain" start="1340" end="1823"/>
     <ddomain name="Signal peptide" start="2424" end="2495"/>
     <exon name="Exon 1" start="2186" end="2227"/>
     <intron name="Intron 1" start="2228" end="2406"/>
    </gene>
   </locus>
   <locus>
   ...
   </locus>
  </chromosome>
 </organism>
</bioml>
```

Figure 1.3 A BioML document showing some of the information about the human insulin gene. Boxes were drawn around each XML element so that the hierarchical structure is more apparent. XML documents normally indicate the hierarchical structure by successive indentation, as in this example.

What makes XML powerful is the ability to organize data hierarchically. XML elements are much more than just self-describing records; each element can contain other elements, which can contain other elements, to arbitrary depth. Figure 1.3 shows a small part of the genomic data for the insulin gene represented using the Biopolymer Markup Language (BioML) (BioML 2003). This XML document consists of a `bioml` element containing an `organism` element. The `organism` element, in turn, contains `chromosome` elements, which contain `locus` elements, which contain `genes`, which contain the DNA sequence, domains, exons, introns, and so on. Along the way, the elements also contain references to database entries that furnish the source material for the genomic information. This example shows the organization of information about biopolymers starting at the organism level and successively elaborating until one sees individual DNA bases.

Because of the hierarchical nature of an XML document, there is always a "top" of the hierarchy called the *root*. In figure 1.3 the root is the `bioml` element. The root is split into a series of branches, which in turn split into branches, and so on, like the branching of a tree. The terminology of family trees is commonly used for the relationships within the hierarchy. The elements contained in an element are called its *child* elements, and the containing element is the *parent*. The children of the same parent are *siblings*. Note that this family tree is asexual: each element (except for the root) has exactly one parent.

The tags and attributes occurring in an XML DTD constitute the *vocabulary* of the ontology. When one is creating an ontology it is important to choose the tags and attributes so that they correspond to how people use the words. Ontologies should facilitate communication between people as well as between computer systems. Because of this emphasis on communication, ontologies are often referred to as *languages*. Ontologies are also important for information retrieval from databases, and the terminology in an ontology is called a *controlled vocabulary* in this context. An ontology is a specialized language for communication in a particular domain. The communication can be between people, between people and computers, or between computers. Ontologies based on XML are more specifically called *markup languages* because of the historical origin of XML as a means of marking up text for the purpose of typesetting documents. Thus the "ML" in BioML and CML both stand for "Markup Language" even though neither of these ontologies is concerned with typesetting.

Summary

- Classification is one way in which people organize a domain in order to understand it more easily.

- Classifications are frequently organized in the form of a hierarchy.
- XML elements are hierarchical: each element can contain other elements, that in turn can contain other elements, and so on.

1.4 Creating and Updating XML

This little example illustrates how statements can be much worse than just being false: they can be meaningless. One of the main functions of a good ontology is that it limits what can be said, so that statements using the ontology always make sense to a member of the community served by the ontology. This is done by using *constraints*. Some constraints have already been discussed in section 1.2 where we saw that one can specify what attributes are allowed for each kind of element. One can also specify which elements can be contained in other elements as well as how many are allowed. These constraints are especially useful when one is creating and updating XML documents, and that is the topic of this section.

Viewing and updating an XML document may seem to be a formidable task, but one rarely looks directly at an XML document any more than one would look at the page source of an HTML (Hypertext Markup Language) document. One uses an XML tool for creating, viewing, and updating. The single term "editing" is used for all three of these activities. An XML editor is a tool that supports the editing of an XML document. XML editors automatically take care of routine tasks such as escaping special characters and making sure that the document is consistent. There are many such tools available. The examples in this book used Xerlin (Xerlin 2003), an open source XML editor that is available from the Apache project. XML viewers and editors make good use of the hierarchical structure of an XML document. This structure is analogous to a file folder or directory structure: The XML document is viewed and updated in much the same way as files in a directory. In figure 1.4 one can see a typical file manager compared with an XML document editor showing the BioML insulin gene document.

The DTD of an XML document specifies more than just the attributes of each element. For example, in CML, a `molecule` contains an `atomArray` and a `bondArray`, and they must occur in this order: the `atomArray` must occur first, and the comma indicates that the `bondArray` must occur second. An `atomArray` element contains one or more `atom` elements, and a `bondArray` consists of one or more bond elements. The DTD would specify this as follows:

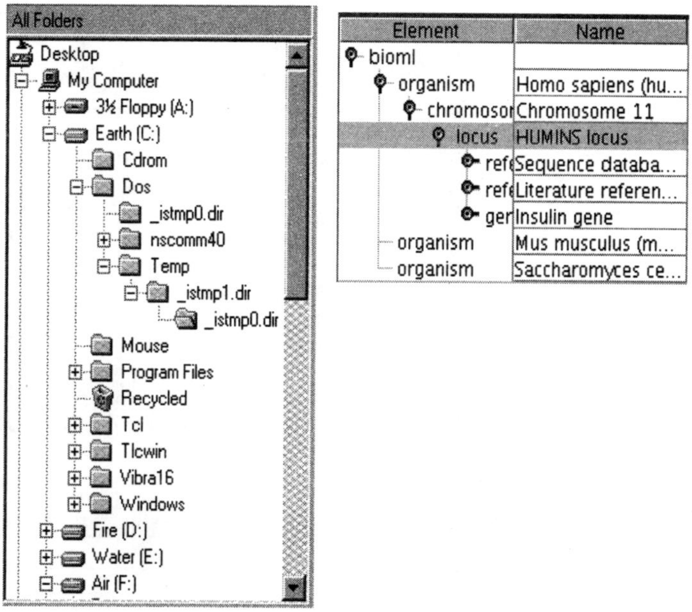

Figure 1.4 File management vs. XML document management. The image on the left used the Windows file manager. It shows disk drives, folders, and files on a PC. The image on the right used the Xerlin XML document editor. It shows the elements of a single XML document.

```
<!ELEMENT molecule (atomArray,bondArray)>
<!ELEMENT atomArray (atom+)>
<!ELEMENT bondArray (bond+)>
```

The `ELEMENT` statements above determine the *content* of these elements. A specification such as `(atom+)` is called a content model. The `ATTLIST` statement for `molecule` given earlier determines the attributes that can be in an element. A DTD will normally have one `ELEMENT` statement and one `ATTLIST` statement for each kind of element that can be in the document. A more complete DTD for molecules is shown in figure 1.6. Because the same attributes are allowed in many elements, a DTD can be very long. `ENTITY` statements are a method for simplifying the writing of DTDs, by allowing one to specify content and lists of attributes just once. In figure 1.7 two entities were defined and then used several times. Large DTDs such as CML use a large number of entities. These are just two of the entities in CML.

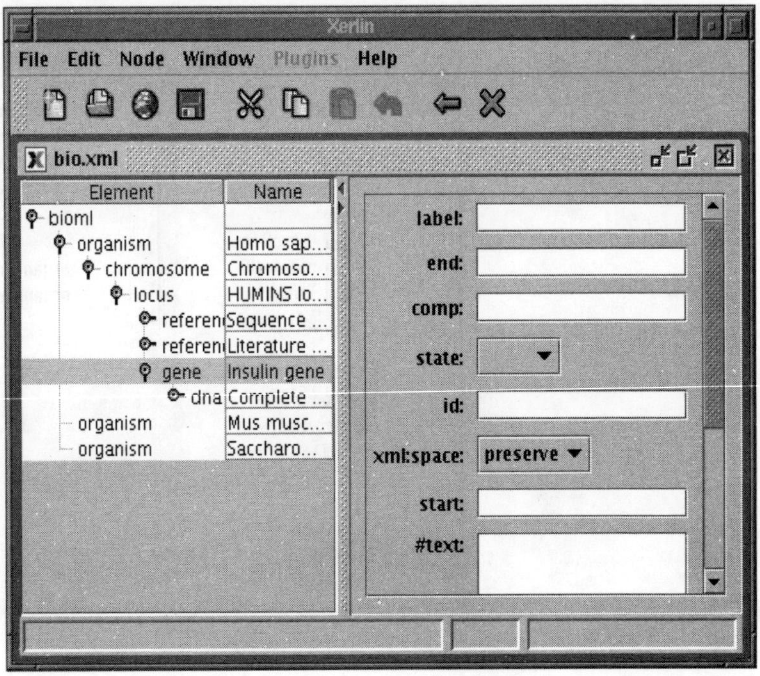

Figure 1.5 Data entry screen for an element of an XML document. The window on the left shows the hierarchical structure of the XML document in the same manner as a file manager. A gene element is highlighted, indicating that this is the currently open element. The attributes for the gene element are shown in the right window. The window on the right acts like a data entry screen for viewing and updating the attributes of the element.

```
<!ELEMENT molecule (atomArray, bondArray)>
<!ATTLIST molecule
      title           CDATA   #IMPLIED
      id              CDATA   #IMPLIED
      convention      CDATA   "CML"
      dictRef         CDATA   #IMPLIED
      count           CDATA   "1"
>
<!ELEMENT atomArray (atom+)>
<!ATTLIST atomArray
      title           CDATA   #IMPLIED
      id              CDATA   #IMPLIED
      convention      CDATA   "CML"
```

```
>
<!ELEMENT atom EMPTY>
<!ATTLIST atom
        elementType     CDATA    #IMPLIED
        title           CDATA    #IMPLIED
        id              CDATA    #IMPLIED
        convention      CDATA    "CML"
        dictRef         CDATA    #IMPLIED
        count           CDATA    "1"
>
<!ELEMENT bondArray (bond+)>
<!ATTLIST bondArray
        title           CDATA    #IMPLIED
        id              CDATA    #IMPLIED
        convention      CDATA    "CML"
>
<!ELEMENT bond EMPTY>
<!ATTLIST bond
        title           CDATA    #IMPLIED
        id              CDATA    #IMPLIED
        convention      CDATA    "CML"
        dictRef         CDATA    #IMPLIED
        atomRefs        CDATA    #IMPLIED
>
```

Figure 1.6 Part of the Chemical Markup Language DTD. This part defines the content and some of the attributes of a `molecule` element as well as the content and some of the attributes of elements that can be contained in a `molecule` element.

```
<!ENTITY % title_id_conv '
        title           CDATA    #IMPLIED
        id              CDATA    #IMPLIED
        convention      CDATA    "CML"     '>
<!ENTITY % title_id_conv_dict
%title_id_conv;
```

```
        'dictRef          CDATA    #IMPLIED'>

<!ELEMENT molecule (atomArray, bondArray)>
<!ATTLIST molecule
%title_id_conv_dict;
        count            CDATA    "1"
>
<!ELEMENT atomArray (atom+)>
<!ATTLIST atomArray
%title_id_conv;
>
<!ELEMENT atom EMPTY>
<!ATTLIST atom
        elementType      CDATA    #IMPLIED
%title_id_conv_dict;
        count            CDATA    "1"
>
<!ELEMENT bondArray (bond+)>
<!ATTLIST bondArray
%title_id_conv;
>
<!ELEMENT bond EMPTY>
<!ATTLIST bond
%title_id_conv_dict;
        dictRef          CDATA    #IMPLIED
        atomRefs         CDATA    #IMPLIED
>
```

Figure 1.7 Part of the Chemical Markup Language DTD. This DTD uses entities to simplify the DTD in figure 1.6.

In addition to simplifying a DTD, there are other uses of XML entities:

- Entities can be used to build a large DTD from smaller files. The entities in this case refer to the files being incorporated rather than to the actual value of the entity. Such an entity would be defined like this:

```
<!ENTITY % dtd1 SYSTEM "ml.dtd">
```

To include the contents of the file `ml.dtd`, one uses `%dtd1;` in your DTD. One can use a URL instead of a filename, in which case the DTD information will be obtained from an external source.

- Entities can be used to build a large document from smaller documents. The smaller documents can be local files or files obtained from external sources using URLs. For example, suppose that an experiment is contained in five XML document files. One can merge these files into a single XML document as follows:

```
<?xml version="1.0"?>
<!DOCTYPE ExperimentSet SYSTEM "experiment.dtd"
[
  <!ENTITY experiment1 SYSTEM "experiment1.xml">
  <!ENTITY experiment2 SYSTEM "experiment2.xml">
  <!ENTITY experiment3 SYSTEM "experiment3.xml">
  <!ENTITY experiment4 SYSTEM "experiment4.xml">
  <!ENTITY experiment5 SYSTEM "experiment5.xml">
]>
<ExperimentSet>
  &experiment1;
  &experiment2;
  &experiment3;
  &experiment4;
  &experiment5;
</ExperimentSet>
```

  Note that entities used within documents use the ampersand rather than the percent sign. This example is considered again in section 11.6 where it is discussed in more detail.

When one is editing an XML document, the DTD assists one to identify the attributes and elements that need to be provided. Figure 1.5 shows the BioML insulin gene document. The "directory" structure is on the left, and the attributes are on the right. In this case a `gene` element is open, and so the attributes for the gene element are displayed. To enter or update an attribute, click on the appropriate attribute and use the keyboard to enter or modify the attribute's value. When an attribute has only a limited list of possible values, then one chooses the desired value from a menu. Attributes are specified in the same manner as one specifies fields of a record in a database using a traditional "data entry" screen. An XML document is effectively an entire database with one table for every kind of element.

In addition to attributes, an XML element can have text. This is often referred to as its *text content* to distinguish it from the elements it can contain. In an XML editor, the text content is shown as if it were another child element, but labeled with `#text`. It is also shown as if it were another attribute, also labeled with `#text`.

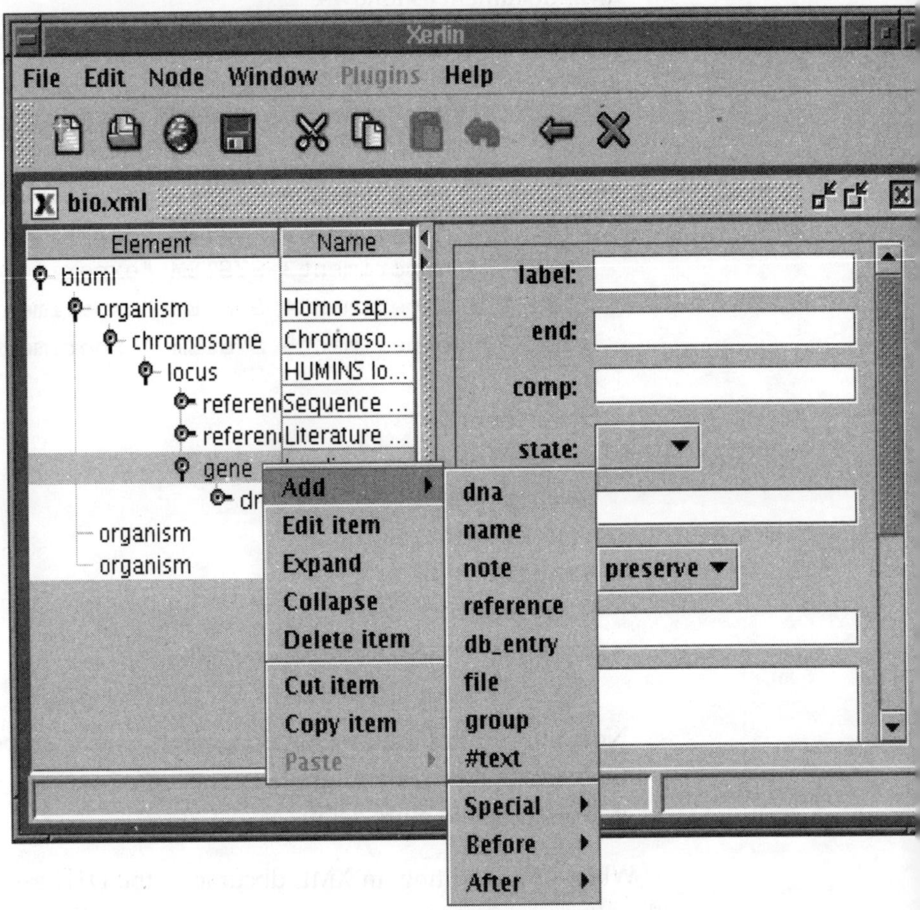

Figure 1.8 The process of adding a new element to an XML document. The menus shown were obtained by right-clicking on the gene element and then selecting the Add choice. The menu containing dna, name, and so on shows the elements that are allowed in the gene element.

The hierarchical structure of an XML document is manipulated by opening and closing each element just as one opens and closes directories (folders) when managing a collection of files. Unlike file management, the DTD of an XML document can be very precise about what elements are allowed to be in another element, as well as the order in which they must appear. Figure 1.8 shows how to add an element to the `gene` element of a BioML document. Click (with the right mouse button) on the `gene` element and a menu appears that allows one to `Add` another element. Selecting this choice in the menu displays another menu. This second menu shows all the kinds of child elements that are allowed at this time in the parent element. In figure 1.9 a `note` element was chosen. Figure 1.10 shows the result of making this selection: the `gene` element now has a new `note` child element. The right window now displays the attributes for the newly created element.

The elements that are allowed in an element depend on the editing context. The `molecule` element must have exactly two child elements: `atomArray` and `bondArray`. As a result, when a `molecule` is selected the first time, the only choice for a child element will be `atomArray`. After adding such an element, the only choice will be `bondArray`.

The file directory metaphor is a compelling one, but it is important to note how XML differs from a directory structure. The differences are explained in table 1.1. Just as a database schema is an example of an ontology, an XML DTD is also an example of an ontology. However, unlike a database, an XML document is self-describing. One can understand much of the meaning of the data without recourse to the ontology. Indeed, there are tools which can guess the DTD for an XML document that does not have one.

Summary

- XML documents are examined and updated by taking advantage of the hierarchical structure.

- The XML DTD assists in updating a document by giving clues about what attributes need to be entered as well as what elements need to be added.

1.5 The Meaning of a Hierarchy

Each kind of element in an XML document represents a *concept*. Concepts are the means by which people understand the world around them. They classify the world into units that allow comprehension. They also make it

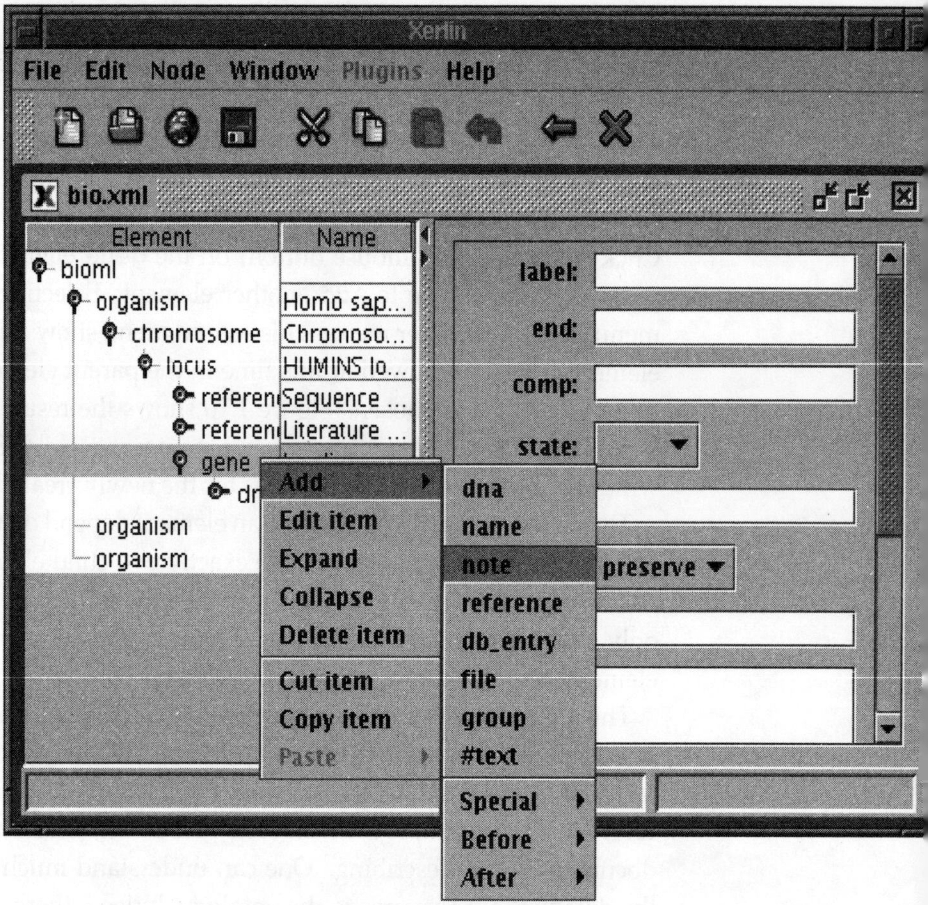

Figure 1.9 Adding a new element to an XML document. A `note` element has been chosen to be added to a `gene` element.

possible for people to communicate with each other. Individuals must have a shared conceptual framework in order to communicate, but communication requires more than just a shared conceptualization; it is also necessary for the concepts to have names, and these names must be known to the two individuals who are communicating.

Biochemistry has a rich set of concepts ranging from very generic notions such as *chemical* to exquisitely precise notions such as *Tumor necrosis factor alpha-induced protein 3*. Concepts are typically organized into hierarchies to capture at least some of the relationships between them. XML document hierarchies are a means by which one can represent such hierarchical organi-

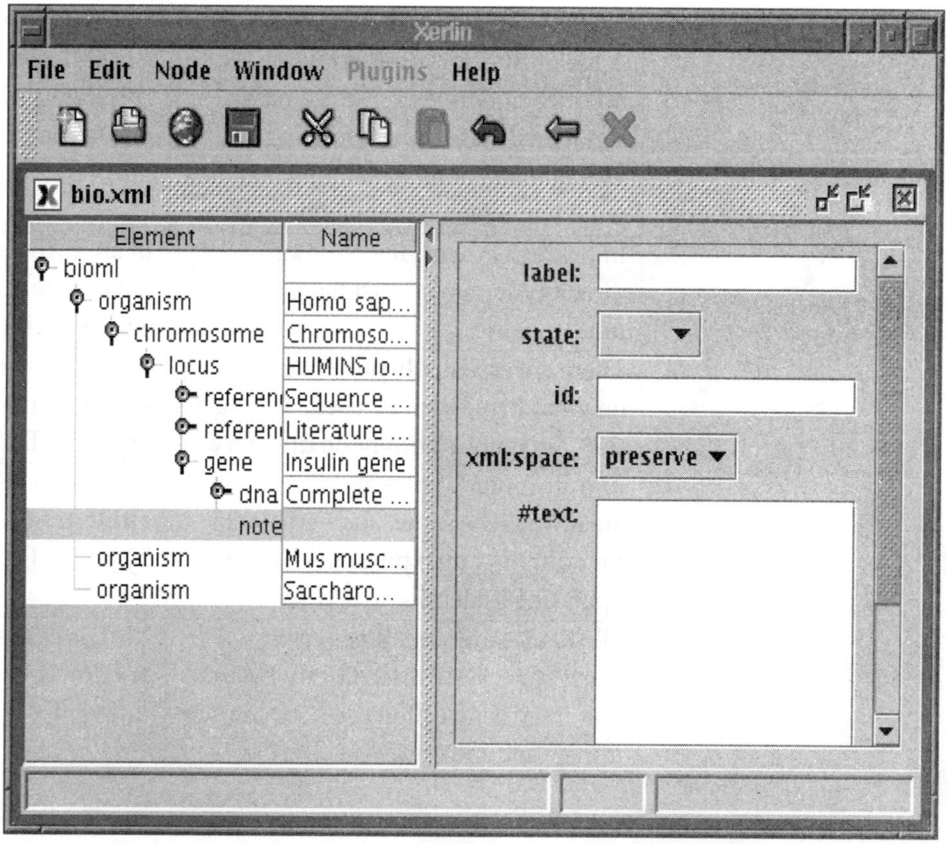

Figure 1.10 Result of adding a new element to an XML document. A `note` element has been added to a `gene` element. The active element is now the `note` element, and its attributes appear in the window on the right side.

zations of knowledge.

Aristotle (384-322 BC) was the first who understood the difficulty of categorizing living organisms into classes according to their anatomical and physiological characteristics (Asimov 1964). Since then, this tradition of classification has been one of the major themes in science. Figure 1.11 illustrates a hierarchy of chemicals taken from EcoCyc (EcoCyc 2003). For example, *protein* is more specific than *chemical*, and *enzyme* is more specific than *protein*. Classifications that organize concepts according to whether concepts are more general or more specific are called *taxonomies* by analogy with biological classifications into species, genera, families, and so on.

Hierarchies are traditionally obtained by starting with a single all-inclu-

File Manager	XML Editor
A file or directory (folder) name uniquely identifies it. Such a name is not well suited for describing the contents, or for specifying what it means.	An XML element tag is for specifying what the element means, not how to obtain it.
In a directory each file or other directory within it must have a unique name.	In an XML element one can have more than one child element with the same tag.
There are essentially no constraints on what names can be used, as long as they are unique within the directory.	XML elements can only have tags that are allowed by the DTD.
The attributes of files and directories are always the same, and serve only for administrative purposes by the operating system.	XML elements can have any attributes that are allowed by the DTD.
File names are sometimes case insensitive. Case insensitivity means that there is no difference between upper- and lower-case letters.	XML is case sensitive. Upper- and lower-case letters are different.

Table 1.1 Comparison of directory/file management with XML document editing

sive class, such as "living being," and then subdividing into more specific subclasses based on one or more common characteristics shared by the members of a subclass. These subclasses are, in turn, subdivided into still more specialized classes, and so on, until the most specific subclasses are identified. We use this technique when we use an outline to organize a task: the most general topic appears first, at the top of the hierarchy, with the more specialized topics below it. Constructing a hierarchy by subdivision is often called a "top-down" classification.

An alternative to the top-down technique is to start with the most specific classes. Collections of the classes that have features in common are grouped together to form larger, more general, classes. This is continued until one collects all of the classes together into a single, most general, class. This approach is called "bottom-up" classification. This is the approach that has been used in the classification of genes (see figure 1.12). Whether one uses a

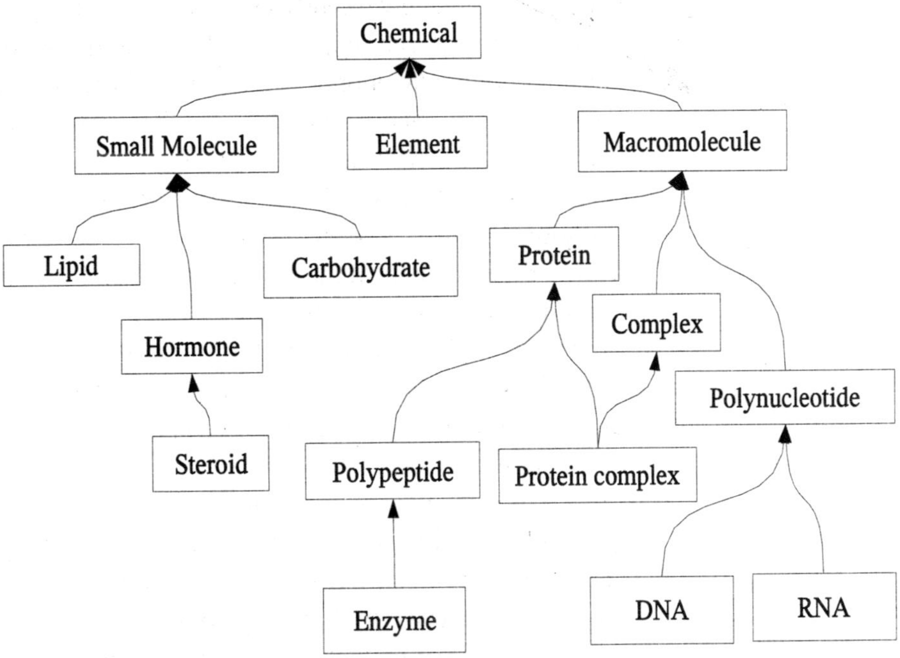

Figure 1.11 Chemical hierarchy (EcoCyc 2003).

top-down or bottom-up technique, it is always presumed that one can define every class using shared common characteristics of the members.

There are many algorithms for constructing hierarchical classifications, especially taxonomies, based on attributes of entities (Jain and Dubes 1988). Such algorithms are usually referred to as *data-clustering* algorithms. An example of a hierarchy constructed by a data-clustering algorithm is shown in figure 5.3. The entities being clustered in this case are a set of genes, and the hierarchy appears on the left side of the figure. Such automated classifications have become so routine and common that many tools construct them by default.

The notions of taxonomy and hierarchy have been an accepted part of Western civilization since the time of Aristotle. They have been a part of this culture for so long that they have the status of being completely obvious and natural. Aristotle already emphasized that classifications must be "correct," as if they had the status of a law of nature rather than being a means for understanding the world. This attitude toward classification was not questioned until relatively recently, and is still commonly accepted. By the middle of the nineteenth century, scholars began to question the implicit assumptions underlying taxonomic classification. Whewell, for example, discussed classification in science, and observed that categories are not usually specifiable by shared characteristics, but rather by resemblance to what he called "paradigms." (Whewell 1847) This theory of categorization is now called "prototype theory." A *prototype* is an ideal representative of a category from which other members of the category may be derived by some form of modification. One can see this idea in the classification of genes, since they evolve via mutation, duplication, and translocation (see figure 1.13). Wittgenstein further elaborated on this idea, pointing out that various items included in a category may not have one set of characteristics shared by all, yet given any two items in the category one can easily see their common characteristics and understand why they belong to the same category (Wittgenstein 1953). Wittgenstein referred to such common characteristics as "family resemblances," because in a family any two members will have some resemblance, such as the nose or the eyes, so that it is easy to see that they are related, but there may be no one feature that is shared by all members of the family. Such a categorization is neither top-down nor bottom-up, but rather starts somewhere in the middle and goes up and down from there.

This is especially evident in modern genetics. Genes are classified both by function and by sequence. The two approaches interact with one another in complex ways, and the classification is continually changing as more is learned about gene function. Figure 1.12 shows some examples of the classification of genes into families and superfamilies. The superfamily is used to describe a group of gene families whose members have a common evolutionary origin but differ with respect to other features between families. A gene family is a group of related genes encoding proteins differing at fewer than half their amino acid positions. Within each family there is a structure that indicates how closely related the genes are to one another. For example figure 1.13 shows the evolutionary structure of the nuclear receptor gene family. The relationships among the various concepts is complex, including evolution, duplication and translocation.

The hierarchies shown in figure 1.11, 1.12, and 1.13 are very different from one another due to the variety of purposes represented in each case. The chemical hierarchy in figure 1.11 is a specialization/generalization hierarchy. The relationship here is called *subclass* because mathematically it represents a subset relationship between the two concepts. The gene families and superfamilies in figure 1.12 are also related by the subclass relationship, but the

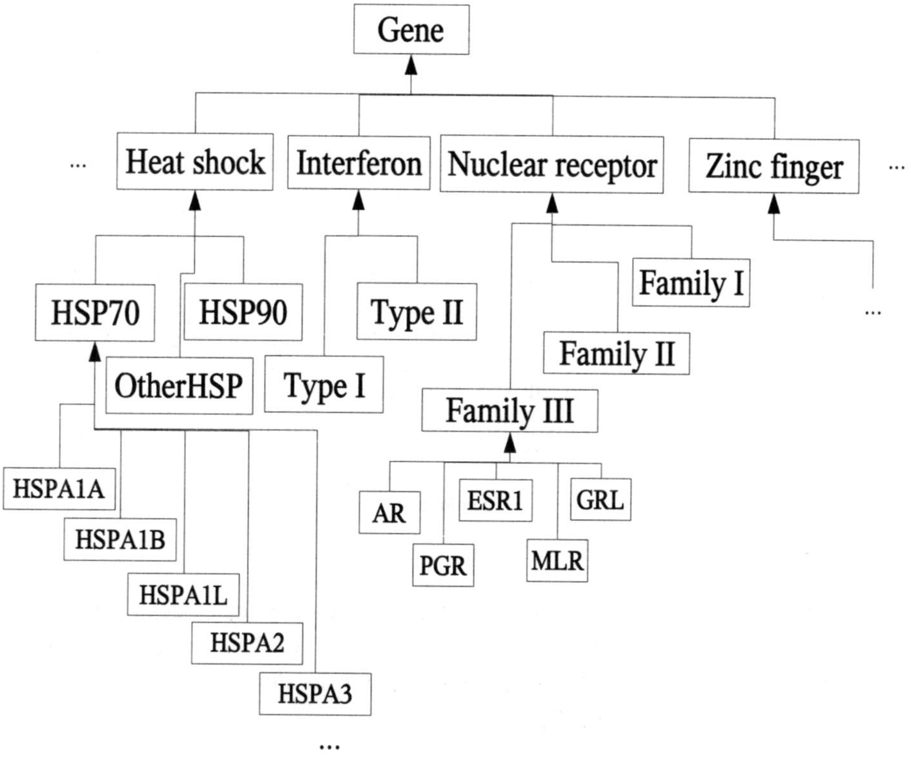

Figure 1.12 Some gene families. The first row below `Gene` in this classification consists of superfamilies. The row below that contains families. Below the families are some individual genes. See (Cooper 1999), Chapter 4.

individual genes shown in the diagram are members (also called *instances*) of their respective families rather than being subsets. However, the nuclear receptor gene diagram in figure 1.13 illustrates that the distinction between subclass and instance is not very clear-cut, as the entire superfamily evolved from a single ancestral gene. In any case, the relationships in this last diagram are neither subclass nor instance relationships but rather more complex relationships such as: *evolves by mutation*, *duplicates*, and *translocates*.

Although hierarchical classification is an important method for organizing complex information, it is not the only one in common use. Two other techniques are partitioning and self-organizing maps. Both of these can be regarded as classification using attribute values rather than hierarchical structures. In *partitioning*, a set of entities is split into a specified number of subsets (MacQueen 1967). A *self-organizing map* is mainly used when a large num-

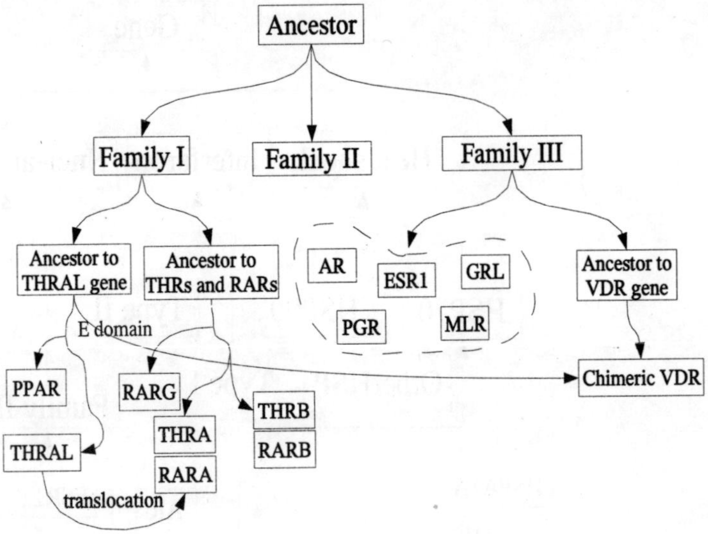

Figure 1.13 The human nuclear receptor gene superfamily. A common ancestor evolved into the three gene families. Unlabeled arrows represent evolution over time. Labeled arrows indicate translocation between families or subfamilies. See (Cooper 1999), Figure 4.28.

ber of attributes for a set of entities is to be reduced to a small number of attributes, usually two or three (Kohonen 1997). The attributes are then displayed using visual techniques that make the clusters easy for a person to see. There are also many clustering techniques that combine some or all of these techniques.

Summary

- Classifications can be constructed top-down, bottom-up, or from the middle.
- Classifications can be based on many principles: subclass (subset), instance (member), or more complex relationships.
- It is even possible for a classification to be based on several relationships at the same time.

1.6 Relationships

Of course, most titles are like this, and the abstract quickly clears up the confusion. However, it does point out how important such connecting phrases can be to the meaning of a document. These are called *relationships*, and they are the subject of this section.

The organization of concepts into hierarchies can capture at least some of the relationships between them, and such a hierarchy can be represented using an XML document hierarchy. The relationship in an XML document between a parent element and one of its child elements is called *containment* because elements contain each other in the document. However, the actual relationship between the parent element and child element need not be a containment. For example, it is reasonable to regard a `chromosome` as containing a set of `locus` elements because a real chromosome actually does contain loci. Similarly, a gene really does contain exons, introns, and domains. However, the relationship between a `gene` and a `reference` is not one of containment, but rather the *referral* or *citation* relationship.

One of the disadvantages of XML is that containment is the only way to relate one element to another explicitly. The problem is that all the various kinds of hierarchy and various forms of relationship have to be represented using containment. The hierarchy in figure 1.13 does not use any relationships that could reasonably be regarded as being containment. Yet, one must use the containment relationship to represent this hierarchy. The actual relationship is therefore necessarily implicit, and some auxiliary, informal technique must be used to elucidate which relationship is intended.

Unfortunately, this is not a small problem. One could not communicate very much if all one had were concepts and a single kind of relationship. Relating concepts to each other is fundamental. Linguistically, concepts are usually represented by nouns and relationships by verbs. Because relationships relate concepts to concepts, the linguistic notion of a simple sentence, with its subject, predicate, and object, represents a basic fact. The subject and object are the concepts and the predicate is the relationship that links them.

One can specify relationships in XML, but there are two rather different ways that this can be done, and neither one is completely satisfactory. The first technique is to add another "layer" between elements that specifies the relationship. This is called *striping*. A BioML document could be represented using striping, as in figure 1.14. If one consistently inserts a relationship element between parent and child concept elements, then one can unambiguously distinguish the concept elements from the relationship elements.

Striping was first introduced in the Resource Description Framework (RDF) (Lassila and Swick 1999).

```
...
<locus name="HUMINS locus">
 <contains>
  <gene name="Insulin gene">
   <isStoredIn>
    <db_entry name="Genbank sequence" entry="v00565"
              format="GENBANK"/>
    <db_entry name="EMBL sequence" format="EMBL"
              entry="V00565"/>
   </isStoredIn>
   <isCitedBy>
    <db_entry name="Insulin gene sequence" format="MEDLINE"
              entry="80120725"/>
    <db_entry name="Insulin mRNA sequence" format="MEDLINE"
              entry="80236313"/>
    <db_entry name="Localization to Chromosome 11" format="MEDLINE"
              entry="93364428"/>
   </isCitedBy>
   <hasSequence>
    <dna name="Complete HUMINS sequence" start="1" end="4992">
     1 ctcgaggggc ctagacattg ccctccagag agagcaccca acaccctcca ggcttgaccg
     ...
    </dna>
   </hasSequence>
  </gene>
 </contains>
</locus>
...
```

Figure 1.14 Using striping to represent relationships involving the human insulin gene. The shaded elements in the figure are the relationships that link a parent element to its child elements.

Another way to specify a relationship is to use a *reference*. A reference is an attribute of an XML element that refers to some other data. The referenced data can be anything and anywhere, not just XML elements and not just in the same XML document. This technique is much more flexible and powerful than striping. An example of a molecule with two atoms bound to each other is shown in figure 1.15. The two atoms in the `atomArray` are referenced by the bond in the `bondArray`. In general, a reference could be to anything that has a URI.

Striping and references can be used in the same document. In RDF, the two techniques can be used interchangeably, and they have exactly the same

```
<molecule id="m1">
  <atomArray>
    <atom id="a1"/>
    <atom id="a2"/>
  </atomArray>
  <bondArray>
    <bond atomRefs2="a1 a2"/>
  </bondArray>
</molecule>
```

Figure 1.15 The use of references to specify a bond between two atoms in a molecule. The arrows show the atoms that are being referenced by the bond element.

meaning. A relationship specified with either striping or a reference forms a statement. For example, in figure 1.14 there is the statement "The human insulin gene is cited by db entry 80129725." Both striping and references help organize XML documents so that relationships are explicit. They contribute to the goal of ensuring that data are self-describing. References are commonly used in bioinformatics ontologies, but striping is seldom used outside of RDF ontologies.

One feature of RDF that makes it especially attractive is that its semantics have been formalized using mathematical logic. There are now a number of ontology languages that extend RDF and that also have formal semantics. The DARPA Agent Markup Language (DAML) is a DARPA project that produced the DAML+OIL language. This language has recently been superseded by the Web Ontology Language (OWL). OWL is a standard of the World Wide Web Consortium (W3C). The RDF and OWL standards are available on the W3C website (`www.w3c.org`). Both RDF and OWL will be discussed in much more detail in the rest of this book.

Summary

- Relationships connect concepts to each other.
- XML has only one explicit kind of relationship: containment.
- Relationships can be specified in XML in two ways:
  1. adding a new layer (striping),
  2. using references.

- RDF and languages based on it allow one to use either striping or references interchangeably.

1.7 Namespaces

So far, all of the examples of XML documents used a single DTD. It is becoming much more common to use several DTDs in a single document. This has the important advantage that markup vocabulary that is already available can be reused rather than being invented again. However, simply merging the vocabularies of multiple DTDs can have undesirable consequences, such as:

- The same term can be used in different ways. For example, "locus" is an attribute in the Bioinformatic Sequence Markup Language (BSML), but it is an element in BioML.
- The same term can have different meanings. This is especially true of commonly occurring terms such as "value" and "label."
- The same term might have the same use and meaning, but it may be constrained differently. For example, the "Sequence" element occurs in several DTDs and has the same meaning, but the content and attributes that are allowed will vary.

Namespaces were introduced to XML to allow one to use multiple DTDs or XML schemas without confusing the names of elements and attributes that have more than one meaning. A *namespace* is a URI that serves as means of distinguishing a set of terms. For example, `reaction` is used both in the Systems Biology Markup Language (SBML) (SBML 2003) and in CML. The SBML namespace is `http://www.sbml.org/sbml/level2`. The CML namespace dealing with chemical reaction terminology is `http://www.xml-cml.org/schema/cml2/react`. By using the namespaces one can ensure that any use of `reaction` is unambiguous.

Within an XML document namespaces are specified using an abbreviation called the *namespace prefix*. For example, if one wishes to use both CML and SBML reactions in the same document, then one must declare prefixes as follows:

```
xmlns:cmlr="http://www.xml-cml.org/schema/cml2/react"
xmlns:sbml="http://www.sbml.org/sbml/level2"
```

These declarations are attributes that can be added to any element, but they are most commonly added to the root element. Once the prefixes have been declared, one can use the prefixes for elements and for attributes. For example, the following document mixes CML, BioML and SBML terminology:

```
<bioml:organism
 xmlns:cml="http://www.xml-cml.org/schema/cml2/core"
 xmlns:cmlr="http://www.xml-cml.org/schema/cml2/react"
 xmlns:bioml="http://xml.coverpages.org/bioMLDTD-19990324.txt"
 xmlns:sbml="http://www.sbml.org/sbml/level2"
>
  <bioml:species>Homo sapiens</bioml:species>
  <sbml:reaction sbml:id="reaction_1" sbml:reversible="false">
    <sbml:listOfReactants>
      <sbml:speciesReference sbml:species="X0"/>
    </sbml:listOfReactants>
    <sbml:listOfProducts>
      <sbml:speciesReference sbml:species="S1"/>
    </sbml:listOfProducts>
  </sbml:reaction>
  <cmlr:reaction>
    <cmlr:reactantList>
      <cml:molecule cml:id="r1"/>
    </cmlr:reactantList>
    <cmlr:productList>
      <cml:molecule cml:id="p1"/>
    </cmlr:productList>
  </cmlr:reaction>
  ...
</bioml:organism>
```

There are several ambiguities in the document above. As we have already noted, CML and SBML both use `reaction`. The meanings are the same, but they are specified differently. For example, CML uses `reactantList` for what SBML calls `listOfReactants`. A more subtle ambiguity is the use of `species` by both SBML and BioML. Here the the meanings are different. In SBML a species is a chemical species. In BioML it is an organism species.

One can use any prefix to designate a namespace within an XML element. For example, one could have used `xyz` instead of `bioml` in the document above. However, it is better to use prefixes that clearly abbreviate the namespace URI. When an element name or attribute has a namespace prefix, it is said to be *qualified* by the namespace.

One can also declare a namespace to be the *default* namespace. When there is a default namespace, then unqualified element names belong to the default namespace. The example above could be simplified somewhat by using a default namespace as follows:

```
<organism
 xmlns="http://xml.coverpages.org/bioMLDTD-19990324.txt"
 xmlns:sbml="http://www.sbml.org/sbml/level2"
 xmlns:cml="http://www.xml-cml.org/schema/cml2/core"
 xmlns:cmlr="http://www.xml-cml.org/schema/cml2/react"
>
  <species>Homo sapiens</species>
  <sbml:reaction sbml:id="reaction_1" sbml:reversible="false">
    <sbml:listOfReactants>
      <sbml:speciesReference sbml:species="X0"/>
    </sbml:listOfReactants>
    <sbml:listOfProducts>
      <sbml:speciesReference sbml:species="S1"/>
    </sbml:listOfProducts>
  </sbml:reaction>
  ...
</organism>
```

It is important to note that the default namespace applies only to element names, not to attributes. Because of this limitation, many authors have chosen to avoid using default namespaces altogether and to explicitly qualify every element and attribute. This has the advantage that such documents are somewhat easier to read, especially when one is using more than two or three namespaces.

The namespace URI need not be the same as the location of the DTD or schema. For example, the CML core has the namespace `http://www.xml-cml.org/schema/cml2/core`, but the actual location of the schema is `www.xml-cml.org/dtdschema/cmlCore.xsd`. Consequently, for each namespace one needs to know the URI, the location and the most commonly used abbreviation. The namespaces that are the most important for ontologies are

bioml: Biopolymer Markup Language
`http://xml.coverpages.org/bioMLDTD-19990324.txt`

cellml: Cell Markup Language
`http://www.cellml.org/cellml/1.0`

cmeta: Cell Meta Language
`http://www.cellml.org/metadata/1.0`

cml: Chemical Markup Language
`http://www.xml-cml.org/schema/cml2/core`

dc: Dublin Core Elements
`http://purl.org/dc/elements/1.1/`

dcterms: Dublin Core Terms
`http://purl.org/dc/terms/`

go: Gene Ontology
`http://ftp://ftp.geneontology.org/pub/go/xml/dtd/go.dtd`

mathml: Mathematics Markup Language
`http://www.w3.org/1998/Math/MathML`

owl: Web Ontology Language
`http://www.w3.org/2002/07/owl`

rdf: RDF
`http://www.w3.org/1999/02/22-rdf-syntax-ns`

rdfs: RDF Schema
`http://www.w3.org/2000/01/rdf-schema`

sbml: Systems Biology Markup Language
`http://www.sbml.org/sbml/level2`

stm: Technical Markup Language
`http://www.xml-cml.org/schema/stmml`

xmlns: XML Namespaces
`http://www.w3.org/XML/1998/namespace`

xsd: XML Schema (original)
`http://www.w3.org/2000/10/XMLSchema`

xsd: XML Schema (proposed)
`http://www.w3.org/2001/XMLSchema`

xsi: XML Schema instances
`http://www.w3.org/2001/XMLSchema-instance`

xsl: XML Transform
`http://www.w3.org/1999/XSL/Transform`

xtm: Topic Maps
`http://www.topicMaps.org/xtm/1.0/`

Summary

- Namespaces organize multiple vocabularies so that they may be used at the same time.
- Namespaces are URIs that are declared in an XML document.
- Each namespace is either the default namespace or it has an abbreviation called a prefix.
- Element names and attributes may be qualified by using the namespace prefix.
- The default namespace applies to all unqualified elements. It does not apply to unqualified attributes.

1.8 Exercises

1. A spreadsheet was exported in comma-delimited format. The first few lines look like this:

```
element_id,sequence_id,organism_name,seq_length,type
U83302,MICR83302,Colaptes rupicola,1047,DNA
U83303,HSU83303,Homo sapiens,3460,DNA
U83304,MMU83304,Mus musculus,51,RNA
U83305,MIASSU833,Accipiter striatus,1143,DNA
```

   Show how these records would be written as XML elements using the `bio_sequence` tag.

2. For the spreadsheet in exercise 1.1 above, show the corresponding XML DTD. The `element_id` attribute is a unique key for the element. Assume that all attributes are optional. The molecule type is restricted to the biologically significant types of biopolymer.

3. Here is a relational database table that defines some physical units:

name	prefix	unit	exponent
millisecond	milli	second	1
per_millisecond	milli	second	-1
millivolt	milli	volt	1
microA_per_mm2	micro	ampere	1
microA_per_mm2	milli	meter	-2
microF_per_mm2	micro	farad	1
microF_per_mm2	milli	meter	-2

A physical unit is, in general, composed of several factors. This was encoded in the relational table by using several records, one for each factor. The `microF_per_mm2` unit, for example, is the ratio of microfarads by square millimeters.

This relational database table illustrates how several distinct concepts can be encoded in a single relational table. In general, information in a relational database about a single concept can be spread around several records, and a single record can include information about several concepts. This can make it difficult to understand the meaning of a relational table, even when the relational schema is available.

Show how to design an XML document so that the information about the two concepts (i.e, the physical units and the factors) in the table above are separated.

4. This next relational database table defines some of the variables used in the Fitzhugh-Nagumo model (Fitzhugh 1961; Nagumo 1962) for the transmission of signals between nerve axons:

component	variable	initial	physical_unit	interface
membrane	u	-85.0	millivolt	out
membrane	Vr	-75.0	millivolt	out
membrane	Cm	0.01	microF_per_mm2	
membrane	time		millisecond	in
ionic_current	I_ion		microA_per_mm2	out
ionic_current	v			in
ionic_current	Vth		millivolt	in

The physical units are the ones defined in exercise 3 above. Extend the solution of that exercise to include the data in the table above. Note that once again, multiple concepts have been encoded in a single relational database table. This exercise is based on an example on the CellML website (CellML 2003).

5. Use an XML editor (such as Xerlin or XML Spy) to construct the examples in the previous two exercises. Follow these steps:

   (a) Cut and paste the following DTD into a file:

```
<?xml version="1.0">
<!DOCTYPE model [
  <!ELEMENT model (physical_unit*,component*)>
  <!ELEMENT physical_unit (factor)*>
  <!ATTLIST physical_unit name ID #REQUIRED>
  <!ELEMENT factor EMPTY>
  <!ATTLIST factor
      prefix   CDATA #IMPLIED
      unit     CDATA #REQUIRED
      exponent CDATA "1">
  <!ELEMENT component (variable)*>
  <!ATTLIST component name ID #REQUIRED>
  <!ELEMENT variable EMPTY>
  <!ATTLIST variable
     name          CDATA   #REQUIRED
     initial       CDATA   #IMPLIED
     physical_unit IDREF   "dimensionless"
     interface     (in|out) #IMPLIED>
]>
<model/>
```

   (b) Open the file with your XML editor.

   (c) Create the elements and enter the attributes shown in the two database tables in the two previous exercises.

   (d) Save the file, and open it with an ordinary text editor.

   (e) Verify that the resulting file has the data as shown in the answers to the exercises above.

2 *XML Semantics*

2.1 The Meaning of Meaning

Semantics is a surprisingly simple notion. It is concerned with when two terms or statements are the *same*. For example, one can write the number 6 in many ways such as: 3 + 3, 06, six and VI. All of these ways to write 6 look different, but they all have the same meaning. The field of semantics is concerned with extracting a single abstract concept from the many ways that the concept can be represented, such as words, phrases, sounds, and pictures. Semantics is part of a more general field called *semiotics* which studies the relationship between concrete representations and the phenomena in the world they signify.

Meaning is always relative to a context. For example, "lumbar puncture" and "spinal tap" are synonymous in the context of medical procedures. However, they are not synonymous in the context of movies (the latter being the name of a movie, while the former has never been used in the name of a movie).

Semantics is often contrasted with syntax. The *syntax* of a language defines what statements can be expressed in the language. Syntax is concerned with the *grammar* of the language. However, there can be many ways to say the same thing. The common concept behind the syntactic variations is the semantics.

The usual method for defining the sameness relationship is to use mathematics. Terminology and statements are then mapped to an abstract mathematical structure, usually called the *model*. Two terms or statements are the same when they map to the same model. Integers, for example, have an abstract formal model which is used to define their semantics. Medical terminology can be defined using a standard vocabulary such as the Unified

Medical Language System (UMLS). In the UMLS, "spinal tap" has concept identifier C0553794. All terms with this same concept identifier are synonymous.

An ontology is a means by which the language of a domain can be formalized (Heflin et al. 1999; Opdahl and Barbier 2000; Heflin et al. 2000; McGuinness et al. 2000). As such, an ontology is a context within which the semantics of terminology and of statements using the terminology are defined. Ontologies define the syntax and semantics of concepts and of relationships between concepts. Concepts are used to define the vocabulary of the domain, and relationships are used to construct statements using the vocabulary. Such statements express known or at least possible knowledge whose meaning can be understood by individuals in the domain. Representing knowledge is therefore one of the fundamental purposes of an ontology. Classic ontologies in philosophy are informally described in natural language. Modern ontologies differ in having the ability to express knowledge in machine-readable form. Expressing knowledge in this way requires that it be represented as data. So it is not surprising that ontology languages and data languages have much in common, and both kinds of language have borrowed concepts from each other. As we saw in section 1.1, a database schema can be regarded as a kind of ontology. Modern ontology languages were derived from corresponding notions in philosophy. See the classic work (Bunge 1977, 1979), as well as more recent work such as (Wand 1989; Guarino and Giaretta 1995; Uschold and Gruninger 1996). Ontologies are fundamental for communication between individuals in a community. They make it possible for individuals to share information in a meaningful way. Formal ontologies adapt this idea to automated entities (such as programs, agents, or databases). Formal ontologies are useful even for people, because informal and implicit assumptions often result in misunderstandings. Sharing of information between disparate entities (whether people or programs) is another fundamental purpose of an ontology.

It would be nice if there were just one way to define ontologies, but at the present time there is not yet a universal ontology language. Perhaps there will be one someday, but in the meantime, one must accept that there will be some diversity of approaches. In this chapter and in chapter 4, we introduce the diverse mechanisms that are currently available, and we compare their features. The ontology languages discussed in chapter 4 make use of logic and rules, so we introduce them in chapter 3.

Two examples are used throughout this chapter as well as chapter 4. The first one is a simplified Medline document, and the second is the specification for nitrous oxide using CML. The document-type definitions were highly simplified in both cases. The simplified Medline document is figure 2.1. The original Medline citation is to (Kuter 1999). The DTD being used is given in figure 2.2. The nitrous oxide document is figure 2.3. The simplified CML DTD being used is given in figure 1.6.

```
<MedlineCitation Owner="NLM" Status="Completed">
  <MedlineID>99405456</MedlineID>
  <PMID>10476541</PMID>
  <DateCreated>
    <Year>1999</Year>
    <Month>10</Month>
    <Day>21</Day>
  </DateCreated>
  <ArticleTitle>Breast cancer highlights.</ArticleTitle>
</MedlineCitation>
```

Figure 2.1 Example of part of a Medline citation using the Medline DTD.

```
<!ELEMENT MedlineCitation
        (MedlineID, PMID, DateCreated, ArticleTitle?)>
<!ATTLIST MedlineCitation
        Owner   CDATA                        "NLM"
        Status  (Incomplete|Completed)   #REQUIRED>
<!ELEMENT MedlineID       (#PCDATA)>
<!ELEMENT PMID            (#PCDATA)>
<!ELEMENT DateCreated     (Year, Month, Day)>
<!ELEMENT Year            (#PCDATA)>
<!ELEMENT Month           (#PCDATA)>
<!ELEMENT Day             (#PCDATA)>
<!ELEMENT ArticleTitle    (#PCDATA)>
```

Figure 2.2 Simplification of the Medline DTD.

```
<molecule id="m1" title="nitrous oxide">
  <atomArray>
    <atom id="n1" elementType="N"/>
    <atom id="o1" elementType="O"/>
  </atomArray>
  <bondArray>
     <bond atomRefs="n1 o1"/>
  </bondArray>
</molecule>
```

Figure 2.3 The representation of nitrous oxide using CML.

2.2 Infosets

Although XML is not usually regarded as being an ontology language, it is formally defined, so it certainly can be used to define ontologies. In fact, it is currently the most commonly used and supported approach to ontologies among all of the approaches considered in this book.

The syntax for XML is defined in (W3C 2001b). The structure of a document is specified using a DTD as discussed in section 1.2. A DTD can be regarded as being an ontology. A DTD defines concepts (using element types) and relationships (using the parent-child relationship and attributes). The concept of a DTD was originally introduced in 1971 at IBM as a means of specifying the structure of technical documents, and for two decades it was seldom used for any other purpose. However, when XML was introduced, there was considerable interest in using it for other kinds of data, and XML has now become the preferred interchange format for any kind of data.

The formal semantics for XML documents is defined in (W3C 2004b). The mathematical model is called an *infoset*. The mathematical model for the XML document in figure 2.1 is shown in figure 2.4. The infoset model consists of *nodes* (shown as rectangles or ovals) and *relationship links* (shown as arrows). There are various types of nodes, but the two most common types are *element nodes* and *text nodes*. There are two kinds of relationship link: *parent-child link* and *attribute link*. Every infoset model has a *root node*. For an XML document, the root node has exactly one child node, but infosets in general can have more than one child node of the root, as, for example, when the infoset represents a fragment of an XML document or the result of a query.

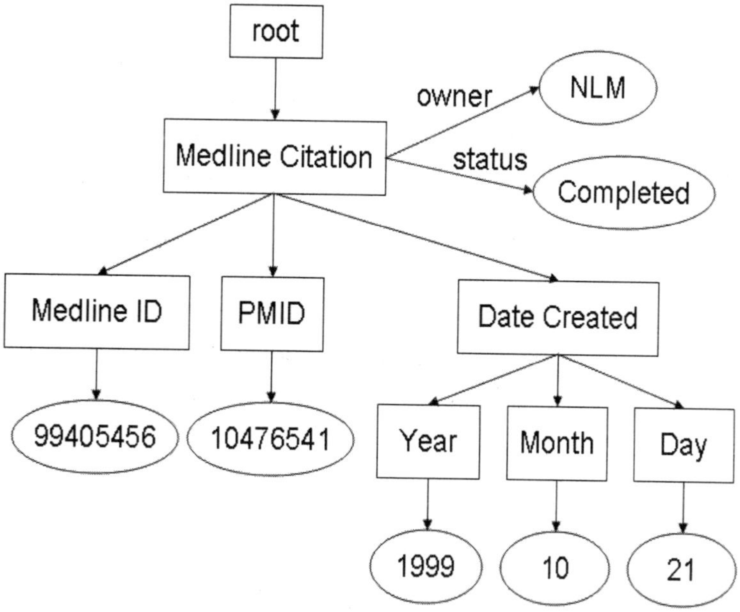

Figure 2.4 XML data model for a typical Medline citation. Element nodes are shown using rectangles, text nodes are shown using ovals, child links are unlabeled, and attributes are labeled with the attribute name.

The formal semantics of XML documents is concerned with whether two infoset models are the same. The most obvious requirement for two infosets to be the same is that the nodes and links correspond with one another. The more subtle requirement is concerned with the arrangement of the nodes. For two infosets to be the same, the children of each node must be in the same order. For example, suppose that the two first two child elements of the Medline citation were reversed as follows:

```
<MedlineCitation Owner="NLM" Status="Completed">
  <PMID>10476541</PMID>
  <MedlineID>99405456</MedlineID>
  <DateCreated>
    <Year>1999</Year>
    <Month>10</Month>
    <Day>21</Day>
  </DateCreated>
</MedlineCitation>
```

The corresponding infoset is shown in figure 2.5 and differs from that in figure 2.4 in only one way: the MedlineID and PMID child nodes have been reversed. These two infosets are *different*.

By contrast, the attribute links can be in any order. For example suppose that the attributes of the MedlineCitation were reversed as follows:

```
<MedlineCitation Status="Completed" Owner="NLM">
```

The corresponding infoset is shown in figure 2.6 and differs from that in figure 2.4 in only one way: the owner and status links have been reversed. These two infosets are *the same*.

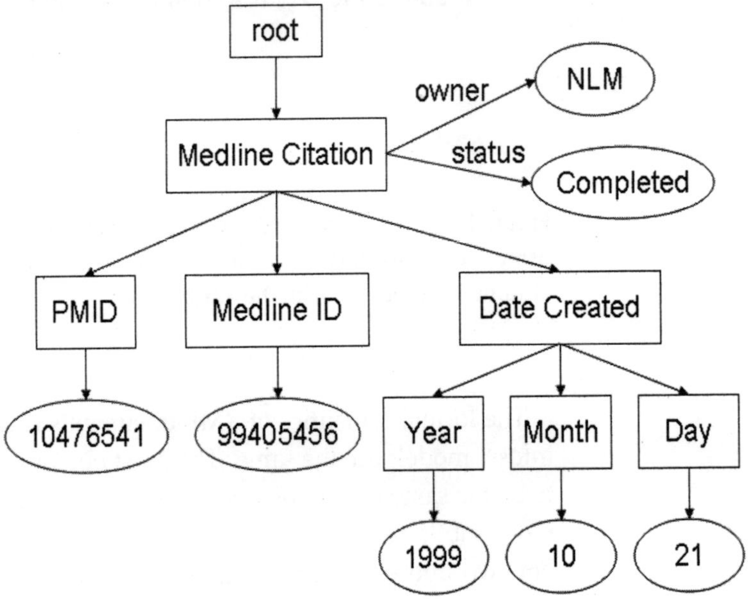

Figure 2.5 XML data model for a Medline citation in which the MedlineID and PMID nodes are in the opposite order.

This example illustrates that the semantics of XML does not always correctly capture the semantics of the domain. In this case, the XML documents in which the PMID and MedlineID elements have been reversed have a different meaning in XML but are obviously conveying the same information from the point of view of a bibliographic citation. One can deal with this problem by specifying in the DTD that these two elements must always appear in one order. In this case, the MedlineID element must occur before the PMID element.

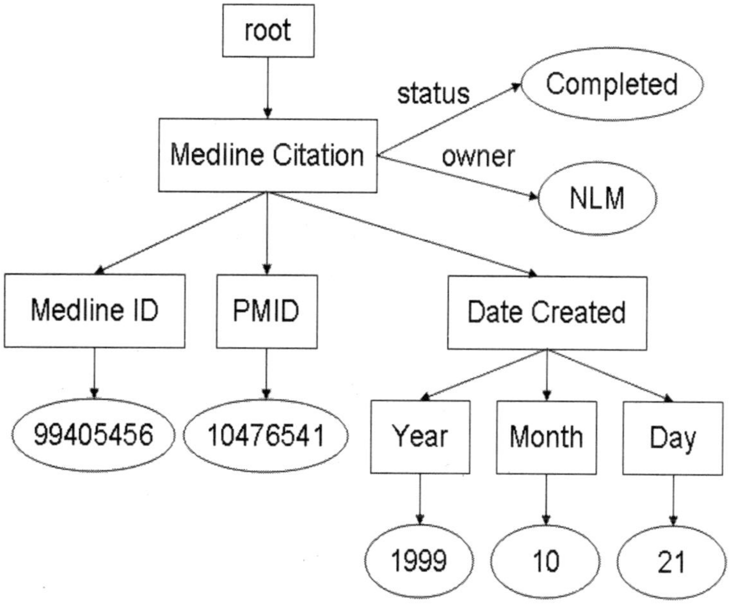

Figure 2.6 XML data model for a Medline citation in which the status and owner attributes are in the opposite order.

The infoset for the nitrous oxide document in figure 2.3 is shown in figure 2.7. If the first two atom elements were reversed the infoset would be as in figure 2.8. These two infosets are *different*. However, from a chemical point of view, the molecules are the same. This is another example of a clash between the semantics of XML and the semantics of the domain. Unlike the previous example, there is no mechanism in XML for dealing with this example because all of the child elements have the same name (i.e., they are all atom elements). So one cannot specify in the DTD that they must be in a particular order. One also cannot specify that the order does not matter.

Summary

- An XML DTD can be regarded as an ontology language.
- The formal semantics of an XML document is defined by its infoset.
- The order of attributes does not matter.
- The order of elements does matter.

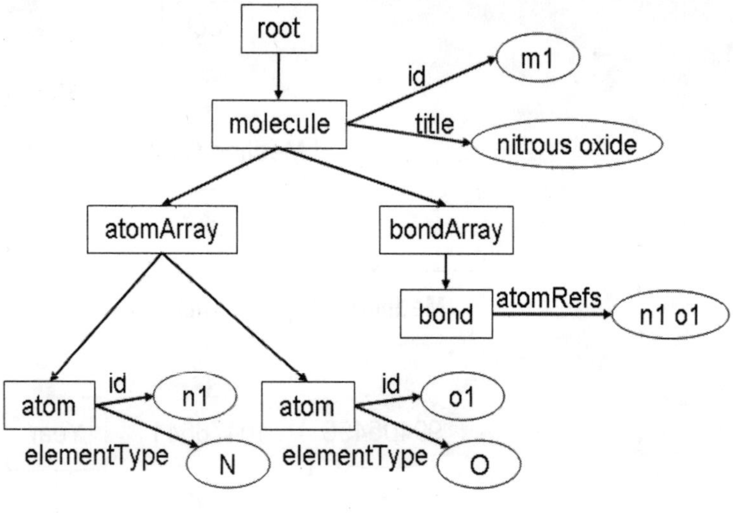

Figure 2.7 XML data model for a molecule represented using CML.

2.3 XML Schema

As XML began to be used for applications other than technical documents, the limitations and flaws of DTDs soon became apparent.

1. The DTD uses a very different syntax from that of XML.
2. There are only a few data types. Because of this limitation, nearly all attributes are defined to have type CDATA, that is, ordinary text, more commonly known as *strings*. Important types such as numbers, times, and dates cannot be specified.
3. Techniques that are common in modern databases and programming languages such as customized types and inheritance are not available.

As a result, an effort was started to express DTDs in XML and to support more general data-structuring mechanisms. The end result was XML Schema (W3C 2001c), often abbreviated as XSD. There are two parts to XSD:

1. Part 1: Complex data types. These can be used to define all of the commonly used data structures used in computer science, such as lists and records.
2. Part 2: Simple data types. These include numbers, times, durations of time, dates, and web addresses. One can also introduce customized sim-

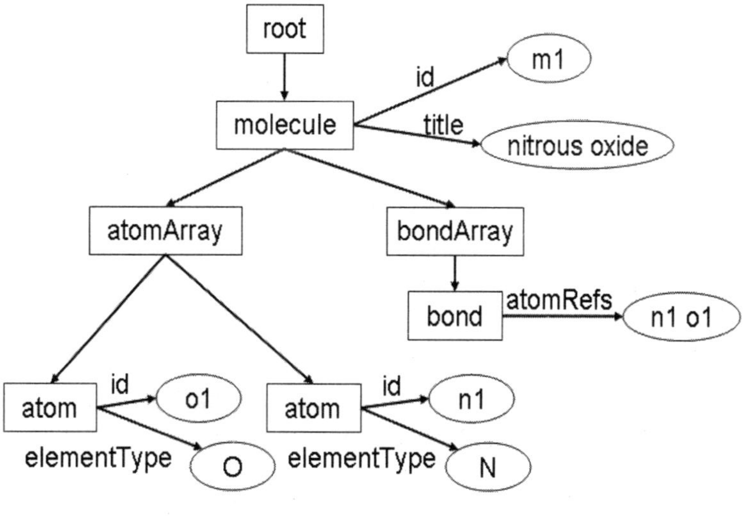

Figure 2.8 XML data model for the same molecule as in figure 2.7 except that the two atoms have been reversed.

ple data types for specialized needs. For example, one could define a DNA sequence to be text containing only the letters A, C, G, and T.

There is a tool written in Perl, called `dtd2xsd.pl` (W3C 2001a) that translates DTDs to XML schemas. However, one must be cautious when using this tool. It does not support all of the features of XML. For example, conditional sections are not supported. As one of the authors pointed out, "It is worth pointing out that this tool does not produce terribly high quality schemas, but it is a decent starting point if you have existing DTDs." When one is using this tool one must manually check that the translation is correct. One can then enhance the schema to improve the semantics using features of XSD that are not available in DTDs.

Applying the `dtd2xsd.pl` program to figure 2.2 gives the XML schema shown below. The XML schema is considerably longer than the DTD. We leave it as an exercise to do the same for the molecule DTD in figure 1.6.

```
<schema
  xmlns='http://www.w3.org/2000/10/XMLSchema'
  targetNamespace='http://www.w3.org/namespace/'
  xmlns:t='http://www.w3.org/namespace/'>

 <element name='MedlineCitation'>
```

```
   <complexType>
    <sequence>
     <element ref='t:MedlineID'/>
     <element ref='t:PMID'/>
     <element ref='t:DateCreated'/>
     <element ref='t:ArticleTitle'
       minOccurs='0' maxOccurs='1'/>
    </sequence>
    <attribute name='Owner' type='string'
       use='default' value='NLM'/>
    <attribute name='Status' use='required'>
     <simpleType>
      <restriction base='string'>
       <enumeration value='Incomplete'/>
       <enumeration value='Completed'/>
      </restriction>
     </simpleType>
    </attribute>
   </complexType>
  </element>

  <element name='MedlineID'>
   <complexType mixed='true'>
   </complexType>
  </element>

  <element name='PMID'>
   <complexType mixed='true'>
   </complexType>
  </element>

  <element name='DateCreated'>
   <complexType>
    <sequence>
     <element ref='t:Year'/>
     <element ref='t:Month'/>
     <element ref='t:Day'/>
    </sequence>
   </complexType>
```

```
 </element>

 <element name='Year'>
  <complexType mixed='true'>
  </complexType>
 </element>

 <element name='Month'>
  <complexType mixed='true'>
  </complexType>
 </element>

 <element name='Day'>
  <complexType mixed='true'>
  </complexType>
 </element>

 <element name='ArticleTitle'>
  <complexType mixed='true'>
  </complexType>
 </element>
</schema>
```

The XML schema shown above has exactly the same meaning as the DTD. Having translated this DTD to XSD, one can make use of features of XSD that are not available in a DTD. Some examples of these features are shown in the next section.

Abstract Syntax Notation One (ASN.1) is another mechanism for encoding hierarchically structured data. The development of ASN.1 goes back to 1984, and it was a mature standard by 1987. It is mainly used in telecommunications, but it is also being used in other areas, including biomedical databases. ASN.1 and XSD have similar capabilities and semantics. The main difference is that ASN.1 allows for much more efficient encoding than XML. XER is an encoding of XSD using ASN.1 and the `xsdasn1` script translates from XSD to ASN.1. Both XER and `xsdasn1` are available at `asn1.elibel.tm.fr`.

Summary

- XSD adds additional data-type and data-structuring features to XML.

- XSD does not change the semantics of XML documents.
- An XML DTD can be converted to XSD using `dtd2xsd.pl`.

2.4 XML Data

An XML DTD has a rich language for specifying the kinds of element that can be contained within each element. By contrast there is very little one can say about attributes and text content with an XML DTD. In practice, nearly all attributes are defined to be of type `CDATA` which allows any kind of text data except for XML elements. When an element has text, then its content is defined to be of type `#PCDATA` which allows any kind of text data. Many important types of data are routinely used by computer and database systems, such as numbers, times, dates, telephone numbers, product codes, and so on. The limitations of XML DTDs have prevented XML processors from properly validating these types. The result has been that individual application writers have had to implement type checking in an ad hoc manner. The XSD datatype recommendation addresses the need of both document authors and applications writers for a robust, extensible datatype system for XML. This standard has been very successful, and it has now been incorporated into other XML-related standards such as RDF and OWL, to be discussed in chapter 4.

For example, in the Medline schema part of which was shown in section 2.3, the `Day` element specifies the day of the month, but this schema allows one to use any text whatsoever as a day. At the very least, one should limit the values to positive numbers. To do this one should change the specification for the Day element to the following:

```
<element name='Day' type='xsd:positiveInteger'/>
```

An even better specification would further restrict the possible numbers to be positive numbers no larger than 31 as in

```
<element name='Day'>
 <simpleType>
   <xsd:restriction base='xsd:positiveInteger'>
     <xsd:maxInclusive value='31'/>
   </xsd:restriction>
 </simpleType>
</element>
```

One can make similar restrictions for the Year and Month elements. However, this still does not entirely capture all possible restrictions. For example, it would allow February to have 31 days. As it happens, there is an XML datatype for a date which includes all restrictions required for an arbitrary calendar date. To use this datatype, replace the Year, Month, and Day elements with the following:

```
<element name='DateCreated' type='xsd:date'/>
```

Using this approach, the Medline citation in figure 2.1 would look like this:

```
<MedlineCitation Owner="NLM" Status="Completed">
  <MedlineID>99405456</MedlineID>
  <PMID>10476541</PMID>
  <DateCreated>1999-10-21</DateCreated>
  <ArticleTitle>Breast cancer highlights.</ArticleTitle>
</MedlineCitation>
```

The semantics of an XML datatype is given in three parts:

1. The *lexical space* is the set of strings that are allowed by the datatype. In other words, the kind of text that can appear in an attribute or element that has this type.

2. The *value space* is the set of abstract values being represented by the strings. Each string represents exactly one value, but one value may be represented by more than one string. For example, 6.3200 and 6.32 are different *strings* but they represent the same *value*. In other words, two strings have the same *meaning* when they represent the same value.

3. A set of *facets* that determine what operations can be performed on the datatype. For example, a set of values can be sorted only if the datatype has the *ordered* facet.

For some datatypes, the lexical space and value space coincide, so what one sees is what it means. However, for most datatypes there will be multiple representations of the same value. When this is the case, each value will have a *canonical* representation. Since values and canonical representations correspond exactly to each other, in a one-to-one fashion, it is reasonable to think of the canonical representation as being the meaning.

XSD includes over 40 built-in datatypes. In addition one can construct datatypes based on the built-in ones. The built-in datatypes that are the most useful to bioinformatics applications are:

1. **string**. Arbitrary text without embedded elements.
2. **decimal**. A decimal number of any length and precision.
3. **integer**. An integer of any length. This is a special case of **decimal**. There are many special cases of integer, such as **positiveInteger** and **nonNegativeInteger**.
4. **date**. A Gregorian calendar date.
5. **time**. An instant of time during the day, for example, 10:00.
6. **dateTime**. A date and a time instance during that date.
7. **duration**. A duration of time.
8. **gYear**. A Gregorian year.
9. **gYearMonth**. A Gregorian year and month in that year.
10. **boolean**. Either true or false.
11. **anyURI**. A web resource.

There are three ways to construct a new datatype from other datatypes:

1. **Restriction**. The most common way to define a datatype is to restrict another datatype. For example, to define a telephone number, start with string and restrict to those strings that have an acceptable pattern of digits, and plus and minus signs. One can restrict using any combination of the following techniques:
   (a) Bounds. The maximum and minimum, either inclusive or exclusive.
   (b) Length. The number of characters of a string (minimum or maximum or both), the number of digits of a number, or the number of digits after the decimal point.
   (c) Pattern. A pattern that must be matched. The XML pattern language is similar to the one used by Perl; see subsection 10.1.4
   (d) Enumeration. An explicit list of all possibilities.
2. **Union**. One can combine the set of values of several datatypes. This is handy for adding special cases to another datatype.
3. **List**. A sequence of values.

For example, suppose that it is possible for the `dateCreated` attribute to be not applicable to a citation. Simply omitting this attribute would mean that it has a value which is unknown rather than that the attribute does not apply to the citation. To allow for such special cases, one can add additional values to a datatype by using a *union* as follows:

```
<element name='DateCreated'>
  <simpleType>
    <xsd:union memberTypes='xsd:date'>
      <xsd:simpleType>
        <enumeration value='N/A'/>
      </xsd:simpleType>
    </xsd:union>
  </simpleType>
</element>
```

Summary

- XSD provides built-in datatypes for the most commonly used purposes, such as strings, numbers, dates, times, and resource references (URIs).
- New datatypes can be defined by restricting another datatype, combining several datatypes (union), or allowing a sequence of values (list).

2.5 Exercises

1. Convert the molecule DTD shown in figure 1.6 to an XML schema.
2. Revise the molecule schema in exercise 2.1 above so that the `elementType` attribute can only be one of the standard abbreviations of the 118 currently known elements in the periodic table.
3. Define a simple datatype for a single DNA base. Hint: Use an enumeration as in exercise 2.2 above.
4. Define a simple datatype for a DNA sequence.
5. Define a more realistic datatype for a DNA sequence. It is a common practice to break up amino acid and DNA sequences into more manageable pieces. For example, the following is a sequence in the European Molecular Biology Laboratory (EMBL) database:

```
cctggacctc ctgtgcaaga acatgaaaca nctgtggttc ttccttctcc tggtggcagc    60
tcccagatgg gtcctgtccc aggtgcacct gcaggagtcg ggcccaggac tggggaagcc   120
tccagagctc aaaaccccac ttggtgacac aactcacaca tgcccacggt gcccagagcc   180
caaatcttgt gacacacctc ccccgtgccc acggtgccca gagcccaaat cttgtgacac   240
acctccccca tgcccacggt gcccagagcc caaatcttgt gacacacctc ccccgtgccc   300
nnngtgccca gcacctgaac tcttgggagg accgtcagtc ttcctcttcc ccccaaaacc   360
caaggatacc cttatgattt cccggacccc tgaggtcacg tgcgtggtgg tggacgtgag   420
ccacgaagac ccnnnngtcc agttcaagtg gtacgtggac ggcgtggagg tgcataatgc   480
caagacaaag ctgcgggagg agcagtacaa cagcacgttc cgtgtggtca gcgtcctcac   540
cgtcctgcac caggactggc tgaacggcaa ggagtacaag tgcaaggtct ccaacaaagc   600
cctcccagcc cccatcgaga aaaccatctc caaagccaaa ggacagcccn nnnnnnnnnn   660
nnnnnnnnnn nnnnnnnnnn nnnnngagga gatgaccaag aaccaagtca gcctgacctg   720
cctggtcaaa ggcttctacc ccagcgacat cgccgtggag tgggagagca atgggcagcc   780
ggagaacaac tacaacacca cgcctcccat gctggactcc gacggctcct tcttcctcta   840
cagcaagctc accgtggaca agagcaggtg gcagcagggg aacatcttct catgctccgt   900
gatgcatgag gctctgcaca accgctacac gcagaagagc ctctccctgt ctccgggtaa   960
atgagtgcca tggccggcaa gccccgctc cccgggctct cggggtcgcg cgaggatgct  1020
tggcacgtac cccgtgtaca tacttcccag gcacccagca tggaaataaa gcacccagcg  1080
ctgccctgg                                                         1089
```

The sequence is divided into groups of 60 bases, and these groups are divided into subgroups of 10 bases. A number follows each group of 60 bases. The letter n is used when a base is not known.

6. Define a datatype for an amino acid sequence (protein). Here is an example of such a sequence:

```
  1 meepqsdpsv epplsqetfs dlwkllpenn vlsplpsqam ddlmlspddi eqwftedpgp
 61 deaprmpeaa ppvapapaap tpaapapaps wplsssvpsq ktyqgsygfr lgflhsgtak
121 svtctyspal nkmfcqlakt cpvqlwvdst pppgtrvram aiykqsqhmt evvrrcphhe
181 rcsdsdglap pqhlirvegn lrveylddrn tfrhsvvvpy eppevgsdct tihynymcns
241 scmggmnrrp iltiitleds sgnllgrnsf evrvcacpgr drrteeenlr kkgephhelp
301 pgstkralpn ntssspqpkk kpldgeyftl qirgrerfem frelnealel kdaqagkepg
361 gsrahsshlk skkgqstsrh kklmfktegp dsd
```

Like DNA sequences, it is divided into groups of 60 amino acids, and these groups are divided into subgroups of 10 amino acids. A number precedes each group. The letter x is used for an unknown amino acid. The letters j, o, and u are not used for amino acids.

3 *Rules and Inference*

Rules and inference are important for tackling the challenges in bioinformatics. For example, consider the Biomolecular Interaction Network (BIND). The problem of defining interactions is very complex, and interactions must be obtained from several sources, such as the Protein Data Bank (PDB), metabolic/regulative pathways, or networks. Rules can be used to model and query these interaction networks.

3.1 Introduction to Rule-Based Systems

An *enzyme* is a protein that exerts an effect on a specific molecule called its *substrate*. The enzyme and its substrate combine to form the *enzyme-substrate complex*. The active site of the enzyme and the corresponding part of the substrate have shapes that match each other in a complementary manner, known as the lock-and-key model, proposed by Emil Fischer in 1890. The substrate fits into the enzyme just as a key fits into a lock. This is a good model for visualizing how enzymes catalyze reactions. More recently, in 1958, Daniel E. Koshland, Jr. described how the active site of an enzyme can change when the substrate binds to the enzyme. Thus enzymes can take an active role in creating a shape into which the substrate fits. This process is known as the induced-fit model.

A *regulatory transcription factor* is a protein that exerts an effect on the rate at which a gene is transcribed. Such a protein forms a complex with a chromosome at specific binding sites determined by the sequence of bases at these sites. One may consider the binding site of a transcription factor to be a "word" encoded in the DNA. The "words" are the DNA binding motifs for their respective transcription factors. The binding sites for a gene encode the mechanism by which cells control such important biological functions as cell growth, proliferation, and differentiation. It is common for this binding to be somewhat imprecise, so that the resulting binding is not too tight. This allows genes to be turned on and off dynamically in response to chemical signals.

Both of these situations illustrate a common theme in protein-mediated biological processes. Proteins match targets, and when a match takes place (i.e., the protein forms a complex with its target), then the protein exerts a biological effect (e.g., catalysis or gene regulation). The match can be exquisitely precise, as in the classic Fischer lock-and-key model that characterizes many enzymes, or it can be less precise, as in the Koshland induced-fit model or in transcription factor bindings.

Not surprisingly, a similar theme is used in many computer systems. This style of programming goes by various names, such as the "pattern-action paradigm," "expert systems," "rule-based inferencing," or "declarative programming." When programming in this style, one specifies a collection of *rules*. Each rule has two parts:

1. The *pattern*, also called the *antecedent* or the *hypothesis*. This part of the rule specifies the match condition.

2. The *action, consequent,* or *conclusion*. This part of the rule specifies the effect that is exerted when the match condition holds.

A rule can be regarded as a logical statement of the form "if the match condition holds, then perform the action." When considered from this point of view, the match condition is a "hypothesis," and the action is a "conclusion." Just as in organisms, match conditions range from being very precise to being very generic.

The condition of a rule is a Boolean combination of elementary facts, each of which may include constants as well as one or more variables. A query is essentially a rule with no conclusion, just a condition. At the other extreme, a fact is a rule with no condition, just a conclusion. The result of a query is the set of assignments to the variables that cause the rule to fire. From the point of view of relational databases, a query can be regarded as a combination of selections, projections, and joins. The variables in a rule engine query correspond to the output variables (i.e., the projection) and join conditions of a relational query. The constants occurring in a rule engine query correspond to the selection criteria of a relational query. Both rule engines and relational databases support complex Boolean selection criteria.

When the match condition of a rule is found to hold and the consequent action is performed, the rule is said to have been "invoked" or "fired." The firing of a rule affects the environment, and this can result in the firing of other rules. The resulting cascade of rule firings is what gives rule-based systems their power. By contrast, the most common programming style (the so-called procedural programming or imperative style) does not typically have such a cascading effect.

Rule-based inferencing has another benefit. Rules express the meaning of a program in a manner that can be much easier to understand. Each rule should stand by itself, expressing exactly the action that should be performed in a particular situation. In principle, each rule can be developed and verified independently, and the overall system will function correctly provided only that it covers all situations. Unfortunately, rules can interact in unexpected ways, so that building a rule-based system is not a simple as one might suppose. The same is true in organisms, and it is one of the reasons why it is so difficult to understand how they function.

Rules have been used as the basis for computer software development for a long time. Rule-based systems have gone by many names over the years. About a decade ago they were called "expert systems," and they attracted a great deal of interest. While expert systems are still in use, they are no longer as popular today. The concept is certainly a good one, but the field suffered from an excess of hubris. The extravagantly optimistic promises led to equally extreme disappointment when the promises could not be fulfilled. Today it is recognized that rules are only one part of any knowledge-based system, and it is important to integrate rules with many other techniques. The idea that rules can do everything is simply unreasonable.

The process of using rules to deduce facts is called *inference* or *reasoning*, although these terms have many other meanings. Systems that claim to use reasoning can use precise (i.e., logical reasoning), or various degrees of imprecise reasoning (such as "heuristic" reasoning, case-based reasoning, probabilistic reasoning, and many others). This chapter focuses on logical reasoning. In chapter 13 logical inference is compared and contrasted with scientific inference.

Logical reasoners act upon a collection of *facts* and *logical constraints* (usually called *axioms*) stored in a *knowledge base*. Rules cause additional facts to be *inferred* and stored in the knowledge base. Storing a new fact in the knowledge base is called *assertion*. The most common action of a rule is to assert one or more facts, but any other action can be performed.

Many kinds of systems attempt automated human reasoning. A system that evaluates and fires rules is called a *rule engine*, but there are may other kinds of automated reasoners. Including rule-based systems, the main kinds of automated reasoners are

1. forward-chaining rule engines,
2. backward-chaining rule engines,
3. theorem provers,
4. constraint solvers,
5. description logic reasoners,
6. business rule systems,
7. translators,
8. miscellaneous systems.

These are not mutually exclusive categories, and some systems support more than one style of reasoning. We now discuss each of these categories in detail, and then give a list of some of the available software for automated reasoning.

Summary

- Rule-based programming is a distinct style from the more common procedural programming style.
- A rule consists of an antecedent and a consequent. When the antecedent is satisfied, the consequent is invoked (fired).
- Rule engines logically infer facts from other facts, and so are a form of automated reasoning system.
- There are many other kinds of reasoning system such as theorem provers, constraint solvers, and business rule systems.

3.2 Forward- and Backward-Chaining Rule Engines

Forward-chaining rule engines are the easiest rule engines to understand. One simply specifies a set of rules and some initial facts. The engine then fires rules as long as there are rules whose match condition is true. Of course, there must be a mechanism to prevent rules from firing endlessly on the same facts. A rule is normally only invoked once on a particular set of facts that match the rule. When the rule engine finds that no new facts can be inferred, it stops. At that point one can query the knowledge base.

Backward-chaining rule engines are much harder to understand. They maintain a knowledge base of facts, but they do not perform all possible inferences that a forward-chaining rule engine would perform. Rather, a backward-chaining engine starts with the query to be answered. The engine then tries to determine whether it is already known (i.e., it can be answered with known facts in the knowledge base). If so, then it simply retrieves the facts. If the query cannot be answered with known facts, then it examines the rules to determine whether any one of them could be used to deduce the answer to the query. If there are some, then it tries each one. For each such rule, the rule engine tries to determine whether the hypothesis of the rule is true. It does this the same way as it does for answering any query: the engine first looks in the knowledge base and then the engine tries to deduce it by using a rule.

Thus a backward-chaining rule engine is arguing backward from the desired conclusion (sometimes called the "goal") to the known facts in the knowledge base. In contrast with the forward-chaining technique that matches the hypothesis and then performs the corresponding action, a backward-chaining engine will match the conclusion and then proceed backward to the hypothesis. Actions are performed only if the hypothesis is eventually verified. Rules are invoked only if they are relevant to the goal. Thus actions that would be performed by a forward-chaining engine might not be performed by a backward-chaining engine. On the other hand, actions that would be performed just once by a forward-chaining engine could be performed more than once by a backward-chaining engine.

The best-known example of a backward-chaining rule engine is the Prolog programming language (Clocksin et al. 2003). However, there are many others, especially commercial business rule engines, which are discussed later in this chapter.

Backward chainers have some nice features. Because of their strong focus on a goal, they only consider relevant rules. This can make them very fast. However, they also have disadvantages. They are much more prone to infinite loops than forward-chaining engines, and it is difficult to support some forms of reasoning such as paramodulation, which is needed by OWL ontologies (see section 4.4). Programming in backward-chaining mode is also counterintuitive. As a result it takes considerable skill to do it well compared with programming in forward-chaining mode or programming with a procedural language.

Forward-chaining engines behave much more like the protein-substrate binding process introduced at the beginning of this chapter. Modern forward-chaining rule engines are nearly always based on the Rete algorithm introduced by (Forgy 1982). The Rete algorithm processes the rules and facts with a network of interrelationships that captures all of the ways that rules can be fired, either fully or in part. By capturing partially fired rules, it is much easier to introduce new facts. The Rete network matches the new fact against all of the partially fired rules to determine whether it can be extended to form a rule that fires or at least is closer to firing.

Summary

- Both forward- and backward-chaining rule engines require a set of rules and an initial knowledge base of facts.
- Forward-chaining rule engines apply rules which cause more facts to be asserted until no more rules apply. One can then query the knowledge base.
- Backward-chaining rule engines begin with a query and attempt to satisfy it, proceeding backward from the query to the knowledge base.

3.3 Theorem Provers and Other Reasoners

As the name suggests, a *theorem prover* attempts to prove theorems. A program consists of a theory expressed using axioms and facts. A conjecture is presented to the system, and the theorem prover attempts to find a proof. A *proof* consists of a sequence of rule invocations which start with the axioms and end with the conjecture. A proved conjecture is called a theorem. Conjectures can be regarded as queries, and the theorem-proving process is a mechanism for answering the queries. However, this is not quite the same as the query mechanism supported by rule engines or relational databases. In the latter systems, the result of a query is the set of ways that it can be satisfied, not whether it can be satisfied. Theorem provers are not usually capable of dealing with queries in the same way as a relational database system.

To illustrate this distinction between theorem provers and rule engines, consider the case of consistency checking. This is an important problem for logical theories that corresponds to type checking in modern programming languages. If one uses a theorem prover for consistency checking, the result will normally be only one of these three possibilities: consistent, inconsistent, or unknown. By contrast, with a rule engine the result will also show, in some detail, the reasons why the theory is inconsistent or why the theory could not be shown to be either consistent or inconsistent. Needless to say, this is much more useful. Of course, some theorem provers are much better at explaining the reason for their conclusion, but the most popular ones give up this capability for the sake of efficiency.

Theorem provers use a variety of strategies, and many systems offer several strategies and considerable customization of those strategies. Theorem proving in general is undecidable, so there is always a possibility that an attempt to prove a theorem will fail. One popular strategy is the use of "tableaux." When attempting to prove a theorem, one must try many possibilities. When one attempt fails, it is necessary to backtrack to an earlier point in the attempt and then try again. From this point of view, theorem proving is a search in a very large (potentially infinite) space of proofs, and tableaux are a means of controlling the search.

Because theorem provers and rule engines can be inefficient, a large number of specialized automated reasoners have been introduced that are limited to specific kinds of reasoning problems but are much more efficient than the general reasoners. We discuss some of the most popular of these in this section.

Constraint solvers allow one to specify a collection of constraints on variables. The constraints are most commonly linear equalities or inequalities, and the variables are usually real-valued. A solution is an assignment of real values to the variables so that all of the constraints hold. Sometimes one also asks for the solution that maximizes some linear function. Constraint solvers are not reasoning about the variables in the same way that rule engines and theorem provers are reasoning about their variables. However, constraint solvers are sometimes sufficient for some classes of reasoning problems, and they are much more efficient at finding solutions than a rule engine or theorem prover would be on the same problem.

Although constraint solvers are excellent for multivariate constraint problems, they are rarely compatible with the reasoning needed for ontologies. So they will not be discussed in any more detail.

Description logic (DL) reasoners are a form of theorem prover that is optimized for a special class of theories. Biological information captured using DLs is classified in a rich hierarchical lattice of concepts and their interrelationships. Most biological concepts can be defined using DLs, and they allow limited forms of reasoning about biological knowledge (Baker et al. 1999). However, not all concepts can be defined using DLs, and many forms of reasoning cannot be expressed in this framework. Database joins, for example, cannot be expressed using DLs. A DL reasoner will be very efficient, but the limitations of DL reasoners can be too severe for many application domains. This efficiency leads to another problem: it is difficult to extract the reasons for conclusions made by the reasoner. Consequently, DL reasoners provide little feedback in tasks such as consistency checking.

The Web Ontology Language (OWL) was discussed in section 1.6 and will be covered in detail in section 4.4. OWL has three language levels, depending on what features are supported. The lowest level, OWL Lite, has the minimum number of features that are necessary for specifying ontologies. The intermediate level, OWL-DL, has more features than OWL Lite, but still has some restrictions. The restrictions were chosen so that OWL-DL ontologies could be processed using a DL reasoner. The highest level, OWL Full, has no restrictions. The OWL Full level cannot be processed by a DL reasoner, and one must use a theorem prover.

Business rule systems can be classified as rule engines (and some of them are excellent in this regard). However, they tend to emphasize ease of use via graphical user interfaces (GUIs) rather than support for underlying functionality. They are intended to be used by individuals who do not have a background in logic or reasoning systems. Business rule systems are nearly always proprietary, and their performance is usually relatively poor, although there are exceptions. Typically the rule system is only part of a larger system, so the poor performance is effectively masked by the other activities occurring at the same time. Web portal servers often contain a business rule system. Some business rule systems have full support for ontologies, most commonly ontologies expressed in RDF or OWL.

Many systems simply translate from one language to another one, perform the reasoning using a different system, and then translate back to the original language. The advantage is flexibility. The disadvantage is that they can be much less efficient than systems that are optimized for the target language. Translators are commonly used for processing ontologies.

Many other kinds of reasoning system exist, such as Boolean constraint solvers and decision support systems. These may be regarded as optimized reasoners (just as a DL reasoner is an optimized specialization of a theorem prover). However, such reasoners are generally much too limited for processing ontologies.

Summary

- Theorem provers prove theorems.
- Queries to a theorem-proving system are expressed as conjectures.
- Theorem proving is very difficult, so numerous strategies have been developed.
- Specialized theorem provers limit the kind of theory so as to improve performance:

  1. Constraint solvers
  2. Description logic
  3. Business rule systems

3.4 Performance of Automated Reasoners

Any particular reasoning operation (whether it is an inference or a query) is always performed within a fixed, known context. If the reasoning context (also known as the knowledge base) changes over time, then steps must be taken to ensure that the operation is not affected by the changes. This is the same as query processing in a relational database which uses transactions (most commonly implemented by using a locking mechanism) to ensure that updates do not interfere with the results. Because it can be inefficient for each query to be performed by scanning all of the data, relational databases usually maintain indexes on the data. However, there is a cost, both in time and in storage space, to maintain the indexes. An index is useful only if the cost of maintaining the index is compensated by the improvement in performance that it provides. If most uses of the database require examining most of the data, then an index should not be maintained.

A similar situation occurs in automated reasoners. Specialized indexing structures, such as the Rete network maintained by a Rete engine, requires additional computer time and storage space to maintain. This overhead must be compensated by an improvement in performance for the indexing structure to be worthwhile. The way that the knowledge base is used will determine whether the Rete network provides an overall improvement in performance. Generally speaking, the theorem provers (including description logic systems) are suitable primarily for static knowledge bases. Backward chainers and non-Rete forward chainers are intermediate. They have some support for dynamic knowledge bases, but not as much as a Rete forward chainer. The Rete-based systems are especially well suited for knowledge bases that both add data (and rules) and retract them.

It is important to bear in mind that a system in one class can be used to perform reasoning that is normally associated with another class. Prolog, for example, is a general-purpose programming language, so one can, in principle, implement a theorem prover or a DL reasoning system in Prolog (and this is commonly done). However, by itself, Prolog is not a theorem prover.

Summary

- Automated reasoners use specialized indexes to improve performance.
- A Rete network is a specialized index that improves the performance of forward-chaining rule engines.
- Most reasoning systems require an unchanging knowledge base. Only forward-chaining engines and some backward-chaining engines allow for dynamically changing knowledge bases.

4 *The Semantic Web and Bioinformatics Applications*

Many people have had the experience of suddenly realizing that two of their acquaintances are actually the same person, although it usually is not as dramatic as it was for the main characters in the movie *You've Got Mail*. The other kind of identity confusion is considerably more sinister: two persons having the same identity. This is a serious problem, known as *identify theft*. The issue of whether two entities are the same or different is fundamental to semantics.

Addressing logical issues such as whether two entities are the same requires substantially more powerful reasoning capabilities than XML DTDs or schemas provide. Someday, automated reasoners and expert systems may be ubiquitous on the web, but at the moment they are uncommon. The web is a powerful medium, but it does not yet have any mechanism for rules and inference. Tim Berners-Lee, the director of the World Wide Web Consortium, has proposed a new layer above the web which would make all of this possible. He calls this the *Semantic Web* (Miller et al. 2001).

The World Wide Web is defined by a language, the Hypertext Markup Language (HTML), and an Internet protocol for using the language, the Hypertext Transfer Protocol (HTTP). In the same way, the Semantic Web is defined by languages and protocols. In this chapter, we introduce the languages of the Semantic Web and explain what they mean.

4.1 The Semantic Web in Bioinformatics

Biologists use the web heavily, but the web is geared much more toward human interaction than automated processing. While the web gives biologists access to information, it does not allow users to easily integrate different data sources or to incorporate additional analysis tools. The Semantic Web addresses these problems by annotating web resources and by providing rea soning and retrieval facilities from heterogeneous sources.

To illustrate a possible use of the Semantic Web, consider the following hy pothetical scenario. A scientist would like to determine whether a novel pro tein (protein Y) interacts with p21-activated kinase 1 (PAK1). To answer this question, the scientist first goes to the kinase pathway database `kinasedb.ontology.ims.u-tokyo.ac.jp` to obtain a list of all known proteins that interact with PAK1 (e.g., MYLK and BMX). The scientist then writes a set of rules to determine whether the protein Y is structurally similar to any PAK1-interacting proteins. After applying the rules using a Semantic Web-enabled protein interaction server, one hit, protein X, is found. This leads to the prediction that protein Y will interact with PAK1, as in figure 4.1. Next, the scientist wishes to relate this interacting pair to a particular signaling pathway. As all the tools used refer to the same ontologies and terminology defined through the Gene Ontology (GO), the researcher can easily map this interacting pair to a relevant signaling pathway obtained from a Semantic Web–enabled pathway server. During the information foraging described above, the scientist constantly used literature databases to read relevant articles. Despite the tremendous growth of more than 5000 articles each week, the biologist still managed to quickly find the relevant articles by using an ontology-based search facility.

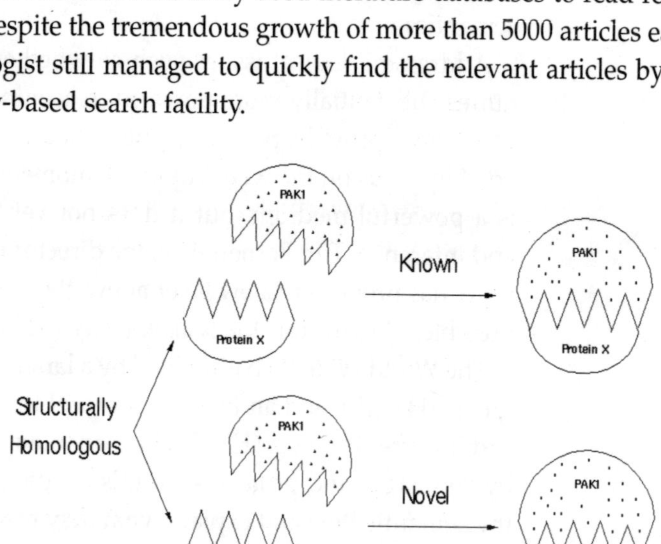

Figure 4.1 Based on a set of user-defined rules of 3D structural similarity, alignments of protein X (known) and protein Y (novel protein), a novel PAX-protein Y interaction can be predicted in silico using the Semantic Web.

As another example, consider a biologist who has just found a novel DNA sequence from an *Anopheles* clone which may be important in the developmental process. To find related sequences, the biologist runs a `blastn` search based on a set of requirements (e.g., the sequence identities must be over 60% and the E-value must be less than 10^{-10}). These requirements can be captured as rules and constraints which could be taken into account by an online Semantic Web–enabled sequence comparison service. If the researcher found a number of significantly similar sequences in *Drosophila*, the scientist could then obtain gene expression data for the relevant genes from a Semantic Web–enabled gene expression database. Rules can then be specified which capture the interesting expression profiles, such as genes which are highly expressed at specified time points in the developmental process.

In both of these examples, the activities can, in principle, be carried out manually by the researcher. The researcher reads material, selects the relevant data, copies and pastes from the web browser, and then struggles with diverse formats, protocols, and applications. Constraints and rules can be enforced informally by manually selecting the desired data. All of this is tedious and error-prone, and the amount of data that can be processed this way is limited. The Semantic Web offers the prospect of addressing these problems.

Summary

The Semantic Web addresses two important problems in Bioinformatics:

1. The dramatic increase of bioinformatics data available in web-based systems and databases calls for novel processing methods.

2. The high degree of complexity and heterogeneity of bioinformatics data and analysis requires semantic-based integration methods.

4.2 The Resource Description Framework

One of the most basic mechanisms of computing systems and networks is the ability to obtain information identified by an address of some kind. To open a web page using a browser one must enter a URL either manually or by clicking on a hypertext link. A URL identifies a *resource* which can be a whole web page, an item within a webpage or a query for information that can be obtained at a website. The ability to access resources using a URL is a fundamental service provided by the Internet.

Hypertext links are one of the most important features that the Worl Wide Web adds to the underlying Internet. If one regards hypertext link as defining relationships between resources, then the World Wide Web wa responsible for adding relationships to the resources that were already avai able on the Internet prior to the introduction of the web. Indeed, the nam "World Wide Web" was chosen because its purpose was to link together th resources of the Internet into an enormous web of knowledge. However, a we discussed in section 1.6, for relationships to be meaningful, they must b explicit. As stated by Wittgenstein in Proposition 3.3 of (Wittgenstein 1922) "Only the proposition has sense; only in the context of a proposition has name meaning." Unfortunately, hypertext links by themselves do not co vey any meaning. They do not explicitly specify the relationship between th two resources that are linked.

The Semantic Web is a layer above the World Wide Web that adds meanir to hypertext links. In other words, the Semantic Web makes hypertext linl into ontological relationships. The Semantic Web is a means for introducir formal semantics to the World Wide Web. All reasoning in the Semantic We is formal and rigorous. The Semantic Web is defined by a series of progre sively more expressive languages and recommendations of the World Wic Web Consortium. The first of these is the Resource Description Framewo (RDF) (Lassila and Swick 1999) which is introduced in this section. RDF developing quickly (Decker et al. 1998), and there are now many tools ar products that can process RDF. In section 4.4 we introduce the Web Ontolo Language (OWL) which adds many new semantic features to RDF.

As the name suggests, RDF is a language for representing informatic about resources in the World Wide Web. It is particularly intended for repr senting annotations about web resources, such as the title, author, and mo ification date of a webpage. However, RDF can also be used to represe information about anything that can be identified on the web, even wh it cannot be directly retrieved. Thus one could use URIs to represent di eases, genes, universities, and hospitals, even though none of these are w resources in the original sense.

The following is the beginning and end of the GO database, as express in RDF:

```
<go:go
   xmlns:go="http://www.geneontology.org/dtds/go.dtd#"
   xmlns:rdf="http://www.w3.org/1999/02/22-rdf-syntax-ns#">
  <rdf:RDF>
   <go:term rdf:about="http://www.geneontology.org/go#GO:0003673"
```

```
          n_associations="149784">
      <go:accession>GO:0003673</go:accession>
      <go:name>Gene_Ontology</go:name>
    </go:term>
    <go:term rdf:about="http://www.geneontology.org/go#GO:0003674"
          n_associations="101079">
      <go:accession>GO:0003674</go:accession>
      <go:name>molecular_function</go:name>
      <go:definition>Elemental activities, such as catalysis or
        binding, describing the actions of a gene product at the
        molecular level.  A given gene product may exhibit one or
        more molecular functions.
      </go:definition>
      <go:part_of
        rdf:resource="http://www.geneontology.org/go#GO:0003673"/>
    </go:term>
       ...
  </rdf:RDF>
</go:go>
```

The entire GO database is currently over 350 MB. The root element (named `go:go`), defines the two namespaces that are used by the database: RDF and the GO. RDF statements are always contained in an element named `rdf:RDF`. Within the `rdf:RDF` elements look like ordinary XML elements, except that they are organized in alternating layers, or *stripes*, as discussed in section 1.6. The first layer defines instances belonging to classes. In this case, the GO database defines two instances of type `go:term`. The second layer makes statements about these instances, such as the `go:accession` identifier and the `go:name`. The `rdf:about` attribute is special: it gives the resource identifier (URI) of the resource *about* which one is making statements. The `rdf:resource` attribute is also special: it *refers* to another resource. Such a reference is analogous to a web link used for navigating from one page to another page on the web. If there were a third layer, then it would define instances, and so on. For an example of deeper layers, see figure 1.14.

XML, especially when using XML Schema (XSD), is certainly capable of expressing annotations about URIs, so it is natural to wonder what RDF adds to XSD. Tim Berners-Lee wrote an article in 1998 attempting to answer this question (Berners-Lee 2000b). The essence of the article is that RDF semantics can be closer to the semantics of the domain being represented. As we discussed in section 2.2, there are many features of the semantics of a domain that are difficult to capture using DTDs or XML schemas. Another way of putting this is that XML documents will make distinctions (such as the order of child elements) that are semantically irrelevant to the information being represented. Recall that semantics is concerned with the abstraction that emerges from syntactic variation. Tim Berners-Lee claims that RDF is much better at abstracting semantics from syntax than ordinary XML.

RDF differs from ordinary XML in several important ways:

1. *Explicit relationships.* In XML there is only one relationship between elements: the unlabeled parent-child relationship (i.e., the relationship between an element and the elements it directly contains). In RDF a resource can participate in many different relationships. RDF relationships are expressed by adding intermediate layers (stripes) or by using attributes that refer to another resource.

2. *Order does not matter.* The order of the child elements is semantically significant in an XML document. In RDF, on the other hand, one can choose whether order matters or not. By default, the order in which RDF statements are asserted does not affect the meaning.

3. *Many-to-many relationships.* An XML element can have many child elements, but each child element can only be contained in one parent element. In RDF a resource can be related to any number of other resources, in either direction. In other words, all RDF relationships are many-to-many, whereas XML relationships can only be one-to-many.

4. *Syntactic flexibility.* RDF allows a larger variety of ways to express facts than XML. For example, a relationship can be specified using either an XML attribute or a child element. As another example, one can specify facts about a resource in several places in the same document or even in documents anywhere on the web. One can mention a URI in an XML document, but this is just another piece of data, and it is has no semantic significance for XML.

5. *Inference.* RDF has a number of built-in rules that define important notions such as inheritance. These built-in rules are part of the semantics of RDF.

6. *Uniform notation.* An important distinction between XML DTDs and XSD is that DTDs use a different language than the one used for XML documents. XSD, on the other hand, uses XML for both the schemas and the documents. RDF also uses the same language for both ontologies and data, but it goes even further than XSD in erasing all distinctions between ontologies and data. One can freely intermix RDF ontological and data statements. By contrast, in XSD the schemas are distinct from any XML data conforming to the schemas.

7. *Open vs. closed worlds.* When evaluating queries it is important to know whether the database is complete or "closed." Whether based on the relational model or on XML, databases differ markedly from RDF in this regard: RDF is open while databases are closed.

RDF refers to relationships as *properties*, which includes both attributes and the relationships specified using striping. When a property has a simple data value, one can use either an attribute or a child element to express it. Both of these have the same meaning in RDF, while they have very different meanings in XML. When one is designing a DTD or schema, it is necessary to make a choice about how relationships will be expressed. The question of whether one should use an attribute or a child element is one of the most common decisions one must make. Some DTDs, such as the one for Medline, primarily use child elements. Other DTDs, such as CML, prefer to use attributes. RDF eliminates the need for this choice.

The mathematical model that defines the semantics of RDF is a graph (W3C 2004a), consisting of nodes and links, much like the one used by XML infosets. However, RDF graphs can be arbitrary graphs, while XML infosets are strictly hierarchical, starting from the root. In addition, as mentioned before, all RDF links are labeled, while XML infosets only label the attribute links. An RDF node is a resource, and an RDF link is labeled by a property. Resources are classified using RDF classes, and every resource is an instance of at least one RDF class. Classes can be related to one another by the RDF property named `rdfs:subClassOf`. When an RDF class is a subclass of another, then the instances of the subclass are automatically also instances of the other class. The `subClassOf` relationship defines the class hierarchy of an RDF ontology. Properties can be related to one another by `rdfs:subPropertyOf`, and it has a similar meaning to `subClassOf`, but it is much less commonly used.

Classes and properties are the fundamental organizational entities of RDF. Consider the first part of the example in figure 1.14:

```
<locus name="HUMINS locus">
  <contains>
    <gene name="Insulin gene">
      <isStoredIn>
        <db_entry name="Genbank sequence" entry="v00565"
                  format="GENBANK"/>
        <db_entry name="EMBL sequence" format="EMBL"
                  entry="V00565"/>
```

```
      </isStoredIn>
    </gene>
  </contains>
</locus>
```

The corresponding RDF graph is shown in figure 4.2. The element names alternate between names of classes and names of properties, depending on the "striping" level. Thus `locus` is a class, `contains` is a property, `gene` is a class, and so on. Attributes are always names of properties. The nodes with no label (i.e., the empty ovals in the graph) are called *blank* or *anonymous* resources. They are important for conveying meaning, but they do not have explicit URIs. RDF processors generate URIs for blank nodes, but these generated URIs have no significance. The use of blank nodes in RDF complicates query processing, compared with XML. However, high-performance graph-matching systems have been developed that are efficient and scalable. This will be discussed in section 6.6.

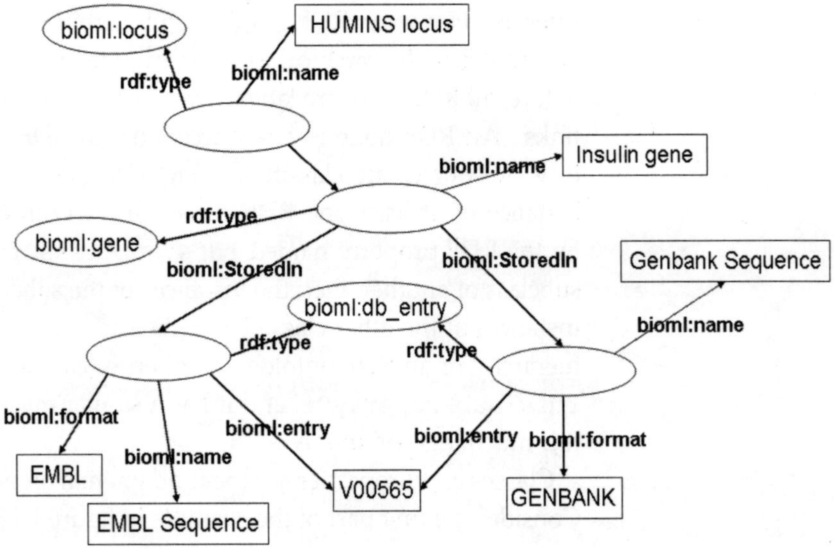

Figure 4.2 RDF graph for an XML document. Resources are represented using ovals, and rectangles contain data values.

Every link in an RDF graph has three components: the two resources being linked and the property that links them. Properties are themselves resources, so a link consists of three resources. The two resources being linked are called the *subject* and *object*, while the property that does the linking is called the *predicate*. Together the three resources form a *statement*, analogous to a statement in natural language. It is a good practice to use verbs for the names of predicates so that each RDF statement looks just like a sentence, and means essentially the same. RDF statements are also called *triples*. Some of the triples of the RDF graph in figure 4.2 include the following:

```
_:1 rdf:type bioml:locus
_:1 bioml:name "HUMINS locus"
_:1 bioml:contains _:2
_:2 rdf:type bioml:gene
_:2 bioml:name "Insulin gene"
_:2 bioml:isStoredIn _:3
...
```

The underscore means that the resource is a blank node so it does not have a URI. The other resources are part of either the BioML ontology or are part of the RDF language. When expressed in English the triples above might look like the following:

```
Anonymous node #1 is of type locus.
Anonymous node #1 has name "HUMINS locus".
Anonymous node #1 contains anonymous node #2.
Anonymous node #2 is of type gene.
Anonymous node #2 has name "Insulin gene".
Anonymous node #2 is stored in anonymous node #3.
...
```

Simple data values, such as the text string "HUMINS locus" are formally defined by XSD datatypes as in section 2.4.

Unlike the conversion from DTDs to XSD, it is not possible to automate the conversion from DTDs to RDF. The problem is that relationships are not explicitly represented in either DTDs and XSD. In the Medline DTD shown in figure 2.2, some of the elements correspond to RDF classes while others correspond to RDF properties. A person who is familiar with the terminology can usually recognize the distinction, but because the necessary information is not available in the DTD or schema, the conversion cannot be automated. The `MedlineCitation` element, for example, probably corresponds to an RDF class, and each particular Medline citation is an instances of this RDF class. After a little thought, it seems likely that all of the other elements in the Medline DTD correspond to RDF properties. However, these choices are speculative, and one could certainly make other choices, all of which would result in a consistent conversion to RDF. Converting from a DTD to RDF is further complicated by implicit classes. When converting the Medline DTD to RDF, it is necessary to introduce an RDF class for the date, yet there is no such element in the Medline DTD. In general, XML element types can correspond to either RDF classes or RDF properties, and both RDF classes and RDF properties can be implicit. In other words, XML DTDs and schemas are missing important information about the concepts being represented.

One specifies an RDF ontology using RDF itself. The fact that a resource is an RDF class, for example, is stated using an ordinary RDF. For example, one possibility for the classes and properties of the RDF ontology corresponding to the Medline DTD is shown in figure 4.3. There are two namespaces used by RDF. The first one is RDF, and the second is RDF Schema (RDFS). The RDF namespace is sufficient for specifying ordinary facts, while RDFS is necessary for specifying an RDF ontology.

```
<rdf:RDF
  xmlns:rdf="http://www.w3.org/1999/02/22-rdf-syntax-ns#"
  xmlns:rdfs="http://www.w3.org/2000/01/rdf-schema#">
 <rdfs:Class rdf:ID="MedlineCitation"/>
 <rdf:Property rdf:ID="Owner"/>
 <rdf:Property rdf:ID="Status"/>
 <rdf:Property rdf:ID="MedlineID"/>
 <rdf:Property rdf:ID="PMID"/>
 <rdf:Property rdf:ID="DateCreated"/>
 <rdfs:Class rdf:ID="Date"/>
 <rdf:Property rdf:ID="Year"/>
 <rdf:Property rdf:ID="Month"/>
 <rdf:Property rdf:ID="Day"/>
 <rdf:Property rdf:ID="ArticleTitle"/>
</rdf:RDF>
```

Figure 4.3 One possible way to represent the Medline DTD of figure 2.2 using an RDF ontology.

The Medline citation in figure 2.1 is already almost in a form that is compatible with RDF. All that is needed is to add a `Date` element as shown in figure 4.4. However, RDF gives one the freedom to represent the same information in many other ways. The document shown in figure 4.5 is equivalent. Both representations have the same RDF graph, shown in figure 4.6.

```
<MedlineCitation Owner="NLM" Status="Completed">
  <MedlineID>99405456</MedlineID>
  <PMID>10476541</PMID>
  <DateCreated>
    <Date>
      <Year>1999</Year>
      <Month>10</Month>
      <Day>21</Day>
    </Date>
  </DateCreated>
  <ArticleTitle>Breast cancer highlights.</ArticleTitle>
</MedlineCitation>
```

Figure 4.4 Part of a Medline citation written using RDF.

```
<MedlineCitation Owner="NLM" Status="Completed"
    MedlineID="99405456" PMID="10476541"
    ArticleTitle="Breast cancer highlights">
  <DateCreated>
    <Date Year="1999" Month="10" Day="21"/>
  </DateCreated>
</MedlineCitation>
```

Figure 4.5 Part of a Medline citation written using RDF. Although it looks different, the information is the same as that in figure 4.4.

In an XML DTD or schema one can restrict the content of an element. The analogous kinds of restriction in RDF are called domain and range constraints. A *domain constraint* of a property restricts the kinds of resource that can be the subjects of statements using that property. A *range constraint* restricts the kinds of resource that can be objects of that property. For example, the `DateCreated` property is only allowed to link a `MedlineCitation` to a `Date`. An RDF ontology would state this as follows:

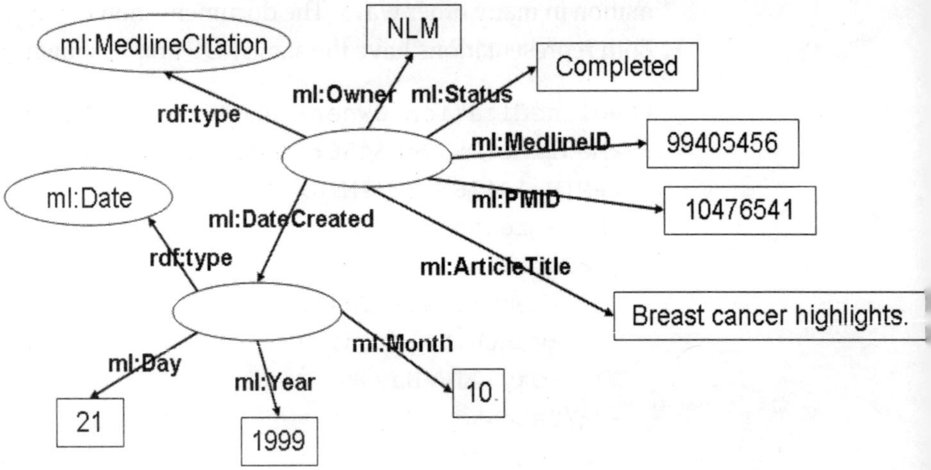

Figure 4.6 RDF graph for a typical Medline citation. Resource nodes are shown using ovals, text nodes are shown using rectangles. All links are labeled with the property. The `ml:` prefix stands for the Medline ontology.

```
<rdf:Property rdf:ID="DateCreated">
  <rdfs:domain rdf:resource="#MedlineCitation"/>
  <rdfs:range rdf:resource="#Date"/>
</rdf:Property>
```

The `rdf:ID` attribute is used for *defining* a resource. Each resource is defined exactly once. At this point one can also annotate the resource with additional property values, as was done above. The `rdf:about` attribute is used for *annotating* a resource. Use this when one is adding property values to a resource that has been defined elsewhere. The `rdf:resource` attribute is used for *referring* to a resource. In terms of statements, use `rdf:ID` and `rdf:about` for a resource that is to be the subject of the statement, and use `rdf:resource` when the resource is to be the object of the statement. We leave it as an exercise to the reader to restate the molecule DTD as an RDF ontology and to write the nitrous oxide molecule document in RDF.

The last important feature that distinguishes RDF from XML is its incorporation of built-in inference rules. The most important built-in rule is the *subClass rule* because this is the rule that implements inheritance and taxonomic classification of concepts. Although there are many notions of hierarchy, as discussed in section 1.5, the most commonly used is the notion of taxonomy which is based on the mathematical notion of set containment.

The fundamental relationship between classes in a taxonomy is the *subclass* relationship. For example, in the chemical hierarchy in figure 1.11, Macromolecule is a subclass of Chemical, and Protein is a subclass of Macromolecule.

In RDF, the subclass relationship is called `rdfs:subClassOf`. To specify that Protein is a subclass of Macromolecule, which is a subclass of Chemical, one would use these RDF statements:

```
<Class rdf:about="#Protein">
  <subClassOf rdf:resource="#Macromolecule"/>
</Class>
<Class rdf:about="#Macromolecule">
  <subClassOf rdf:resource="#Chemical"/>
</Class>
```

Now suppose that one is describing a particular protein using RDF:

```
<Protein rdf:ID="rhodopsin"/>
```

RDF will automatically infer that rhodopsin is also a macromolecule and a chemical. In other words, the fact that rhodopsin is a macromolecule is inherited from the fact that it is a protein. This can be important for information retrieval and information transformation. Without this inference, a query for a chemical would not recognize rhodopsin as being a chemical.

As discussed in chapter 3, a rule has an antecedent and a consequent. The antecedents for RDF rules consist of one or more RDF statements. The conclusion consists of exactly one RDF statement. If the antecedent statements have been previously asserted (i.e., either explicitly stated or previously inferred), then the consequent statement is inferred. The most important inference rules that are built into RDF are the following:

1. **Subclass rule.** If class A is a `subClassOf` class B, and if the resource R has `rdf:type` A, then the resource R also has `rdf:type` B. In other words, class A is a subset of class B.

2. **Subproperty rule.** If property P is a `subPropertyOf` property Q, and if the resource R is linked to resource S using property P, then the resource R is also linked to resource S using property Q. In other words, the links (statements) using property P are a subset of the links using Q.

3. **Domain rule.** If property P has `domain` D, and if the resource R is linked to resource S using property P, then the resource R has `rdf:type` D.

4. **Range rule.** If property *P* has `range` *C*, and if the resource *R* is linked to resource *S* using property *P*, then the resource *R* has `rdf:type` *C*.

The full list of all RDF inference rules is in (Hayes 2004).

The meaning of an RDF graph is intimately connected with the RDF rules. Any query or other use of an RDF graph always presumes that all applicable inference rules have been applied. This is very different from XML where "what you see is what you get," that is, the meaning of XML is entirely determined by what is explicitly stated. Rules can be a powerful mechanism for expressing meaning, but this power comes at a price. The subClass rule, for example, is very powerful and useful, but because it does not allow for the possibility of any exceptions, it is not applicable to many taxonomies in the real world. As another example, suppose one makes a mistake and states that a Medline citation is owned by rhodopsin. This does not make sense: rhodopsin is a protein not an institution. An XML processor would immediately give an error message, but an RDF processor would not. The RDF processor would just infer that rhodopsin is both a protein and an institution!

One flaw of XML is that the order of child elements is semantically significant whether one wants it to be significant or not. RDF allows one to choose whether the order should matter or not, and it provides two mechanisms for this. The older mechanism is the notion of a *container*. For example, the Krebs cycle is an ordered list of eight enzymes.[1] Using ordinary XML, order matters, so one could define the Krebs cycle enzymes with the following:

```
<Pathway name="Krebs Cycle">
  <Protein name="Citrate synthase"/>
  <Protein name="Aconitase"/>
  <Protein name="Isocitrate dehydrogenase"/>
  <Protein name="a-Ketoglutarate dehydrogenase complex"/>
  <Protein name="Succinyl-CoA synthetase"/>
  <Protein name="Succinate dehydrogenase"/>
  <Protein name="Fumerase"/>
  <Protein name="Malate dehydrogenase"/>
</Pathway>
```

In RDF the relationship must be explicitly specified so it would look something like this:

1. It is actually a cycle. The order is significant, but one can start with any one of the enzymes.

```
<Pathway name="Krebs Cycle">
 <usesEnzyme>
  <Protein name="Citrate synthase"/>
  <Protein name="Aconitase"/>
  <Protein name="Isocitrate dehydrogenase"/>
  <Protein name="a-Ketoglutarate dehydrogenase complex"/>
  <Protein name="Succinyl-CoA synthetase"/>
  <Protein name="Succinate dehydrogenase"/>
  <Protein name="Fumerase"/>
  <Protein name="Malate dehydrogenase"/>
 </usesEnzyme>
</Pathway>
```

However, now the order is lost. The fact that the statements are in the right order does not matter. An RDF processor will not maintain this order, and one cannot make any use of it. Fortunately, there are two mechanisms for retaining the ordering. The older method is to place the enzymes in a *sequence container* as follows:

```
<Pathway name="Krebs Cycle">
  <usesEnzyme>
    <rdf:Seq>
      <rdf:li>
        <Protein name="Citrate synthase"/>
      </rdf:li>
      <rdf:li>
        <Protein name="Aconitase"/>
      </rdf:li>
      <rdf:li>
        <Protein name="Isocitrate dehydrogenase"/>
      </rdf:li>
      <rdf:li>
        <Protein name="a-Ketoglutarate dehydrogenase complex"/>
      </rdf:li>
      <rdf:li>
        <Protein name="Succinyl-CoA synthetase"/>
      </rdf:li>
      <rdf:li>
        <Protein name="Succinate dehydrogenase"/>
      </rdf:li>
      <rdf:li>
```

```
      <Protein name="Fumerase"/>
    </rdf:li>
    <rdf:li>
      <Protein name="Malate dehydrogenase"/>
    </rdf:li>
  </rdf:Seq>
 </usesEnzyme>
</Pathway>
```

The sequence is itself a resource as well as being the container of the other resources. Notice the use of the `rdf:li` property for the members of the container. This name was borrowed from HTML where it is used for the members of lists. There are three kinds of container:

1. **rdf:Seq.** A sequence contains an ordered list of resources. In HTML one uses the `ol` tag for such a list.

2. **rdf:Bag.** A bag is an unordered container of resources. A bag can have no resources at all, and it can contain the same resource more than once. In HTML one uses the `ul` tag for such a container.

3. **rdf:Alt.** An Alt container is intended to represent a set of alternatives. The first resource is the default or preferred alternative, and there is no preference among the others. So an Alt container is a set with one distinguished member. This corresponds to a "drop-down menu" in HTML, and is specified by using the `select` tag.

More recently, a second mechanism for ordered lists was added to RDF, called a `collection`. The Krebs cycle can now be expressed as follows:

```
<Pathway name="Krebs Cycle">
 <usesEnzyme rdf:parseType="Collection">
  <Protein name="Citrate synthase"/>
  <Protein name="Aconitase"/>
  <Protein name="Isocitrate dehydrogenase"/>
  <Protein name="a-Ketoglutarate dehydrogenase complex"/>
  <Protein name="Succinyl-CoA synthetase"/>
  <Protein name="Succinate dehydrogenase"/>
  <Protein name="Fumerase"/>
  <Protein name="Malate dehydrogenase"/>
 </usesEnzyme>
</Pathway>
```

To get this kind of list, one only needs to say that the `rdf:ParseType` if the property value is `Collection`. This is much simpler than using a container.

Summary

- RDF is a framework for representing explicit many-to-many relationships (called properties) between web-based resources and data.
- The semantics of RDF is defined by RDF graphs.
- RDF has built-in inference rules for subclasses, subproperties, domains and ranges.
- Inference is a powerful feature, but one must be careful when using it.
- Conversion from XML DTDs or schemas to RDF cannot be automated.
- Ordered structures can be defined using RDF containers and collections.

4.3 XML Topic Maps

The XML Topic Maps (XTM) language is another XML-based ontology language that has a very different history and semantics than any of the other XML-based ontology languages (XTM 2000). XTM provides a model and grammar for representing the structure of information resources used to define topics, and the associations (relationships) between topics. Names, resources, and relationships are characteristics of abstract subjects, which are called *topics*. Topics have their characteristics within scopes, that is, the limited contexts within which the characteristics of topics apply.

Consider the Medline citation shown in figure 4.5. When written in XTM it would look something like this:

```
<topic id="PMID10476541">
  <instanceOf><topicRef xlink:href="#MedlineCitation"/></instanceOf>
  <baseName>
    <baseNameString>Breast cancer highlights</baseNameString>
  </baseName>
  <occurrence>
    <instanceOf><topicRef xlink:href="#html-format"/></instanceOf>
    <resourceRef
      xlink:href="http://www.ncbi.nlm.nih.gov/entrez/query.fcgi?..."/>
  </occurrence>
</topic>
```

```
<association>
  <instanceOf>
    <topicRef xlink:href="#citation-attributes"/>
  </instanceOf>
  <member>
    <roleSpec><topicRef xlink:href="#owner"/></roleSpec>
    <topicRef xlink:href="#NLM"/>
  </member>
  <member>
    <roleSpec><topicRef xlink:href="#status"/></roleSpec>
    <topicRef xlink:href="#completed"/>
  </member>
  <member>
    <roleSpec><topicRef xlink:href="#date-created"/></roleSpec>
    <topicRef xlink:href="#date991021"/>
  </member>
</association>

<topic id="date991021">
  <baseName>
    <scope>
      <topicRef
        xlink:href="http://kmi.open.ac.uk/psi/datatypes.xtm#date"/>
    </scope>
    <baseNameString>1999-10-21</baseNameString>
  </baseName>
</topic>
```

Except for some syntactic details such as striping and built-in attributes in the RDF namespace, RDF documents can be very similar to general XML documents. As the example above illustrates, XTM documents have no such advantage.

XTM is a graph-based language that has much in common with RDF. Both of them are intended to be a mechanism for annotating web resources. The web resources that are being annotated occur within documents which furnish the "primary structure" defining the resources. The annotations are a "secondary structure" known as *metadata* or "data about data."

Although XTM and RDF have many similarities, they also differ in some important respects:

- XTM relationships (called *associations*) can have any number of roles. By contrast, the RDF languages only support binary relationships.

- XTM has a notion of scope or context that the RDF languages lack.

- The RDF languages have a formal semantics. XTM only has a formal metamodel.
- XTM makes a clear distinction between metadata and data, while RDF does not. In RDF one can annotate anything, including annotations.

4.4 The Web Ontology Language

We began this chapter with a story about the experience people sometimes have of suddenly realizing that two of their acquaintances are actually the same person. Although RDF introduces some important prerequisites for reasoning about whether two entities are the same or different, it does not have the ability to deal with this issue. OWL differs from RDF in a number of ways, but one of the most important is how it deals with identity. One can explicitly state that two resources are the same or that they are different. It is also possible to infer one or the other of these two cases. However, it is also possible that one will not be able to infer either one. This has a profound impact on the semantics of OWL. Unlike XML or RDF, one must now consider many "possible worlds," differing from one another with respect to whether resources are the same or different, yet each of the possible worlds is compatible with the known facts.

Logicians refer to the process of equating entities as *paramodulation*. The OWL paramodulation properties are `sameAs` and `differentFrom`. In addition to paramodulation, OWL extends the RDF vocabulary to include relations between classes (such as disjointness), cardinality constraints on properties (e.g., "exactly one"), characteristics of properties (e.g., symmetry), and enumerated classes. All of these features of OWL are intimately connected to the issue of identity. For example, suppose that one states in the molecule ontology that every `atom` has exactly one `elementType`. If a particular atom does not have an elementType, then one can infer that this atom has an elementType which is the `sameAs` one of the known element types, but one does not know which one. No inference of this kind is possible in RDF.

The OWL language specification is given in (van Harmelen et al. 2003). There are three distinct OWL languages: OWL Lite, OWL-DL, and OWL Full. They differ from each other primarily with respect to what constructs are allowed in each language. OWL Lite is the most restrictive. OWL-DL is less restrictive than OWL Lite but more restrictive than OWL Full. OWL Full is unrestricted. Syntactically, OWL is nearly identical to RDF. Like RDF, one can intermix ontological and data statements. The only syntactic differences are the additional properties introduced by OWL and the restrictions imposed by OWL Lite and OWL-DL.

Consider the following characterization of a disease syndrome: "The irido-corneal endothelial syndrome (ICE) is characterized by corneal endothelium proliferation and migration, iris atrophy, corneal oedema and/or pigmentary iris nevi" (Jurowski et al. 2004). In this statement there are four symptoms, and the ICE syndrome is stated to be characterized by exhibiting one or more of these symptoms. In OWL one can specify an enumeration as follows:

```
<owl:Class rdf:ID="ICE-Symptoms">
  <owl:oneOf parseType="Collection">
    <Symptom name="corneal endothelium
      proliferation and migration"/>
    <Symptom name="iris atrophy"/>
    <Symptom name="corneal oedema"/>
    <Symptom name="pigmentary iris nevi"/>
  </owl:oneOf>
</owl:Class>
```

This defines a class of symptoms consisting of exactly the ones specified. One can then define the ICE syndrome as the subclass of disease for which at least one of these four symptoms occurs:

```
<owl:Class rdf:ID="ICE-Syndrome">
  <owl:intersectionOf parseType="Collection">
    <owl:Class rdf:about="#Disease"/>
    <owl:Restriction>
      <owl:onProperty rdf:resource="#has-symptom"/>
      <owl:someValuesFrom
        rdf:resource="#ICE-Symptoms"/>
    </owl:Restriction>
  </owl:intersectionOf>
</owl:Class>
```

The statements above specify the ICE-Syndrome class as being the intersection of two sets:

1. The set of all diseases

2. The set of things that have at least one of the four ICE symptoms

An OWL restriction is a way of specifying a set of things that satisfy some criterion. It is called a "restriction" because it restricts the set of all things to those that satisfy the criterion. Mathematically, it corresponds to the "set constructor" whereby a set is defined by a condition on its elements. For example, $\{x|x>0\}$ defines the set of positive numbers. Classes are constructed from classes from by using `owl:intersectionOf` (corresponding to the Boolean AND operator), `owl:unionOf` (Boolean OR operator), and `owl:complementOf` (Boolean NOT operator).

OWL has six set constructors. All of them use `owl:Restriction` and `owl:onProperty` as in the ICE-Syndrome example above, together with one of the following:

1. **owl:someValuesFrom.** This is the constructor that was used in the ICE-Syndrome example. It defines the set of resources for which the property has at least one value in the class.

2. **owl:allValuesFrom.** This constructor defines the set of resources for which the property only takes values in the class. In other words, the property does not take values of any other kind.

3. **owl:hasValue.** This constructor defines the set of resources for which the property takes the specified value.

4. **owl:maxCardinality.** In general, a property can have any number of values. This constructor defines the set of resources for which the property is limited to have the specified maximum number of values.

5. **owl:minCardinality.** This is the reverse of `maxCardinality`. It defines the set of resources for which the property is at least the specified minimum number of values.

6. **owl:cardinality.** This constructor defines the set of resources for which the property has exactly the specified number of values.

While classes can be constructed in a large variety of ways, OWL has only one property constructor: `owl:inverseOf`. The inverse of a property is the property for which the roles of subject and object have been reversed. For example, in figure 1.14 there are a number of relationships such as `isStoredIn` and `isCitedBy`. It may be useful to look at these relationships from the other point of view. Here is how one would define the inverse properties using OWL:

```
<owl:ObjectProperty rdf:ID="stores">
  <owl:inverseOf rdf:resource="#isStoredIn"/>
</owl:ObjectProperty>
<owl:ObjectProperty rdf:ID="cites">
  <owl:inverseOf rdf:resource="#isCitedBy"/>
</owl:ObjectProperty>
```

There are no other ways to construct a property in OWL. However, there are some property constraints:

1. **owl:FunctionalProperty.** Such a property may relate a subject to at most one object. If a particular subject is related to two resources, then those two resources must be the same. Mathematically, such a property is a partial function.

2. **owl:InverseFunctionalProperty.** Such a property is allowed to relate an object to at most one subject. This is the same as constraining the inverse property to have the `owl:FunctionalProperty`.

3. **owl:SymmetricProperty.** A symmetric property is the same as its inverse property.

4. **owl:TransitiveProperty.** This imposes the mathematical transitivity condition on the property.

The semantics of OWL is defined in (Patel-Schneider et al. 2004). The semantics is in terms of *interpretations*. The more commonly used term for an interpretation is a *model*, and that is the term that will usually be used in this chapter. The detailed definition of an interpretation is complicated, but it is essentially the same as an RDF graph. An OWL document specifies a collection of statements. One can also specify that it *owl:imports* other OWL documents, and the statements in the imported documents are also regarded as having been stated. The collection of all the statements forms a *theory* about the world. Such a theory is consistent with an infinite collection of models. These models may be regarded as *possible worlds*. As additional facts become known, the collection of possible worlds gets smaller, as the new facts eliminate possibilities. This process is analogous to scientific reasoning, except that in science one attaches probabilities to possible worlds, and observations modify these probabilities (using Bayes' law). This is covered in chapter 14. Logical theories, by contrast, assign no weights to the possible worlds.

Inference for a theory is conceptually simple. If a statement is true in every possible world, then it is a fact. If a statement is true in some worlds, but not in others, then one cannot say that it is either true or false: it is not a fact, at least not yet. A statement that is true in every possible world compatible with a theory is said to be *entailed* by the theory. Although the notion of entailment is simple and intuitive, it is not very practical. After all, there will always be infinitely many possible worlds so it is not possible to examine every one. In practice, some other technique is necessary to determine what statements are entailed.

The most commonly used technique for determining entailment is to use rules. This is similar to how inference is done in RDF, but there is an important difference. Applying RDF rules in any particular RDF document will always eventually terminate. If one does the same for OWL, then it will not terminate because OWL models are always infinite. In other words, one can infer infinitely many facts. To perform inference with OWL one must focus on a particular question that one would like to answer. For example, one might ask for all known diseases that are characterized by the ICE syndrome (i.e., all known instances of the ICE-Syndrome class). A more subtle question would be whether two diseases are the same. By focusing on a particular question, the rule engine can restrict attention to facts and rules that are relevant. The resulting inference process can be very efficient in practice.

Relational databases and XML differ from the Semantic Web with respect to how each interprets the meaning of the known facts about the world. We have already seen some examples of this distinction in the previous section where someone made a mistake and stated that a Medline citation is owned by "rhodopsin." The "owned by" relationship links a citation to the institution that owns it. A relational database or an XML processor would give an error message in this case because "rhodopsin" is not an institution. An RDF processor, on the other hand, would infer that "rhodopsin" is an institution. Databases and XML are said to be assuming a *closed world*. The Semantic Web, on the other hand, is assuming an *open world*.

Logicians refer to this distinction as *monotonicity*. A logical system that assumes an *open world* is monotonic, while a logical system that assumes a *closed world* is nonmonotonic. To understand why one refers to a monotonic system as open, consider another example. Suppose that the "occursIn" property gives the species of an entity such as a gene or chromosome. It is reasonable to require that every gene occurs in at least one species. In OWL this would be written as follows:

```
<owl:Class rdf:ID="Gene">
  <rdfs:subClassOf>
    <owl:Restriction>
      <owl:onProperty rdf:resource="#occursIn"/>
      <owl:someValuesFrom rdf:resource="#Species"/>
    </owl:Restriction>
  </rdfs:subClassOf>
</owl:Class>
```

Now suppose that we state that hemoglobin alpha embryonic-3 (*hbae3*) is a gene but neglect to mention any species that this gene occurs in:

```
<Gene rdf:ID="hbae3">
  <rdfs:label>hemoglobin alpha embryonic-3</rdfs:label>
</Gene>
```

If we are assuming a closed world, then this annotation is inconsistent: one is required to specify at least one species in which each gene occurs. In an open world, on the other hand, one can infer that this gene occurs in some species, we just do not know which ones they are. In an open world, one accepts that one does not know everything, and that some facts have yet to be stated. In a closed world, the world is assumed to be "complete": if something has not been stated then it is not true.

The closed world assumption has been used successfully for many years by database management systems, and there are good reasons for making this assumption. Databases arose in the context of commercial business applications. For example, they are used for storing employee information. If a person does not have a record in the employee database, then the person is not an employee. This may seem unduly harsh, but it makes perfectly good sense for modern businesses where all significant records are stored in databases. Modern storage systems can be made highly reliable and fault tolerant, so much so that they are now more reliable than paper documents. In this case, the database is not simply recording the state of reality, it *is* reality. Being an employee is *defined* by records stored in the computer system.

By contrast, the closed world assumption is not appropriate for the web where information is necessarily incomplete and fragmentary. The web consists of a very large number of independent websites for which there is no central authority. Furthermore, websites can be turned on and off at the whim of the owners.

Reasoning in an open world is sometimes counterintuitive. As an example of this, suppose that we make the additional requirement that every gene belong to exactly one species. This can be specified by adding the following to the specification above:

```
<FunctionalProperty rdf:about="#occursIn"/>
```

If one has not specified that the *hbae3* gene occurs in any species, then one would infer that there is exactly one, as yet unknown, species where this gene occurs. This is shown in figure 4.7.

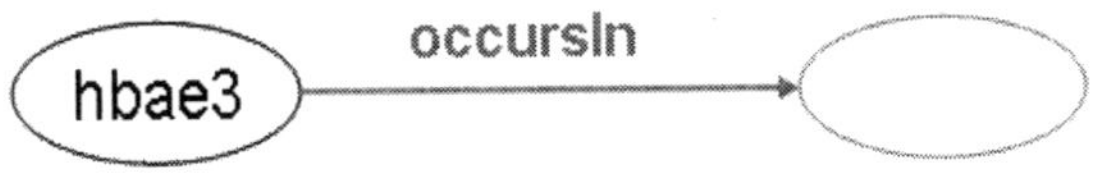

Figure 4.7 An example of an unspecified but mandatory relationship. The blank node represents the anonymous species that is necessary to fulfill the requirement. The inferred resource and relationship are shown in gray.

Now suppose that one specifies that the *hbae3* gene occurs in two species:

```
<Gene rdf:about="hbae3">
  <occursIn rdf:resource="#D.rerio"/>
  <occursIn rdf:resource="#D.danglia"/>
</Gene>
```

In a closed world, this would be inconsistent because the ontology allows a gene to belong to only one species. In an open world, one would infer that *D. rerio* is the same species as *D. danglia*. This is shown graphically in figure 4.8.

The inference that *D. rerio* and *D. danglia* are the same species is clearly incorrect. There are many ways to remedy this erroneous conclusion. One could explicitly specify that the two species are different as follows:

```
<Gene rdf:about="#D.rerio">
  <owl:differentFrom rdf:about="#D.danglia"/>
</Gene>
```

This would cause an OWL processor to signal an inconsistency exactly as in a closed world system. One could also drop the requirement that genes occur in only one species, which is more realistic. Doing this would eliminate the spurious inference in this case, but it would not prevent the inference from

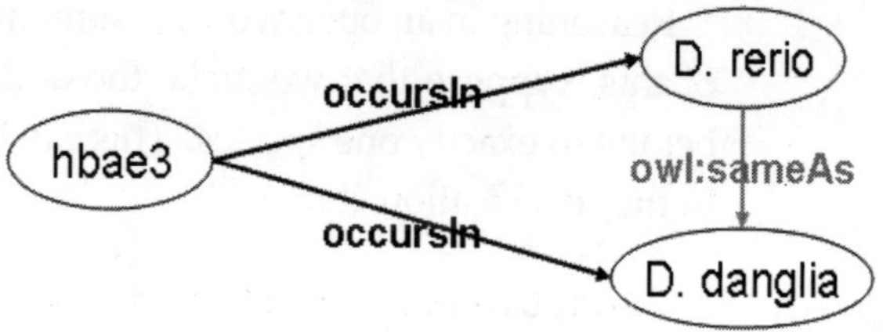

Figure 4.8 An example in which two resources are inferred to be the same. In this case the ontology allows a gene to belong to at most one species. As a result if a gene is linked to more than one species, then all of them must be the same. The inferred relationship is shown in gray.

occurring as a result of other facts and rules. In general, one can reduce spurious inferences in two ways:

1. Never overspecify. Declaring that `occursIn` is a functional property is too strong. It may be true nearly all of the time, but even one exception can result in spurious inferences.

2. Specify distinctions. Explicitly declare that resources are different when they are known to be distinct. However, only specify this if one is really sure that the resources are different. It does sometimes happen that entities such as diseases and proteins that were once thought to be different subsequently turn out to be the same and the consequences can be profound.

Specifying that resources are different can get very tedious if there is a large number of them. To deal with this problem, OWL has a mechanism for specifying that a list of resources are all different from one another. For example, the two species above could have been declared to be different by using the `owl:AllDifferent` resource and `owl:distinctMembers` property as follows:

```
<AllDifferent>
  <distinctMembers parseType="Collection">
    <Gene rdf:about="#D.rerio">
    <Gene rdf:about="#D.danglia">
  </distinctMembers>
</AllDifferent>
```

Summary

- OWL is based on RDF and has three increasingly more general levels: OWL Lite, OWL-DL, and OWL Full.
- An OWL document defines a theory of the world. States of the world that are consistent with the theory are called models of the theory.
- A fact that is true in every model is said to be entailed by the theory. OWL inference is defined by entailment.
- OWL is especially well suited for defining concepts in terms of other concepts using class constructors.
- OWL has only one property constructor, but it has some property constraints.
- Resources can be explicitly stated to be the same or different. It is also possible to infer that two resources are the same or different.
- To perform inference in OWL one must focus on a particular question.
- OWL inference is monotonic, which can limit inferences, but careful design can reduce this problem.

4.5 Exercises

1. Restate the molecule schema in figure 1.6 as an RDF ontology. There will not be a single correct answer to this exercise.
2. Define the nitrous oxide molecule in figure 2.3 using RDF. The answer will depend on the RDF ontology.
3. Rewrite the bio sequence DTD in exercise 4.2 as an OWL ontology.
4. Rewrite the physical units DTD in exercise 4.3 as an OWL ontology.
5. Rewrite the Fitzhugh-Nagumo model and DTD in exercise 4.4 as an OWL ontology.

5 *Survey of Ontologies in Bioinformatics*

There are a large number of biomedical ontologies and databases that are currently available, and more continue to be developed. There is even a site that tracks the publicly available sources. Ontologies have emerged because of the need for a common language to develop effective human and computer communication across scattered, personal sources of data and knowledge. In this chapter, we provide a survey of ontologies and databases used in the bioinformatics community. In the first section we focus on human communication. The ontologies in this section are concerned with medical and biological terminology and with ontologies for organizing other ontologies.

The rest of the chapter shifts the focus to computer communication. In section 5.2 we survey the main XML-based ontologies for bioinformatics. The remaining sections consider some of the many databases that have been developed for biomedical purposes. Each database has its own structure and therefore can be regarded as defining an ontology. However, the focus is on the data contained in the database rather than on the language used for representing the data. These databases differ markedly from one another with respect to how the data are specified and whether they are compatible with the ontologies in the first two sections. Many of the databases are available in several formats. Only databases that can be downloaded were included in the survey.

5.1 Bio-Ontologies

We have seen that ontologies are a versatile mechanism for understanding concepts and relationships. In this section the concern is with the human communication of biomedical concepts as well as with understanding what knowledge is available. We first consider ontologies dealing with terminology. The first one was originally focused on medical terminology but now also includes many other biomedical vocabularies, has grown to be impressively large, but is sometimes incoherent as a result. The second ontology focuses exclusively on terminology for genomics. As a result of its narrow focus, it is very coherent, and a wide variety of tools have been developed that make use of it. Finally, we consider ontologies that organize other ontologies. The number of biomedical ontologies and databases has grown so large that it is necessary to have a framework for organizing them.

5.1.1 Unified Medical Language System

Terminology is the most common denominator of all biomedical literature resources, including the names of organisms, tissues, cell types, genes, proteins, diseases. There are various controlled vocabularies such as the Medical Subject Headings (MeSH) associated with these resources. MeSH was developed by the U.S. National Library of Medicine (NLM). However, having identified terminology as a key integrating factor for biomedical resources does not imply they use standard vocabularies which would make these resources interoperable. In 1986, NLM began a long-term research and development project to build the Unified Medical Language System (UMLS) located at `www.nlm.nih.gov/research/umls`. The UMLS is a repository of biomedical vocabularies and is the NLM's biological ontology (Lindberg et al. 1993; Baclawski et al. 2000; Yandell and Majoros 2002).

The purpose of the UMLS is to improve the ability of computer programs to "understand" the biomedical meaning in user inquiries and to use this understanding to retrieve and integrate relevant machine-readable information for users (Lindberg et al. 1993). The UMLS integrates over 4.5 million names for over 1 million concepts from more than 100 biomedical vocabularies, as well as more than 12 million relations among these concepts. Vocabularies integrated in the UMLS include the the taxonomy of the National Center for Biotechnology Information (NCBI), the Gene Ontology (GO), MeSH and the digital anatomist symbolic knowledge base. UMLS concepts are not only interrelated, but may also be linked to external resources such as GenBank (Bodenreider 2004).

The UMLS is composed of three main components: the Metathesaurus (META), the SPECIALIST lexicon and associated lexical programs, and the Semantic Network (SN) (Denny et al. 2003). We now discuss each of these components in more detail.

META is the main component of the UMLS. This component is a repository of interrelated biomedical concepts that provide metadata, relationships, and semantic information for each concept. However, META is more than a simple concordance of terms. Its developers strive to provide a concept-oriented organization in which synonymous terms from disparate source vocabularies map to the same concepts. The UMLS contains semantic information about terms from various sources, and each concept can be understood and located by its relationships to other concepts. This is a result of the organizing principle of semantic "locality" (Bodenreider et al. 1998). For example, interconcept relationships can be either inherited from the structure of the source vocabularies or generated specifically by the META editors. Relationships can be hierarchical or associative. Statistical relations between concepts from the MeSH vocabulary are also present, derived from the co-occurrence of MeSH indexing terms in Medline citations. Finally, each META concept is broadly categorized by means of semantic types in the SN component of the UMLS. META has been constructed through lexical matching techniques and human review (Tuttle et al. 1989) to minimize inconsistencies of parent-child relationships and to minimize redundancies of multiple META concepts.

The SN is a classification system for the concepts in the META component. As an ontology, the UMLS is an ontology with a class hierarchy containing over 1 million classes, represented by the concepts in META and the semantic types in SN. In this class hierarchy, the semantic types form the top of the hierarchy. The SN serves the additional function of defining part of the property hierarchy of the ontology. However, UMLS concepts can have many other attributes (such as International Classification of Diseases [ICD-9] codes) that implicitly define many other properties. The semantics of the UMLS has yet to be defined precisely, and it has not yet been completely specified using any of the ontology languages.

The SPECIALIST lexicon includes lexical information about a selected core group of biomedical terms, including their parts of speech, inflectional forms, common acronyms, and abbreviations.

In addition to data, the UMLS includes tools such as MetamorphoSys for customizing the META, `lvg` for generating lexical variants of concept names, and MetaMap for extracting UMLS concepts from text. The UMLS knowledge sources are updated quarterly (Bodenreider 2004). MetaMap `ii.nlm.nih.gov/MTI/mmi.shtml` is one of the foundations of NLM's Indexing Initiative System which is being applied to both semiautomatic and fully automatic indexing of the biomedical literature at the library (Aronson 2001). It has been used for mapping text to the UMLS META. For example, MetaMap can be applied to free texts like the title and abstract fields of Medline citations. MetaMap can also be used for constructing a list of ranking concepts by applying the MetaMap indexing ranking function to each UMLS META concept. The UMLS Knowledge Source Server (UMLSKS) umlsks.nlm.nih.gov is a web server that provides access to the knowledge sources and other related resources made available by developers using the UMLS.

The UMLS is a rich source of knowledge in the biomedical domain. The UMLS is used for research and development in a range of different applications, including natural language processing (Baclawski et al. 2000; McCray et al. 2001). UMLS browsers are discussed in section 6.3. Search engines based on the UMLS use it either as a source of keywords or as a means of generating knowledge representations. An example of the keyword approach is the Medical World Search at www.mwsearch.com which is a search engine for medical information in selected medical sites. An example of the knowledge representation approach is the Semantic Knowledge Indexing Platform (SKIP), shown in section 6.6.

5.1.2 The Gene Ontology

The most prominent ontology for bioinformatics is GO. GO is produced by the GO Consortium, which seeks to provide a structured, controlled vocabulary for the description of gene product function, process, and location (GO 2003, 2004). The GO Consortium was initially a collaboration among the Mouse Genome Database, FlyBase, and *Saccharomyces* Genome database efforts. It has since grown to 16 members. GO is now part of the UMLS, and the GO Consortium is a member of the Open Biological Ontologies consortium to be discussed in the next section.

A description of a gene product using the GO terminology is called an *annotation*. One important use of GO is the prediction of gene function based on patterns of annotation. For example, if annotations for two attributes tend to occur together in a database, then a gene holding one attribute is likely to hold for the other as well (King et al. 2003). In this way, functional predictions can be made by applying prior knowledge to infer the function of a novel entity (either a gene or a protein).

GO consists of three distinct ontologies, each of which serves as an *organizing principle* for describing gene products. The intention is that each gene product should be annotated by classifying it three times, once within each ontology (Fraser and Marcotte 2004). The three GO ontologies are:

1. **Molecular function.** The biochemical activity of a gene product. For example, a gene product could be a transcription factor or a DNA helicase. This classifies what kind of molecule the gene product is.

2. **Biological process.** The biological goal to which a gene product contributes. For example, mitosis or purine metabolism. Such a process is accomplished by an ordered assembly of molecular functions. This describes what a molecule does or is involved in doing.

3. **Cellular component.** The location in a cell in which the biological activity of the gene product is performed. Examples include the nucleus, a telomere, or an origin recognition complex. This is where the gene product is located.

The terms within each of the three GO ontologies may be related to other terms in two ways:

1. **is-a.** This is the subclass relationship used by classic hierarchies such as the taxonomy of living beings. For example, *condensed chromosome* `is-a` *chromosome*.

2. **part-of.** This is the containment relationship in which an entity is physically or conceptually contained within another entity. For example, *nucleolus* is `part-of` *nucleus*.

An example of the GO hierarchy for the term "inositol lipid-mediated signaling" is shown in figure 5.1. This shows the series of successively more restrictive concepts to which this concept belongs.

```
GO:0003673 : Gene_Ontology (80972)
 GO:0008150 : biological_process (56741)
  GO:0009987 : cellular process (20309)
   GO:0007154 : cell communication (6336)
    GO:0007165 : signal transduction (4990)
     GO:0007242 : intracellular signaling cascade (1394)
      GO:0019932 : second-messenger-mediated signaling (219)
       GO:0048015 : phosphoinositide-mediated signaling (3)
        GO:0048017 : inositol lipid-mediated signaling (0)
```

Figure 5.1 The GO hierarchy for inositol lipid-mediated signaling. The parentheses show the total number of terms in the category at that level.

Because GO supports two relationships, its ontologies are more expressive than a taxonomy. However, modern ontologies often support many more than just two relationships. An analysis of the names of GO terms suggests that there are many other relationships implicitly contained in the GO terminology. For example, 65.3% of all GO terms contain another GO term as a proper substring. This substring relation often coincides with a derivational relationship between the terms (Ogren et al. 2004). For example, the term *regulation of cell proliferation* (GO:0042127) is derived from the term *cell proliferation* (GO:0008283) by addition of the phrase *regulation of*. The phrase *regulation of* occurs frequently in GO, yet is not itself a GO term. Furthermore, this subterm occurs consistently in different subsets of the GO ontologies. Derivational subterms such as this one indicate interesting semantic relationships between the related terms. Formalizing these relationships would result in a richer representation of the concepts encoded in the ontology, and would assist in the analysis of natural language texts.

Many programs have been developed for profiling gene expression based on GO or the GO file format. These programs have been very useful for translating sets of differentially regulated genes.

DAG-Edit

`sourceforge.net/project/showfiles.php?group_id=36855`

DAG-Edit is an open source tool written in Java for browsing, searching, and modifying structured controlled vocabularies. DAG-Edit was previously called GO-Edit. It is applicable to any kind of structured controlled vocabulary. Three formats are supported: GO flat file, GO serial file, and OBO file.

GenMAPP `www.GenMAPP.org`

This tool visualizes gene expression and other genomic data on maps representing biological pathways and groupings of genes. Integrated with GenMAPP are programs to perform a global analysis of gene expression or genomic data in the context of hundreds of pathway MAPPs and thousands of GO Terms (MAPPFinder), import lists of genes/proteins to build new MAPPs (MAPPBuilder), and export archives of MAPPs as well as expression and genomic data to the web.

GoMiner `discover.nci.nih.gov/gominer`

This program package organizes lists of "interesting" genes (e.g., under- and overexpressed genes from a microarray experiment) for biological interpretation (Zeeberg et al. 2003). GoMiner provides quantitative and statistical output files and two useful visualizations. The first is a treelike structure and the second is a compact, dynamically interactive directed acyclic graph. Genes displayed in GoMiner are linked to major public bioinformatics resources.

NetAffx GO Mining Tool

www.affymetrix.com/analysis/index.affx

This tool permits web-based, interactive traversal of the GO graph in the context of microarray data (Cheng et al. 2004). It accepts a list of Affymetrix probe sets and renders a GO graph as a heat map colored according to significance measurements. The rendered graph is interactive, with nodes linked to public websites and to lists of the relevant probe sets. The GO Mining Tool provides visualization combining biological annotation with expression data, encompassing thousands of genes in one interactive view. An example of using the NetAffx GO Mining Tool in a preterm delivery (PTD) microarray study is shown in figure 5.2. In this figure, it can be seen that the root GO term (level-1 node) "response to external stimulus" has five child GO terms: "response to extracellular stimulus," "response to abiotic stimulus," "taxis," "detection of external stimulus," and "response to biotic stimulus." Three of these level-2 nodes — "response to abiotic stimulus," "detection of external stimulus," and "response to biotic stimulus" — have child GO terms (i.e., level-3 nodes). Also, a child node can have multiple parent nodes in the GO graph. For example, the "detection of abiotic stimulus" GO term (a level-3 node) has two parent nodes: "response to abiotic stimulus" and "detection of external stimulus."

FatiGO fatigo.bioinfo.cnio.es

This tool extracts GO terms that are significantly over or underrepresented in sets of genes within the context of a genome-scale experiment (Al-Shahrour et al. 2004).

GOAL microarrays.unife.it

The GO Automated Lexicon is a web-based application for the automated identification of functions and processes regulated in microarray and serial analysis of gene expression experiments based on GO terms (Volinia et al. 2004).

Onto-Tools vortex.cs.wayne.edu/Projects.html

This is a collection of tools for a variety of tasks all of which involve the use of GO terminology (Draghici et al. 2003),

DAVID david.niaid.nih.gov

The Database for Annotation, Visualization and Integrated Discovery is a web-based tool for rapidly listing genes in GO categories (Dennis, Jr. et al. 2003).

GOTM `genereg.ornl.gov/gotm`
The GOTree Machine is a web-based platform for interpreting microarray data or other interesting gene sets using GO (Zhang et al. 2004).

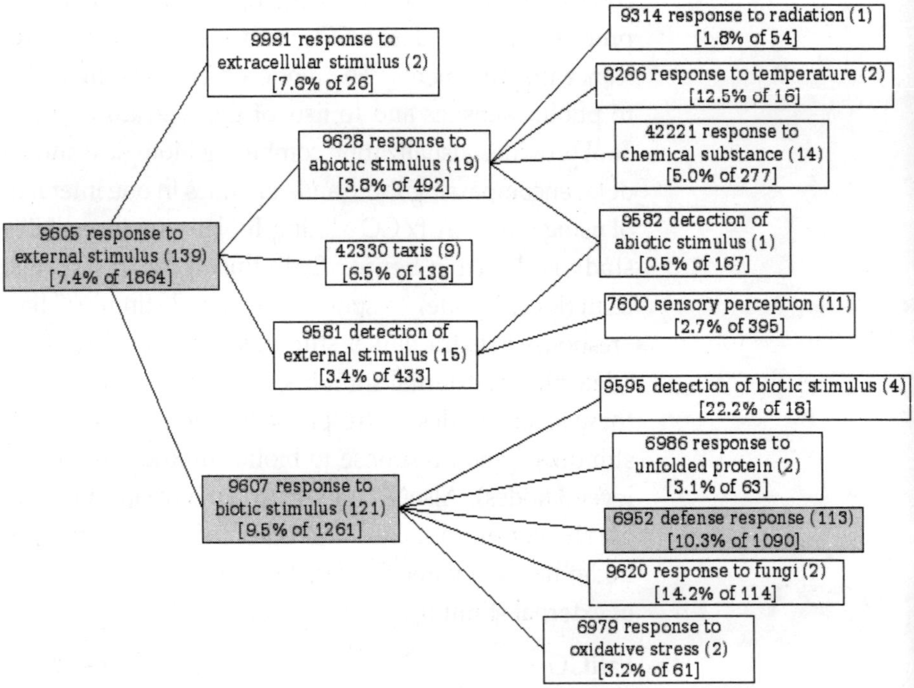

Figure 5.2 A GO network graph generated using the NetAffx Gene Ontology Mining Tool.

As an example of the use of GO, consider PTD, defined as a delivery occurring before the completion of 37 weeks of gestation. PTD is the major determinant of infant mortality, yet the molecular mechanisms of this disorder remain largely unknown. To better understand gene expression changes associated with PTD at the transcriptional level, we extracted total RNA samples using the PAXgene Blood RNA kit from whole-blood samples obtained from black mothers who had PTD (n=8, cases) and ethnicity-matched mothers with term deliveries (n=6, controls) at the Boston Medical Center. Gene expression profiling was carried out using the Affymetrix HU133A GeneChip. Among the 6220 genes that were detected as being expressed, we identified a total of 1559 genes that have significantly different expression patterns between the cases and controls using the t-test at significance level 0.05. This is still a very large number of genes to examine. To focus more precisely on the genes that are most likely to be of importance in PTD, we selected the genes that have been classified as being involved in the "response to external stimulus" GO term of the GO biological process ontology. By using the GO clustering tool of the DNA-Chip Analyzer `www.dchip.org` software, we found 159 genes belonging to this category. The gene expression profile is shown in figure 5.3.

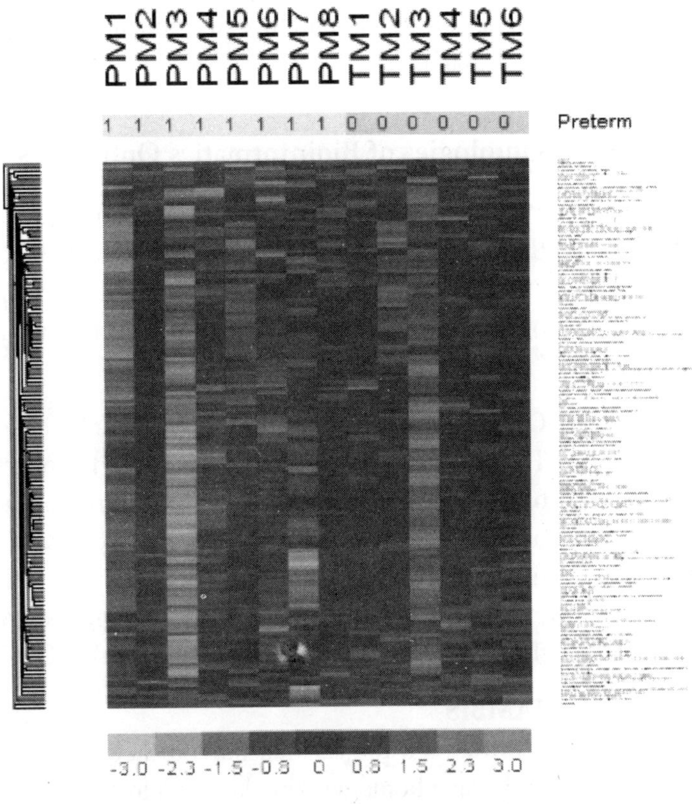

Figure 5.3 A gene expression profiling study of preterm delivery (PTD) of eight mothers with PTDs and six mothers with term deliveries. In this study, 159 genes were found to be significantly belonging to the "Response to External Stimulus" GO term ($P < .0001$).

A number of efforts are underway to enhance and extend GO. The Gene Ontology Annotation (GOA), run by the European Bioinformatics Institute (EBI), is providing assignments of terms from the GO resource to gene products in a number of its databases, such as SWISS-PROT, TrEMBL, and InterP (Camon et al. 2003; GOA 2003).

Although GO is the most prominent of all bio-ontologies, it did not originally make use of a formal ontological framework such as XML or RDF. To remedy this situation, the Gene Ontology Next Generation Project (GONG is developing a staged methodology to evolve the current representation o the GO into the Web Ontology Language (OWL) introduced in section 4.4 OWL allows one to take advantage of the richer formal expressiveness and the reasoning capabilities of the underlying formal logic. Each stage provides a step-level increase in formal explicit semantic content with a view to supporting validation, extension, and multiple classification of GO (Wro et al. 2003).

5.1.3 Ontologies of Bioinformatics Ontologies

With the proliferation of biological ontologies and databases, the ontologies themselves need to be organized and classified. This survey chapter gives an informal classification, but a more formal approach is needed: an ontology of biological ontologies. In this section we review two examples of such "metaontologies."

OBO `obo.sourceforge.net`

The Open Biological Ontologies seeks to collect ontologies for the domains of genomics and proteomics. The criteria for inclusion are that the ontology be open, use either GO or OWL syntax, have definitions and unique identifiers, and complement (rather than compete with) other OBO ontologies. An example of a zebrafish anatomy ontology (development of the zygote from the one-cell stage to the eight-cell stage) in OBO is shown in figure 5.4.

TAMBIS `img.cs.man.ac.uk/tambis`

TAMBIS is a project that aims to help researchers in biological science by building a homogenizing layer on top of various biological information services. The acronym stands for *t*ransparent *a*ccess to *m*ultiple *b*iological *in*formation *s*ources. The TAMBIS ontology is a semantic network that covers a wide range of bioinformatics concepts. It aims to provide transparent information retrieval and filtering from biological information services by building a homogenizing layer on top of the different sources. This layer uses a mediator and many source wrappers to create the illusion of one all-encompassing data source. TAMBIS uses a mediator (information broker) to achieve this goal. This mediator uses an ontology of molecular biology

```
$structures.goff ; ZFIN:0000000
 <001_Zygote\:1-cell\,embryo ; ZFIN:0000004
  <001_Zygote\:1-cell\,blastomere ; ZFIN:0000001
  <001_Zygote\:1-cell\,yolk ; ZFIN:0000012
 <001_Zygote\:1-cell\,extraembryonic ; ZFIN:0000005
  <001_Zygote\:1-cell\,chorion ; ZFIN:0000002
 <002_Cleavage\:2-cell\,embryo ; ZFIN:0000017
  <002_Cleavage\:2-cell\,blastomeres ; ZFIN:0000013
  <002_Cleavage\:2-cell\,yolk ; ZFIN:0000025
 <002_Cleavage\:2-cell\,extraembryonic ; ZFIN:0000018
  <002_Cleavage\:2-cell\,chorion ; ZFIN:0000014
 <003_Cleavage\:4-cell\,embryo ; ZFIN:0000030
  <003_Cleavage\:4-cell\,blastomeres ; ZFIN:0000026
  <003_Cleavage\:4-cell\,yolk ; ZFIN:0000038
 <003_Cleavage\:4-cell\,extraembryonic ; ZFIN:0000031
  <003_Cleavage\:4-cell\,chorion ; ZFIN:0000027
 <004_Cleavage\:8-cell\,embryo ; ZFIN:0000043
  <004_Cleavage\:8-cell\,blastomeres ; ZFIN:0000039
  <004_Cleavage\:8-cell\,yolk ; ZFIN:0000051
 <004_Cleavage\:8-cell\,extraembryonic ; ZFIN:0000044
```

Figure 5.4 An excerpt from `zebrafish_anatomy.ontology` in OBO.

and bioinformatics. This ontology was originally written in a description logic called GRAIL (Baker et al. 1998), but was later changed to OWL. The ontology acts as a universal model to help users form queries that will be understood by the various sources. The wrappers create the illusion of a common query language for all information resources. The latest version of TAMBIS has been translated to to OWL at `imgproj.cs.man.ac.uk/tambis/BabyTao-new.owl`.

5.2 Ontology Languages in Bioinformatics

In this section we survey the main XML-based ontologies that have been developed for bioinformatics. The number of such ontologies is large, and continually increasing, so some of the ontologies will be mentioned only briefly.

BSML www.bsml.org

The Bioinformatic Sequence Markup Language (BSML) is a language that encodes biological sequence information, which encompasses graphical representations of biologically meaningful objects such as nucleotide or protein sequences. The current version (released in 2002) is BSML v3.1. BSML takes advantage of XML features for encoding hierarchically organized information to provide a representation of knowledge about biological sequences.

BSML is useful in capturing the semantics of biological objects (e.g., complete genome, chromosome, regulatory region, gene, transcript, gene product, etc.). BSML can be rendered in the Genomic XML viewer, which greatly facilitates communications among biologists, since biologists are accustomed to visualizing biological objects and to communicating graphically about the these objects and their annotations.

The root element for a BSML document is tagged with Bsml. Consequently, a BSML document should look like the following:

```
<?xml version= "1.0"?>
<!DOCTYPE Bsml PUBLIC
  "http://www.labbook.com/dtd/bsml2_2.dtd">
<Bsml>
  ...
</Bsml>
```

BSML is primarily concerned with DNA, RNA, and protein. Information in a BSML document belongs primarily to one of two broad categories: "sequence data" and "sequence annotation."

1. *Sequence data.* The primary sequence data of the molecule of interest are contained within the sequence element; the information of the sequence is represented using attributes and their associated values, defined in the BSML DTD. figure 5.5 shows an example of using BSML to represent the amino acid sequence of human tumor suppressor p53.

2. *Sequence annotation.* Sequence annotation refers to information for a particular sequence that is beyond the sequence data themselves. Annotations have different types, which include positional annotation, qualitative annotation, quantitative annotation, and referential annotation.

```
<Sequence molecule="aa" length="393">
  <Seq-data>
   1 meepqsdpsv epplsqetfs dlwkllpenn vlsplpsqam ddlmlspddi eqwftedpgp
  61 deaprmpeaa ppvapapaap tpaapapaps wplsssvpsq ktyqgsygfr lgflhsgtak
 121 svtctyspal nkmfcqlakt cpvqlwvdst pppgtrvram aiykqsqhmt evvrrcphhe
 181 rcsdsdglap pqhlirvegn lrveylddrn tfrhsvvvpy eppevgsdct tihynymcns
 241 scmggmnrrp iltiitleds sgnllgrnsf evrvcacpgr drrteeenlr kkgephhelp
 301 pgstkralpn ntssspqpkk kpldgeyftl qirgrerfem frelnealel kdaqagkepg
 361 gsrahsshlk skkgqstsrh kklmfktegp dsd
  </Seq-data>
</Sequence>
```

Figure 5.5 The BSML representation for the SWISS-PROT entry P04637.

BioML `www.rdcormia.com/COIN78/files/XML_Finals/BIOML/Pages/BIOML.htm`

The Biopolymer Markup Language provides an extensible framework for annotating experimental information about molecular entities, such as proteins and genes. Many examples of BioML documents were shown in chapter 1. The four chemical letters of DNA, G, C, A, and T, have their normal meanings as individual nucleotides (case-insensitive). White space (e.g., spaces, tabs, carriage returns) are ignored by the parser, and can be freely added to aid the flow and readability of the file. The parser also ignores any character that cannot be a nucleotide residue, allowing the author to include numbers and other symbols that make reading the file easier. The kinds of element for DNA, RNA, and protein in BioML are presented in table 5.1.

The BioML ontology can also be used to refer to public database entries. For example, one can refer to the GenBank entry for the DNA sequence encoding the human δ-aminolevulinate dehydratase as follows:

```
<bioml>
  <reference>
    <db_entry format="GENBANK" entry="X64467"/>
  </reference>
</bioml>
```

A notable feature of BioML is that it allows for the inclusion of nontextual data, such as binary data (Fenyo 1999). Also, BioML possesses a mechanism

DNA	RNA	Protein
<dna>	<rna>	<protein>
<promoter>	<rdomain>	<subunit>
<gene>	<ra>	<homolog>
<exon>	<rmod>	<peptide>
<intron>	<rvariant>	<domain>
<ddomain>	<rstart>	<aa>
<da>	<rstop>	<amod>
<dmod>		<avariant>
<dvariant>		<aconflict>
<dstart>		
<dstop>		

Table 5.1 The elements for DNA, RNA, and protein in BioML

that accepts information conforming to other standard formats, such as the Protein Data Bank (PDB) format.

SBML `www.sbw-sbml.org`

The Systems Biology Markup Language is an XML-based language for storing biochemical models (Hucka et al. 2003). Formally defined using the Unified Modeling Language (UML) (UML 2004), SBML contains structures for representing compartments, species, and reactions, as well as optional unit definitions, such as parameters and rules (constraints). SBML is still under active development at the California Institute of Technology.

SBML level-1 is aimed at providing a basic representation of biochemical reaction networks. A model in SBML consists of the following components (Hucka et al. 2003):

1. Compartment: a container of finite volume for well-stirred substances where reactions take place

2. Species: a chemical substance or entity that takes part in a reaction

3. Reaction: a statement describing some transformation

4. Parameter: a quantity that has a symbolic name

5. Unit definition: a name for a unit used in the expression of quantities in a model

6. Rule: a mathematical expression that is added to the model equations constructed from the set of reactions

MAGE-ML `www.mged.org`

The MicroArray Gene Expression Markup Language is an XML ontology for microarray data. MAGE-ML aims to create a common data format so that data can be shared easily between projects (Stoeckert, Jr. et al. 2002). The predecessor of MAGE-ML is the Gene Expression Markup Language (GEML), initially developed by Rosetta Inpharmatics (Kohane et al. 2003).

MAGE-ML is a data-exchange syntax for microarray data recently created by the microarray gene expression data group (MGED) (MAGE-ML 2003). In order to standardize the information concerning microarray data, MGED initially introduced the minimal information for the annotation of a microarray experiment (MIAME). MIAME describes the minimum information required to ensure that microarray data can be easily interpreted and that results derived from its analysis can be independently verified (Brazma et al. 2001). Practically speaking, MIAME is a checklist of what should be supplied for publication. MIAME-compliant conceptualization of microarray experiments is then modeled using the UML-based microarray gene expression object model (MAGE-OM). MAGE-OM is then translated into an XML-based data format, MAGE-ML, to facilitate the exchange of data (Spellman et al. 2002).

There is a close relationship between the MAGE-ML and MGED ontologies. The MGED ontology, being developed by the Ontology Working Group of the MGED ontology project, is providing standard controlled vocabularies for microarrays. The goal of the MGED ontology is to create a framework of microarray concepts that reflects the MIAME guidelines and MAGE structure. Therefore, the MGED ontology project has a practical aim to develop standards, and to reduce nonuniform usage of annotation in microarray experiments. Concepts for which existing controlled vocabularies and ontologies can be identified are specified by reference to those external resources, and no new ontologies will be created. Concepts that are microarray-based or tractable (such as experimental conditions) are specified within the MGED ontology. MAGE-ML provides a standard XML format, which supersedes the MicroArray Markup Language (MAML) format, for reporting microarray data and its associated information.

CellML `www.cellml.org`

The CellML ontology is being developed by Physiome Sciences Inc. in Princeton, New Jersey, in conjunction with the Bioengineering Institute at the University of Auckland and affiliated research groups. The purpose of CellML is to store and exchange computer-based biological models. CellML allows scientists to share models even if they are using different model-building software. It also enables them to reuse components from one model in another, thus accelerating model building. CellML includes information about model structure (how the parts of a model are organizationally related to one another), mathematics (equations describing the underlying biological processes), and metadata (additional information about the model that allows scientists to search for specific models or model components in a database or other repository).

CellML is intended to support the definition of models of cellular and subcellular processes. This markup language facilitates the reuse of models and parts of models by employing a component-based architecture. Models are split into logical subparts called components that are connected together to form a model.

CellML separates the specification of the underlying mathematics of a model from a particular implementation of the model's solution. This makes a model independent of a particular operating system or programming language, and allows modelers to easily integrate parts of other peoples' models into their own models. CellML also allows the generation of equations for publishing from the same definition upon which the solution method is based, removing inconsistencies between the model and associated results in academic papers, and allowing others to reliably reproduce these.

RNAML `www-lbit.iro.umontreal.ca/rnaml`

RNAML provides a standard syntax that allows for the storage and exchange of information about RNA sequence as well as secondary and tertiary structures. The syntax permits the description of higher-level information about the data, including, but not restricted to, base pairs, base triples, and pseudoknots (Waugh et al. 2002).

Because of the hierarchical nature of XML, RNAML is a valuable method for structuring the knowledge related to RNA molecules into a nested-structured text document. For example, in RNAML, a "molecule" is an element consisting of the following three lower-level elements: identity (which contains two nested elements, name and taxonomy), sequence (which contains three nested elements, numbering-system, seq-data, and seq-annotation), and structure (which contains one nested element, model). To ensure compatibility with other existing standards of RNA nomenclature, RNAML uses including formats such as the International Union of Pure and Applied Chemistry (IUPAC) lettering and PDB ATOM records. If RNAML needs to depict multiple interacting RNA molecules, the interactions of RNA molecules are presented as character data in an `interaction` element.

AGAVE `www.animorphics.net/lifesci.html`
The Architecture for Genomic Annotation, Visualization and Exchange is an XML language created by DoubleTwist, Inc., for representing genomic annotation data. AGAVE uses XML Schema (XSD) for describing the syntactic structure of the data. A bioperl script can be used to convert data in the European Molecular Biology Laboratory (EMBL) or Genome Annotation Markup Elements (GAME) format into the AGAVE format. The XML EMBL (XEMBL) project of EBI is building a service tool that employs Common Object Request Broker Architecture (CORBA) servers to access EMBL data. The data can then be distributed in XML format via a number of mechanisms (Wang et al. 2002).

CML `www.xml-cml.org`
The Chemical Markup Language was discussed in chapter 1. The purpose of CML is to manage chemical information (e.g., atomic, molecular, crystallographic information). CML is supported by tools such as the popular Jumbo browser. CMLCore retains most of the chemical functionality of CML 1.0, and extends it by adding handlers for chemical substances, extended bonding models, and names (Murray-Rust and Rzepa 2003).

CytometryML
The Cytometry Markup Language is designed for the representation and exchange of cytometry data. CytometryML provides an open, standard XML format, which may replace the Flow Cytometry Standard (Leif et al. 2003).

GAME `www.fruitfly.org/comparative`
GAME is an XML language for curation of DNA, RNA, or protein sequences. GAME uses an XML DTD to specify the syntactic structure of the content of a GAME document. GAME is extensively used within the FlyBase/Berkeley *Drosophila* Genome Project (BDGP). For example, genomic regions for Rhodopsin 1 (ninaE), Rhodopsin 2 (Rh2), Rhodopsin 3 (Rh3), Rhodopsin 4 (Rh4), apterous (ap), even-skipped (eve), fushi-tarazu (ftz) and twist (twi) have been annotated in GAME format (Bergman et al. 2002; GAME 2002) in four *Drosophila* species (*D. erecta*, *D. pseudoobscura*, *D. willistoni*, and *D. littoralis*) covering over 500 kb of the *D. melanogaster* genome.

MML

The Medical Markup Language provides the XML-based standard for medical data exchange/storage (Guo et al. 2003).

MotifML `motifml.org`

MotifML is a language for representing the computationally predicted DNA motifs (often in the regulatory region such as promoters) generated by the Gibbs motif sampler, AlignACE, BioProspector, and CONSENSUS. MotifML was created by the authors of this book and two collaborators (Sui Huang and Jerzy Letkowski). MotifML uses the Web Ontology Language (OWL) to specify the data structure of a MotifML document. MotifML is supported by Java-based visualization tools such as MotifML viewers.

NeuroML `www.neuroml.org/main.html`

The Neural Open Markup Language is an XML language for describing models, methods, and literature for neuroscience. NeuroML uses XSD to specify the syntactic requirements for the model descriptions (Goddard et al. 2001).

ProML

The Protein Markup Language is for specifying protein sequences, structures, and families using an open XML standard. ProML allows machine-readable representations of key protein features (Hanisch et al. 2002).

TML

Taxonomic Markup Language is mainly an XML format for representing the topology of a phylogeny, but also includes a representation for statistical metadata (e.g., branch length, retention index, and consistency index) describing the phylogeny (Gilmour 2000). It is notable that for TML, the hierarchical nature of a phylogeny is readily represented by XML.

5.3 Macromolecular Sequence Databases

The rapid expansion of nucleotide sequence data available in public databases is revolutionizing biomedical research. Sequence databases such as GenBank have a variety of uses, including the discovery of novel genes, identification of homologous genes, analysis of alternative splicing, chromosomal localization of genes, and detection of polymorphisms (Pandey and Lewitter 1999). Macromolecular sequence databases are classified according to whether they deal with nucleotide sequences or protein sequences.

5.3.1 Nucleotide Sequence Databases

GenBank `www.ncbi.nlm.nih.gov/Genbank`

GenBank is a comprehensive database that contains publicly available DNA sequences for more than 140,000 named organisms. The sequences are primarily obtained through submissions from individual laboratories and batch submissions from large-scale sequencing projects (Benson et al. 2004). As of February 2004, GenBank contained over 37 billion bases in over 32 million sequence records. GenBank uses its own non-XML text format.

Most submissions to GenBank are made using the BankIt web service or Sequin program and accession numbers are assigned by GenBank staff upon receipt. Daily data exchange with the EMBL data library in the U.K. and the DNA data bank of Japan (DDBJ) helps ensure worldwide coverage. GenBank is accessible through NCBI's retrieval system, Entrez, which integrates data from the major DNA and protein sequence databases along with taxonomy, genome mapping, protein structure, and domain information, and the biomedical journal literature via PubMed.

EMBL `www.ebi.ac.uk/embl`

The EMBL Nucleotide Sequence Database, maintained at the European Bioinformatics Institute (EBI), incorporates, organizes, and distributes nucleotide sequences from public sources (Kulikova et al. 2004). The database is a part of an international collaboration with DDBJ and GenBank. Data are exchanged between the collaborating databases on a daily basis. The Webin web service is the preferred system for individual submission of nucleotide sequences, including third party annotation (TPA) and alignment data. Automatic submission procedures are used for submission of data from large-scale genome sequencing centers and from the European Patent Office. Database releases are produced quarterly.

EMBL uses its own non-XML text format, but the XEMBL project has made it possible to obtain EMBL data in the AGAVE XML format (Wang et al. 2002). The latest EMBL data collection can be accessed via ftp, email, and web interfaces. The EBI's Sequence Retrieval System (SRS) integrates and links the main nucleotide and protein databases as well as many other specialist molecular biology databases. For sequence similarity searching, a variety of tools (e.g., FASTA and BLAST) are available that allow users to compare their own sequences against the data in EMBL and other databases.

DDBJ `www.ddbj.nig.ac.jp`

DDBJ is maintained at the National Institute of Genetics in Japan (Miyazaki et al. 2004). It is available in several formats, including FASTA and XML. The XML format is defined by the DTD at `ftp://ftp.ddbj.nig.ac.jp/database/ddbj/xml/DDBJXML.dtd`. DDBJ cooperates with both EMBL and GenBank.

5.3.2 Protein Sequence Databases

SWISS-PROT `au.expasy.org/sprot`

SWISS-PROT is the most widely used publicly available protein sequence database. This database aims to be nonredundant, fully annotated, and highly cross-referenced (Jung et al. 2001). SWISS-PROT also includes information on many types of protein modifications. The database is available in both FASTA and XML formats. The XML format is defined both as a DTD and using XSD. The XSD schema is at `www.uniprot.org/support/docs/uniprot.xsd`. The database itself is available at `ftp://ftp.ebi.ac.uk/pub/databases/uniprot/knowledgebase/uniprot_sprot.xml.gz`. Both SWISS-PROT and TrEMBL are available at this site in a variety of formats.

5.4 Structural Databases

Like sequence databases, the structural databases are classified according to whether they deal with nucleotide structure or protein structure.

5.4.1 Nucleotide Structure Databases

NDB `ndbserver.rutgers.edu`

The most prominent nucleotide structure database is the Nucleic Acid Database. NDB was established in 1991 as a resource to assemble and distribute structural information about nucleic acids (both DNA and RNA) (Berman et al. 1992). The core of the NDB has been its relational database of nucleic acid-containing crystal structures. The primary data include the crystallographic coordinate data, structure factors, and information about the experiments used to determine the structures, such as crystallization information, data collection, and refinement statistics. Derived information from experimental data, including valency geometry, torsion angles, and intermolecular contacts, is calculated and stored in the database. Database entries are further annotated to include information about the overall structural features, including conformational classes, special structural features, biological functions, and crystal-packing classifications. The NDB has been used to analyze characteristics of nucleic acids alone as well as complexed with proteins. The NDB database is available in the PDB and mmCIF formats.

5.4.2 Protein Structure Databases

Protein structure databases deal with progressively "higher-order" types of structure: secondary, tertiary, quaternary, and functional. Protein sequence information is also a form of structure: the primary structure. A protein structure database will typically have information about structure on several levels. Accordingly, we have not attempted to perform a strict classification but rather list them approximately by the type of structure, from primary to functional.

Structural classifications range from short motifs and domains to entire protein families, and they derive protein classes based on the molecular similarities in terms of secondary or higher-order structures. Functional classifications range from enzymatic roles to protein interaction networks, and they derive protein classes based on functional similarities in terms of enzyme reaction mechanisms, or participation in biochemical pathways.

Pfam `www.sanger.ac.uk/Software/Pfam`
The Protein Family database is a large collection of protein families and domains (Bateman et al. 2004). The Pfam database is available in FASTA format.

SMART `smart.embl.de`
The Simple Modular Architecture Research Tool is a web tool for the identification and annotation of protein domains, and provides a platform for the comparative study of complex domain architectures in genes and proteins. The January 2004 release of SMART contains 685 protein domains. New developments in SMART are centered on the integration of data from completed metazoan genomes. SMART can be queried using GO terms (Letunic et al. 2004).

PROSITE `www.expasy.org/prosite`
PROSITE is a compilation of sites and patterns found in protein sequences (Sigrist et al. 2002; Hulo et al. 2004). The use of protein sequence patterns (motifs) to determine the protein function has become one of the essential

tools in sequence analysis. PROSITE was developed in 1988 to systematically collect macromolecularly significant patterns (Bairoch 1991). PROSITE is based on multiple sequence alignments (MSAs) which use two kinds of descriptor: patterns and generalized profiles (Hulo et al. 2004). In PROSITE, each PROSITE signature is linked to an annotation document where the user can obtain information regarding the signature. In order to make the three-dimensional (3D) structure more comprehensible, there are links to the representative PDB database. PROSITE is closely related to the SWISS-PROT protein sequence data bank.

The PROSITE descriptors and documentation can also be accessed through InterPro, which uses the detailed family annotation provided by PRINTS (Attwood et al. 2003). InterPro (Mulder et al. 2003) provides an integrated view of several domain databases and offers a large choice of methods to identify conserved regions. ClustalW (Thompson et al. 1994) or T-Coffee (Notredame et al. 2000) are most commonly used to construct the MSAs. However, when the primary sequences are too divergent, it is useful to integrate structural information in the MSAs. In addition, about 3% of profiles in PROSITE are built by using the HMMER hidden Markov model package (Eddy 1998).

The PROSITE database is available as a text file. The format is defined in a separate file and uses a variety of characters (forward slashes, commas, semicolons, etc.) as delimiters.

BLOCKS `blocks.fhcrc.org`

Blocks are defined as ungapped multiple alignments corresponding to the most conserved regions of proteins. Blocks contain "multiple alignment" information, and the use of the BLOCKS database can improve the detection of sequence similarities in searches of sequence databases. The BLOCKS database was introduced to aid in the family classification of proteins (Henikoff and Henikoff 1991). This database turns out to be a very important database, because hits to BLOCKS database entries pinpoint the location of conserved motifs, which are important for further functional characterization (Henikoff et al. 2000). Furthermore, the BLOCKS database can be used for detecting distant relationships (Henikoff et al. 1998). The BLOCKS database is the basis for the BLOSUM substitution tables that are used in amino acid sequence similarity searching, as explained in section 7.1.

The BLOCKS database contains more than 24,294 blocks from nearly 5000 different protein groups (Henikoff et al. 2000). There are a variety of formats for blocks, including the Blocks, FASTA, and Clustal formats. All of the formats are non-XML text formats.

COG `www.ncbi.nlm.nih.gov/COG`

The database of clusters of orthologous groups of proteins (COGs) attempts to give a phylogenetic classification of the proteins encoded in 21 complete genomes of bacteria, archaea, and eukaryotes (Tatusov et al. 2000). The COGs were constructed by applying the criterion of consistency of genome-specific best hits to the results of an exhaustive comparison of all protein sequences from these genomes. The database comprises 2091 COGs that include 56 to 83% of the gene products from each of the complete bacterial and archaeal genomes and approximately 35% of those from the yeast *Saccharomyces cerevisiae* genome. The database is available as a flat file.

PRINTS `umber.sbs.man.ac.uk/dbbrowser/PRINTS`

PRINTS is a compendium of protein fingerprints (Attwood et al. 1999, 2003). It is available in FASTA format.

ProDom `http://protein.toulouse.inra.fr/prodom/current/html/home.php`

ProDom is a comprehensive set of protein domain families automatically generated from the SWISS-PROT and TrEMBL sequence databases (Servant et al. 2002).

TIGRFAMs `http://www.tigr.org/TIGRFAMs/`

The Institute for Genomic Research maintains a database of protein families based on hidden Markov models (Haft et al. 2003). TIGRFAMs currently contains over 1600 protein families. It includes models for both full-length proteins and shorter protein regions grouped at the levels of superfamilies, subfamilies, and "equivalogs," homologous protein sets that are functionally conserved since their last common ancestor. TIGRFAMs is a complementary database to Pfam, whose models typically have a wider coverage across distant homologs. The data can be downloaded as a text file.

PDB `www.rcsb.org/pdb`

The Protein Data Bank is the largest source of publicly available biomolecular 3D structures (Bateman et al. 2004). PDB was established at Brookhaven National Laboratories (BNL) in 1971 as an archive for biological macromolecular crystal structures. According to the PDB holdings list of 9 September 2003, the PDB contains a total of 22,448 structures, 19,062 of which are resolved by X-ray, and the remaining 3386 are resolved by Nuclear Magnetic Resonance (NMR). Generally speaking, NMR structures are more problematic than crystallographic ones, because structures in solution are generally more flexible and less stable than those in a crystal. Indeed, solution structures determined by the NMR data are slightly different from crystal structures. Therefore, NMR is often used to study small and peculiar proteins.

Protein glycosylation is probably the most common and complex type of co- and post-translational modification encountered in proteins (Lutteke et al. 2004). Inspection of the protein databases reveals that 70% of all proteins have potential N-glycosylation sites - Asn-X-Ser/Thr, where X is not Pro (Mellquist et al. 1998). O-glycosylation is even more ubiquitous (Berman et al. 2000). Consequently, PDB entries contain not only protein structures but also pure carbohydrate structures. However, to date, there is no standard nomenclature for carbohydrate residues within the PDB files (Westbrook and Bourne 2000). For example, although many monosaccharide residues are defined in the PDB Het Group Dictionary `pdb.rutgers.edu/het_dictionary.txt`, there is no distinction between the α- and the β-forms. Thus, it is difficult for glycobiologists to find relevant carbohydrate structures from PDB.

The PDB database has two non-XML formats, PDB and mmCIF, that are in use by many other molecular structure databases. Recently an XSD format, PDBML, has been introduced in PDB and automated generation of XML files is driven by the data dictionary infrastructure in use at the PDB. The current XML schema file is located at `deposit.pdb.org/pdbML/pdbx-v1.000.xsd`, and on the PDB mmCIF resource page at `deposit.pdb.org/mmcif/`.

SCOP `scop.mrc-lmb.cam.ac.uk/scop`

The Structural Classification of Proteins database classifies proteins by domains that have a common ancestor based on sequence, structural, and functional evidence (Murzin et al. 1995; Andreeva et al. 2004). In order to understand how multidomain proteins function, it is important to know how they are created during evolution. Duplication is one of the main sources for creating new genes and new domains (Lynch and Conery 2000). For examples of this, see section 1.5. In fact, 98% of human protein domains are duplicates (Gough et al. 2001; Madera et al. 2004; Muller et al. 2002). Once a domain or protein has duplicated, it can evolve a new or modified function.

Access to SCOP requires a license. It is available in a non-XML text format.

CATH `www.biochem.ucl.ac.uk/bsm/cath_new`

This database contains domain structures classified into superfamilies and sequence families (Orengo et al. 1997, 2003). Its name stands for Class/Architecture/Topology/Homology. Each structural family is expanded with domain sequence relatives recruited from GenBank using a variety of efficient sequence search protocols and reliable thresholds. The database is available as a collection of flat files using the fixed-width format, as in section 1.1.

FSSP `www.embl-ebi.ac.uk/dali`

The FSSP database and its new supplement, the Dali Domain Dictionary, present a continuously updated classification of all known 3D protein structures (Holm et al. 1992; Holm and Sander 1998). FSSP stands for the *f*old classification based on *s*tructure-*s*tructure alignment of *p*roteins. The classification is derived using an automatic structure alignment program, called Dali, for the all-against-all comparison of structures in the PDB. From the resulting enumeration of structural neighbors (which form a surprisingly continuous distribution in fold space) a discrete fold classification is derived in three steps: (1) sequence-related families are covered by a representative set of protein chains; (2) protein chains are decomposed into structural domains based on the recurrence of structural motifs; and (3) folds are defined as tight clusters of domains in fold space. The database is available as an SQL dump, using a fixed-width format.

SCOP, CATH, and FSSP are structure classification databases that define, classify, and annotate each domain in the PDB. A systematic comparison of SCOP, CATH, and FSSP found that approximately two thirds of the protein chains are common to all three databases (Hadley and Jones 1999).

REBASE `rebase.neb.com/rebase/rebase.html`

REBASE contains information about restriction enzymes, including their recognition specificities and their sensitivity to DNA methylation (Roberts et al. 2003). There are three major categories of restriction enzymes: type I, type II, and type III. The type II restriction enzymes are among the most valuable tools available to researchers in molecular biology. These enzymes recognize short DNA sequences (four to eight nucleotides) and cleave at, or close to, their recognition sites (Pingoud and Jeltsch 2001). Type II enzymes are widely used not only for molecular cloning and genotyping but also for molecular diagnostics. REBASE contains comprehensive information on all types of restriction enzymes, as well as related kinds of proteins such as methyltransferases, homing endonucleases, and related proteins such as nicking enzymes, specificity subunits of the type I enzymes, control proteins, and methyl-directed restriction enzymes.

The REBASE database is currently available in 39 formats! This extreme heterogeneity is due to the large number of tools, each of which requires its own format. Standard formats would help control this diversity.

MIPS `mips.gsf.de`

The Munich Information Center for Protein Sequences provides protein sequence-related information based on whole-genome analysis (Mewes et al. 2004). The main focus of the work is directed toward the systematic organization of sequence-related attributes as gathered by a variety of algorithms and primary information from experimental data together with information compiled from the scientific literature.

DIP `dip.doe-mbi.ucla.edu`

The Database of Interacting Proteins is a research tool for studying cellular networks of protein interactions (Salwinski et al. 2004). The DIP aims to integrate the diverse body of experimental evidence on protein-protein interactions into a single, easily accessible online database. Because the reliability of experimental evidence varies widely, methods of quality assessment have been developed and utilized to identify the most reliable subset of the interactions. This core set can be used as a reference when evaluating the reliability of high-throughput protein-protein interaction data sets for development of prediction methods, as well as in studies of the properties of protein interaction networks.

Obtaining the DIP database requires registration. The database is available in an XSD format called XIN, as well as in tab-delimited flat files and other formats.

SpiD `http://genome.jouy.inra.fr/cgi-bin/spid/index.cgi`

The Subtilis Protein interaction Database is a protein-protein interaction network database centered on the replication machinery of the gram-positive bacterium *Bacillus subtilis* (Hoebeke et al. 2001). This network was found by using genome-wide yeast two-hybrid screening experiments and systematic specificity assays (Noirot-Gros et al. 2002).

MINT `http://160.80.34.4/mint/`

The Molecular INTeraction database is a relational database containing interaction data between biological molecules (Zanzoni et al. 2002). At present, MINT centers on storing experimentally verified protein-protein interactions with special emphasis on proteomes of mammalian organisms. MINT consists of entries obtained from data mining of the scientific literature. The database is available in either a text format or in XML.

HPID `http://wilab.inha.ac.kr/hpid/`

The Human Protein Interaction Database was designed for the following purposes (Han et al. 2004):

1. Provide human protein interaction data precomputed from existing structural and experimental data using appropriate statistical methods.

2. Provide integrated human protein interactions derived from the Biomolecular Interaction Network Database (BIND) (Bader et al. 2003), DIP (Salwinski et al. 2004), and the Human Protein Reference Database (HPRD) (Peri et al. 2004).

3. Identify potential proteins from the databases that potentially interact with proteins submitted by users. A score composed of three parts is assigned to the predicted interaction data, and interactions with higher scores indicate that the predictions are more reliable.

A set of online software tools has been developed to visualize and analyze protein interaction networks.

5.5 Transcription Factor Databases

In humans, ribosomal RNA genes are transcribed by RNA polymerase I, transfer RNA (tRNA) genes are transcribed by RNA polymerase III, and protein-coding genes are transcribed by RNA polymerase II. Transcription is initiated in the promoter region by a complex of different factors. The following are the main transcription factor databases.

TRANSFAC `transfac.gbf.de`
The most complete transcription factor database is TRANSFAC (Wingender et al. 1996). This database is concerned with eukaryotic transcription regulation. It contains data on transcription factors, their target genes, and regulatory binding sites. The TRANSFAC database requires a license and fee, even for noncommercial use. It uses a flat file format which can be browsed but cannot be downloaded.

TRRD `www.bionet.nsc.ru/trrd`
The Transcription Regulatory Regions Database is a resource containing an integrated description of gene transcription regulation. Each entry of the database is concerned with one gene and contains data on localization and functions of the transcription regulatory regions as well as gene expression patterns (Kolchanov et al. 2002). TRRD contains only experimental data obtained from annotations in scientific publications. TRRD release 6.0 contains

information on 1167 genes, 5537 transcription factor binding sites, 1714 regulatory regions, 14 locus control regions and 5335 expression patterns obtained from 3898 scientific papers.

The TRRD is arranged in seven databases: TRRDGENES (general gene description), TRRDLCR (locus control regions); TRRDUNITS (regulatory regions: promoters, enhancers, silencers, etc.), TRRDSITES (transcription factor binding sites), TRRDFACTORS (transcription factors), TRRDEXP (expression patterns), and TRRDBIB (experimental publications). All of them are relational databases, and the schema consists of a large number of table definitions. SRS is used as a basic tool for navigating and searching TRRD and integrating it with external database and software resources.

COMPEL `compel.bionet.nsc.ru`

COMPEL is a database of composite regulatory elements, the basic structures of combinatorial regulation. Composite regulatory elements are two closely situated binding sites for distinct transcription factors and represent minimal functional units providing combinatorial transcriptional regulation. Both specific factor DNA and factor-factor interactions contribute to the function of composite elements (CEs). Information about the structure of known CEs and specific gene regulation achieved through such CEs appears to be extremely useful for promoter prediction, for gene function prediction, and for applied gene engineering as well.

Access to COMPEL requires registration, but it is free for noncommercial use. The database consists of three relational database tables.

ooTFD `www.ifti.org/ootfd`

The purpose of ooTFD (object-oriented Transcription Factors Database) is to capture information regarding the polypeptide interactions which constitute and define the properties of transcription factors (Ghosh 2000). ooTFD is an object-oriented successor to TFD (Ghosh 1993). The database is currently implemented using ozone, a Java-based object-oriented database system. The schema consists of nine primary Java data structures.

5.6 Species-Specific Databases

SGD `www.yeastgenome.org`

The *Saccharomyces* Genome Database is a database of the molecular biology and genetics of the budding yeast *Saccharomyces cerevisiae* (Dwight et al. 2004). This database collects and organizes biological information about genes and proteins of this yeast from the scientific literature, and presents this information on individual Locus pages for each yeast gene. The Pathway Tools software (Karp et al. 2002a) and the MetaCyc Database of metabolic reactions (Karp et al. 2002b) were used to generate the metabolic pathway information for *S. cerevisiae*. Metabolic pathways are illustrated in graphical format and the information can be viewed at multiple levels, ranging from general summaries to detailed diagrams showing each compound's chemical structure. Enzymatic activities of the proteins shown in each pathway diagram are linked to the corresponding SGD Locus pages.

FlyBase `flybase.bio.indiana.edu`

The fruit fly, *Drosophila melanogaster*, is one of the most studied eukaryotic organisms and a central model for the Human Genome Project (FlyBase 2002). FlyBase is a comprehensive database containing information on the genetics and molecular biology of *Drosophila*. It includes data from the *Drosophila* genome projects and data curated from the literature. FlyBase is a joint project with the Berkeley *Drosophila* Genome Project.

FlyBase is one of the founding participants in the GO consortium. As an example of how FlyBase is related to GO, consider the *D. melanogaster* gene p53 (FlyBase ID: FBgn0039044). Through FlyBase GO annotations, we can learn that p53 is classified by the organization principles as follows:

1. *GO:Molecular function:* The p53 gene encodes a DNA-binding protein product which functions as a transcription factor for RNA polymerase II.
2. *GO:Biological process:* We can also learn that p53 is involved in important molecular processes such as DNA damage response, apoptosis, and response to radiation.
3. *GO:Cellular component:* Lastly, we find that the p53 protein is located in the nucleus.

Besides these GO annotations, we can also learn from the FlyBase report that the p53 gene is expressed not only in adult flies, but also during oogenesis (including nurse cell and oocyte) and during the embryonic stage (including embryonic/larval foregut, embryonic/larval hindgut, embryonic/larval midgut, germ cell, and mesoderm).

MGD `www.informatics.jax.org`

The Mouse Genome Database at the Jackson Laboratory in Bar Harbor, Maine, is a resource for mouse genome information. The human-mouse synteny (i.e., the comparison of the two mammalian organisms) provides important clues regarding gene location, phenotype, and function. Synteny maps are built based on the identification and mapping of conserved human-mouse synteny regions. Comparative mapping is used to pinpoint unknown human homologs of known, mapped mouse genes.

GDB gdbwww.gdb.org
The GDB Human Genome Database is the main repository for all published mapping information generated by the Human Genome Project. This database is specific to *Homo sapiens*. The information stored in GDB includes genetic maps, physical maps (clone, Sequence Tagged Site (STS), and Fluorescence In Situ Hybridization (FISH)-based), cytogenetic maps, physical mapping reagents (clones, STSs), polymorphism information, and citations.

Pathbase www.pathbase.net
Pathbase is a mutant mouse pathology database that stores images of the abnormal histology associated with spontaneous and induced mutations of both embryonic and adult mice (Schofield et al. 2004). The database and the images are publicly accessible and linked by anatomical site, gene, and other identifiers to relevant databases. The database is structured around a novel mouse pathology ontology, called MPATH, and provides high-resolution images of normal and diseased tissues that are searchable through orthogonal taxonomies for pathology, developmental stage, anatomy, and gene attributes. The database is annotated with GO terms, controlled vocabularies for type of genetic manipulation or mutation, genotype, and free text for mouse strain and additional attributes. The MPATH ontology is available in DAG-Edit format.

5.7 Specialized Protein Databases

ORDB senselab.med.yale.edu/senselab/ordb
The Olfactory Receptor Database is a central repository of olfactory receptor (OR) and olfactory receptor-like gene and protein sequences (Crasto et al. 2002). The 2004 Nobel Prize in Physiology or Medicine was awarded jointly to Richard Axel and Linda B. Buck for their discoveries of "odorant receptors and the organization of the olfactory system." Humans detect odorants through ORs, which are located on the olfactory sensory neurons in the olfactory epithelium of the nose (Buck and Axel 1991; Buck 2000).

In building ORDB, relevant HTML files from GenBank and SWISS-PROT and user-supplied text files are parsed to extract relevant data. Upon filtering, an XML-encoded file is then built that is entered into the database via an HTML submission form. The ORDB can be downloaded as an HTML file.

RiboWeb `smi-web.stanford.edu/projects/helix/riboweb.html` RiboWeb is a relational database containing a representation of the primary 3D data relevant to the structure of the ribosome of the prokaryotic 30S ribosomal subunit, which initiates the translation of messenger RNA (mRNA) into protein and is the site of action of numerous antibiotics (Chen et al. 1997). The project has since been expanded to include structural data pertaining to the entire ribosome of prokaryotes (but primarily *Escherichia coli*). The project includes computational modules for constructing and studying structural models,

5.8 Gene Expression Databases

Gene expression profiling includes both transcriptomics and proteomics. The former monitors gene transcription, while the latter monitors gene translation. Proteomics has more restrictive expressions and post-translational modifications. In contrast to transcriptomics, which is an "indirect" measure of gene expression, proteomics provides a more direct measurement of gene expression and is increasingly important in functional genomics. Thus, gene expression databases contain both transcriptomics databases and proteomics databases.

5.8.1 Transcriptomics Databases

It is useful to study the temporal and spatial patterns of gene expression. *Transcriptomics* is defined as the use of quantitative mRNA measurements of gene expression to characterize biological processes and elucidate gene transcription mechanisms. Thus, the goal of gene expression experiments is to quantify mRNA expression, particularly under certain conditions (e.g., drug intervention) or in a disease state. Differential gene expression measurements are performed using a number of high-throughput techniques such as (1) expression sequence tags (ESTs), (2) DNA microarrays (including oligonucleotide microarrays and spotted microarrays), (3) subtractive cloning, (4) differential display, and (5) serial analysis of gene expression (SAGE). Gene expression experiments have as their goal the identification of novel disease genes, drug targets, and coregulated gene groups. Transcriptomics databases provide integrated data management and analysis systems for transcriptional expression. These databases can be used for both hypothesis testing and knowledge discovery.

NCBI's dbEST Database www.ncbi.nlm.nih.gov/dbEST/
The GeneCards Database bioinformatics.weizmann.ac.il/cards
Kidney Development Gene Expression Database organogenesis.ucsd.edu
Gene Expression in Tooth bite-it.helsinki.fi
Mouse Gene Expression Database www.informatics.jax.org
The Cardiac Gene Expression Knowledgebase www.cage.wbmei.jhu.edu
Gene Expression Atlas expression.gnf.org/cgi-bin/index.cgi
NCBI's Gene Expression Omnibus www.ncbi.nlm.nih.gov/entrez/query.fcgi?db=geo
Cancer Gene Expression Database cged.hgc.jp/cgi-bin/input.cgi
***Saccharomyces* Genome Database** www.yeastgenome.org
The Nematode Expression Pattern DataBase nematode.lab.nig.ac.jp
WormBase www.wormbase.org
The *Plasmodium* Genome Resource plasmodb.org
The Zebrafish Information Network zfin.org

5.8.2 Proteomics Databases

Proteomics is defined as the use of quantitative protein-level measurements of gene expression to characterize biological processes and elucidate the mechanisms of gene translation. The goal of proteomics is the quantitative measurement of protein expression in various conditions such as under the influence of a drug or being in a specific disease condition. There are generally two steps in proteomics - protein separation and protein identification. Protein separation is usually performed using 2D polyacrylamide gel electrophoresis (2D-PAGE). Protein identification is usually accomplished using Edman degradation, mass spectrometry, or Western blotting. Protein quantification can be achieved through radiolabeling and scanning or phosphoimaging. Proteomics is important in disease diagnosis and prognosis. For example, human serum contains a spectrum of proteolytically derived peptides (serum peptidome) that may provide a correlate of biological events occurring in the entire organism (Villanueva et al. 2004). Proteomics databases address five biological questions that cannot be answered by DNA analysis: (1) the relative abundance of protein products; (2) post-translational modifications; (3) subcellular localizations; (4) molecular turnover; and (5) protein-protein interactions (Celis et al. 1998).

HEART-2DPAGE userpage.chemie.fu-berlin.de/~pleiss/dhzb.html
Heart High-Performance 2-DE Database www.mdc-berlin.de/~emu/heart
SWISS-2DPAGE au.expasy.org/ch2d
SIENA-2DPAGE www.bio-mol.unisi.it/2d/2d.html
WORLD-2DPAGE us.expasy.org/ch2d/2d-index.html
PMMA-2DPAGE www.pmma.pmfhk.cz
RAT HEART-2DPAGE www.mpiib-berlin.mpg.de/2D-PAGE/RAT-HEART/2d
HSC-2DPAGE www.expasy.org/cgi-bin/dbxref?HSC-2DPAGE
Phosphoprotein Database www-lecb.ncifcrf.gov/phosphoDB
REPRODUCTION-2DPAGE reprod.njmu.edu.cn/cgi-bin/2d/2d.cgi
Toothprint Database biocadmin.otago.ac.nz/tooth/home.htm
COMPLUYEAST-2DPAGE babbage.csc.ucm.es/2d/2d.html
FishProm www.abdn.ac.uk/fishprom/index.shtml

Mining of proteome databases can reveal intrinsic patterns and relationships in proteomics data, for example, protein-protein interactions and protein networks. The identification of patterns in complex proteome data sets can generate new insights into gene translation and post-translational modification conditions and can characterize complex biological networks.

5.9 Pathway Databases

A *pathway* is a system of molecules (especially proteins) that work together. Pathways are also called molecular interaction networks, and include metabolic pathways, regulatory pathways, and molecular complexes.

BioPAX www.biopax.org/
BioPAX is a collaborative effort to create a data exchange format for biological pathway data. The current format is called BioPAX level-1 and represents metabolic pathway information. Future levels are planned for representing

signaling, genetic regulatory, and genetic pathways. BioPAX is an OWL ontology.

KEGG `www.genome.ad.jp/kegg`

The Kyoto Encyclopedia of Genes and Genomes (Kanehisa and Goto 2000; Kanehisa et al. 2002) is the primary database resource of the Japanese GenomeNet service for understanding higher-order functional meanings and utilities of the cell or the organism from its genome information. KEGG consists of the PATHWAY database for the computerized knowledge of molecular interaction networks such as pathways and complexes, the GENES database for information about genes and proteins generated by genome sequencing projects, and the LIGAND database for information about chemical compounds and chemical reactions that are relevant to cellular processes. In addition to these three main databases, limited amounts of experimental data for microarray gene expression profiles and yeast two-hybrid systems are stored in the EXPRESSION and BRITE databases, respectively. Furthermore, a new database, named SSDB, is available for exploring the universe of all protein coding genes in the complete genomes and for identifying functional links and ortholog groups. The data objects in the KEGG databases are all represented as graphs and various computational methods are developed to detect graph features that can be related to biological functions.

The KEGG Markup Language (KGML) is the exchange format of the KEGG graph objects, especially the KEGG pathway maps that are manually drawn and updated. KGML enables automatic drawing of KEGG pathways and provides facilities for computational analysis and modeling of protein networks and chemical networks.

EcoCyc `ecocyc.org`

EcoCyc is an organism-specific pathway database that describes the metabolic and signal transduction pathways of *E. coli* K12 MG1655, its enzymes, and its transport proteins (Karp et al. 2002c). EcoCyc and MetaCyc are part of the BioCyc relational database, which is available as a collection of flat files.

MetaCyc `metacyc.org/`

MetaCyc is a metabolic-pathway database that describes nonredundant, experimentally elucidated metabolic pathways from more than 240 different organisms. (Karp et al. 2002c). Applications of MetaCyc include pathway analysis of genomes, metabolic engineering, and biochemistry education. MetaCyc and EcoCyc are queried using the Pathway Tools GUI, which provides a wide variety of query operations and visualization tools (Karp et al. 2002b).

5.10 Single Nucleotide Polymorphisms

A *single nucleotide polymorphism* (SNP) is defined as a single base change that occurs at a population frequency of at least 1%. SNPs represent the most common form of variation in the human genome. SNPs are important landmarks that can be applied in studies of molecular evolution as well as disease mechanisms.

In contrast to rare Mendelian diseases caused mostly by high-penetrant mutations, low-penetrance SNPs appear to form the most essential component of the heritability of common, complex human diseases. Bioinformatics has provided an unprecedented power and resource for deciphering the enigma of such complex disorders, based on the tremendous amount of data generated by the new, powerful, and high-throughput technologies of genomics and proteomics (Leung and Pang 2002). Several programs have been developed to predict SNP effects in silico on protein functions/gene transcriptional activities (Krishnan and Westhead 2003; Ng and Henikoff 2002, 2003; Conde et al. 2004). Also, there has been a surging interest in studying complex human diseases using SNP-based haplotypes, and a number of haplotype phasing algorithms have been developed (Niu 2004).

This section describes the major SNP and haplotype databases. For a list of the databases in this area, see the HGVbase website at `hgvbase.cgb.ki.se/`.

NCBI dbSNP database `www.ncbi.nlm.nih.gov/SNP`

The NCBI dbSNP database is the central depository for SNPs (Sherry et al. 2001). Because dbSNP entries may contain redundancies, all SNPs contained in dbSNP have been grouped into nonredundant sets of SNPs by clustering SNPs at identical genomic coordinates to create single, representative SNPs, which are called reference SNPs (RefSNPs). These RefSNPs are designated with an *rs* prefix in the ID.

Data are available in a wide variety of formats, including flat files, ASN.1, FASTA, and XSD. The URI for the XSD schema is `ftp://ftp.ncbi.nlm.nih.gov/snp/specs/genoex.xsd`.

HGVbase `hgvbase.cgb.ki.se/`

The objective of the Human Genome Variation Database is to provide an accurate, high-utility, and ultimately fully comprehensive catalog of normal human gene and genome variation, useful as a research tool to help define the genetic component of human phenotypic variation. All records are highly curated and annotated, ensuring maximal utility and data accuracy.

HGVbase is the product of a collaboration between the Karolinska Institute (Sweden), and the European Bioinformatics Institute (U.K.). Recently, a decision was made to develop HGVbase into a phenotype/genotype database. Data exchange with other databases is being maintained, but submissions are not currently being accepted.

Database exchange of core information with dbSNP (Sherry et al. 2001) ensures that HGVbase incorporates data from high-throughput discovery efforts. Release 15 of HGVbase contains information on almost 3 million SNPs, of which 29,000 are found in 10,000 genes and 41,000 have allele frequency information. In HGVbase, the location of each represented variant is presented in the context of available gene predictions, and SNPs within or around genes are described as *exonic*, *intronic*, *utr*, or *flank* (within 2 kb of the gene boundary). HGVbase currently considers only genes with a HUGO nomenclature committee approved definition (Wain et al. 2002), as represented in the Ensembl database (Hubbard et al. 2002). Nonsynonymous SNPs are grouped into three broad classes based on their predicted effect on the protein level: *benign*, *possibly damaging*, and *probably damaging*. The methods used for these functional predictions are described in (Ng and Henikoff 2003; Ramensky et al. 2002).

HGVbase is available in XML, FASTA, MySQL, and flat file formats. The XML format is specified by the XML DTD at `ftp://ftp.ebi.ac.uk/pub/databases/variantdbs/hgbase/hgvbase.dtd`.

Ensembl `www.ensembl.org/`

Ensembl is a a comprehensive source of stable automatic annotation of individual genomes, and of the synteny and orthology relationships between them (Birney et al. 2004). It is also a framework for integration of any biological data that can be mapped onto features derived from the genomic sequence, including SNPs.

Data can be obtained in a variety of formats, including FASTA format, flat files, GenBank format, and MySQL database dump format. The flat file format does not include all the data.

SNP500Cancer `snp500cancer.nci.nih.gov`

The Cancer Genome Anatomy Project (CGAP) was designed to provide public data sets, material resources, and informatics tools to serve as a platform to support the elucidation of the molecular signatures of cancer (Strausberg 2001; Strausberg et al. 2001). The SNP500Cancer Database provides sequence and genotype assay information for candidate SNPs useful in mapping complex diseases such as cancer. The database is an integral component of the Cancer Genome Anatomy Project (Packer et al. 2004) of the National Cancer Institute (NCI). SNP500Cancer provides bidirectional sequencing information on a set of control DNA samples derived from anonymized subjects (102 Coriell samples representing four self-described ethnic groups: African/African-American, White, Hispanic, and Pacific Rim). All SNPs are chosen from public databases and reports, and the choice of genes includes a bias toward nonsynonymous and promoter SNPs in genes that have been implicated in one or more cancers. The website is searchable by gene, chromosome, gene ontology pathway, and by known dbSNP ID. For each analyzed SNP, the database includes the gene location and over 200 bp of surrounding annotated sequence (including nearby SNPs). Other information is also provided such as frequency information in total and per subpopulation and calculation of the Hardy-Weinberg equilibrium for each subpopulation. Sequence validated SNPs with minor allele frequency greater than 5% are entered into a high-throughput pipeline for genotyping analysis to determine concordance for the same 102 samples. The website provides the conditions for validated genotyping assays.

SeattleSNPs Database `pga.mbt.washington.edu`

The SeattleSNPs is a collaboration between the University of Washington and the Fred Hutchinson Cancer Research Center, funded as part of the National Heart Lung and Blood Institute's (NHLBI) Programs for Genomic Applications (PGA). The goal of SeattleSNPs is to discover and model the associations between single nucleotide sequence differences in the genes and pathways that underlie inflammatory responses in humans. In addition to SNP data (location, allele frequency, and function for coding SNPs), haplotypes are presented graphically on the SeattleSNPs website. Haplotype tagging SNPs (htSNPs) information is also provided that will allow fewer SNPs to be genotyped per gene, thereby reducing cost and improving throughput. Data is available in tab-delimited text files.

GeneSNPs `www.genome.utah.edu/genesnps`

The GeneSNPs database is sponsored by the National Institute of Environmental Health Sciences and is being developed by the University of Utah Genome Center. GeneSNPs is a component of the Environmental Genome Project which integrates gene, sequence, and polymorphism data into individually annotated gene models. The human genes included are related to DNA repair, cell cycle control, cell signaling, cell division, homeostasis and metabolism, and are thought to play a role in susceptibility to environmental exposure. Data are available in HTML, FASTA, and XML formats. The

XML format does not use a DTD, and most of the information is encoded as FASTA text within element content.

The SNP Consortium `snp.cshl.org`

The SNP Consortium (TSC) was established in 1999 as a collaboration of several companies and institutions to produce a public resource of SNPs in the human genome (Thorisson and Stein 2003). The initial goal was to discover 300,000 SNPs in 2 years, but the final results exceeded this. For example, at the end of 2001, as many as 1.4 million SNPs had been released into the public domain (ISMWG 2001). The database now contains over 1.8 million SNPs. The data are stored in a relational database and are available in tab-delimited flat files.

International HapMap Project `www.hapmap.org`

The International HapMap project is charting the haplotype structure across the entire human genome in major human ethnic groups (IHMC 2003). The haplotype data of this project are available in XML. The format is specified using XSD in `www.hapmap.org/xml-schema/2003-11-04/hapmap.xsd`.

PART II

Building and Using Ontologies

This part addresses how ontologies are constructed and used. One uses ontologies far more frequently than one creates them, and it is a good idea to have some experience with how ontologies are used before attempting to design new ontologies. Accordingly, this part begins with the many uses for ontologies, and it ends with how one constructs them.

One of the most common uses of ontologies is for querying and retrieval. The first three chapters discuss how query processing works and how to formulate effective queries. Because ontologies have deductive capabilities, the result of a query makes use of inferred information as well as explicitly specified information. There are two main points of view that one can take with respect to retrieval. The first point of view is based on imprecise queries, while the second point of view is based on precise, logical queries. Imprecise bioinformatics queries can be expressed in two ways: natural language or biological sequences. Chapter 6 considers natural language queries, while chapter 7 deals with biological sequence queries. Chapter 8 introduces computer languages for unambiguous queries.

After information retrieval, the most common activity involving ontologies is transformation. The process whereby information is transformed from one format to another is surveyed in chapter 9. Such processes can have many steps and involve many groups of individuals. It is helpful to understand the entire transformation process so that the individual steps can serve the overall process better.

The individual transformation steps use a variety of programming languages and tools. One of the most common is Perl. While Perl is especially well suited for data transformations involving unstructured files, it can also be used for structured data. Chapter 10 is an introduction to Perl that emphasizes its use for data transformations. While Perl can be used effectively on XML documents, there is now a language specifically designed for transforming XML. This language is called XSLT, and it is introduced in chapter 11. As bioinformatics data migrate from flat files to XML structured files, one can expect that XSLT will play an increasing role.

This part ends with a detailed treatment of the process whereby ontologies are built. The ontologies and databases that were surveyed in chapter 5 were substantial endeavors involving many individuals and requiring the agreement of the community being served. While ontologies certainly can be developed in this way, it is also possible for ontologies to serve smaller communities for more limited purposes. Chapter 12 is a practical guide for developing ontologies in a systematic manner, whether the ontology will be used by a large community, a small community, or even a single individual.

6 *Information Retrieval*

6.1 The Search Process

Research is a fundamental activity of knowledge workers, whether they are scientists, engineers, or business executives. While each discipline may have its own interpretation of research, the primary meaning of the word is "a careful and thorough search." In most cases, the thing one is searching for is information. In other words, one of the most important activities of modern educated individuals is searching for information. Whole industries have arisen to meet the need for thorough searching. These include libraries, newspapers, magazines, abstracting services, online search services, and so on.

Biological systems also engage in searches. Enzymes can be highly specific to a particular kind of substrate, and oligonucleotide probes will strongly bind only with complementary DNA. Enzymes and probes diffuse through their medium until they encounter and bind with their matching target. Antibodies are perhaps the most elaborate kind of biological probe. A particular antibody not only continually searches for a particular kind of target but it performs actions when the target is found. The resulting actions can be very elaborate. Antibodies are the research agents of an organism, continually studying their environment and responding to attacks and threats.

Not surprisingly, the search process itself has been studied at least since the 1930s (Saracevic 1975), and a standard model was developed by the mid-1960s (Cleverdon and Keen 1966). In this model, the searcher has an information need which he or she tries to satisfy using a large collection or *corpus* of information objects. The objects that satisfy the searcher's needs are the relevant objects. The searcher expresses an information need using a formal statement called a *query*. Queries may be expressed using topics, categories, or words, individually or severally. The query is then given to a search intermediary. In the past the intermediary was a person who specialized in searching. It is more common today for the intermediary to be a computer system. Such systems are called *information retrieval systems* or *online search engines*. The search intermediary tries to match the topics, categories, and words from the query with information objects in the corpus. The intermediary responds with a set of information objects that, it is hoped, satisfy the searcher's needs.

Queries are certainly not the only way to find information in a corpus. Another very commonly used technique is to follow citations or *references* within the documents in the corpus. This technique is called *browsing*. Online browsing tools are now ubiquitous. Such a tool allows a searcher to follow references contained in information objects by simply clicking on a word or picture within the information object. In the standard model for information retrieval, a sharp distinction is made between searching using queries and searching using references.

In the standard model, the quality of a search is measured using two numbers (Saracevic 1975). The first number represents how thorough the search was. It is the fraction of the total number of relevant information objects that are presented to the searcher. This fraction is called by various names, such as the *sensitivity*, *coverage*, or *recall*. If the coverage is less than 100%, then some relevant information objects have been missed. The second number represents how careful the search was. It is the fraction of the objects presented to the searcher that are judged to be relevant. This number is called the *precision* or *selectivity*. If the precision is less than 100%, then some irrelevant objects were presented to the searcher.

Of course, one can always increase the coverage by adding many more information objects to those already presented, thereby ruining the selectivity. Clearly, one would like to balance the coverage and selectivity so as to achieve a search that is as careful and thorough as possible. In this chapter, a variety of search techniques and services are introduced, and the role that ontologies can play is described.

The queries considered above are expressed using topics, categories, or words, or a combination of these. The assumption is that the query is an imprecise and incomplete specification, and the search engine will make an effort to retrieve documents that are likely, but not guaranteed, to be relevant. Alternatively, one could use precise queries. Such queries must be expressed in a formal language with precise semantics. The best-known example of such a query language is SQL, which is used for retrieving information from relational databases. Formal query languages are programming languages specialized for retrieval. The advantage of using a formal query language is that one always has perfect retrieval: 100% coverage and 100% selectivity. This holds because the criteria for retrieval have no ambiguity. But there are several disadvantages. One must learn to program in the query language, which can require a significant effort, and this technique only applies to a corpus that is highly structured, such as a database or collection of XML documents. Formal query languages for XML documents are discussed in chapter 8.

Summary

- Online search engines are based on the standard model for information retrieval.
- In the standard model, a query is matched against a corpus and the most relevant documents are retrieved.
- The quality of the retrieval is measured by the coverage and selectivity.

6.2 Vector Space Retrieval

The simplest search technique is to look for documents that contain the words specified in a query. From this point of view a document is simply a set of words, and the same is true of a query. Search consists of finding the documents that contain the words of the query. Many retrieval systems use this basic technique, but this is only effective for relatively small repositories. The problem is that the number of matches to a query can be very large, so some mechanism must be provided that selects among the matching documents or arranges the documents so that the best matches appear first.

Simply arranging the matching documents by the number of matching words is not very effective because words differ in their selectivity. A word such as "the" in English has little use in search by word matching because nearly every document that uses English will have this word. For example, PubMed (NIH 2004b) is a very large corpus containing titles, abstracts, and other information about medical research articles. Table 6.1 gives the number of times that the most common words occur in PubMed. The second column of this table gives the number of times that the word occurs in the text parts of the PubMed citations. The third column gives the number of documents that contain the word. Note that "of" occurs in more documents than "the," although the latter occurs more often.

One can deal with the varying selectivity of words in several ways. One could ignore the most commonly occurring words. The list of ignored words is called the "stop word list." One can also weight the matches so that more commonly occurring words have a smaller effect on the choice of documents to be returned. When this technique is used, the documents are arranged in order by how well the documents match the query. Many algorithms have been proposed for how one should rank the selected documents, but the one that has been the most effective is *vector space retrieval*, also called the vector space model. This method was pioneered by (Salton et al. 1983; Salton 1989). In this model, each document and query is represented by a vector in a very high-dimensional vector space. The components of the vector (i.e., the axes or dimensions of the vector space) are all the words that can occur in a document or query and that can be used for searching. Such words are called *terms*. Terms normally do not include stop words, and one commonly maps synonymous words (such as words that differ only by upper- or lower-case distinctions) to the same term.

The vector of a document or query will be very sparse: nearly all entries will be zero for a particular document or query. The entry for a particular term in the vector is a number called the *term weight*. Term weights can be based on many criteria, but the two most important are the following (Salton and McGill 1986):

1. **Term Frequency.** The number of times that the term occurs in a document. The assumption is that if a term occurs more frequently in the document, then it must be more important for that document.

2. **Document Frequency.** The number of documents that make use of the term. When a term occurs in more documents, then it is less important for the purposes of information retrieval. One makes this assumption because terms that occur in more documents are less selective and therefore less useful for distinguishing the relevant documents. For example, "human" occurs in over 8.65 million PubMed documents, while "normetanephrines" only occurs in five documents. Thus "human" is much less selective than "normetanephrines."

The term weight to be assigned to a document should combine the term frequency with the document frequency. The most common way to do this

Word	Number of occurrences	Number of documents
the	61735764	7656676
of	56188095	8838209
and	35936471	7398465
in	33774127	7492125
to	18670139	5482347
a	18257728	5846177
with	12881242	4904841
was	10202515	3683332
for	8961803	4155330
human	8823139	7284714
were	8569580	3348418
by	8312546	3872850
that	7060247	3500824
is	6941721	3340361
metabolism	6459988	2119436
on	5669049	3324379
effects	5383334	2450007
from	5230712	2868785
drug	5141572	2305403
patients	5120897	1680125
as	4631373	2635318
or	4628590	2551277
at	4127714	2236276
cells	4112303	1303540
blood	4040580	1646084
this	3982055	2657101
an	3942049	2647511
male	3908310	3580224
be	3803795	2491683
female	3777242	3496974
support	3751655	2924479
pharmacology	3667686	1550452
analysis	3660441	1964018
are	3650898	2205386

Table 6.1 Number of occurrences of the most common words in PubMed

is based on a probabilistic cost/benefit approach. The two cost factors associated with information retrieval are:

1. The loss associated with the retrieval of an irrelevant document. Such an error is called a Type I error, or a *false positive.* Let c_1 denote the cost of this kind of error.

2. The loss associated with failing to retrieve a relevant document. Such an error is called a Type II error, or a *false negative.* Let c_2 denote the cost of this kind of error.

The cost factors are shown diagrammatically in figure 6.1. This same diagram applies to any situation in which a statistical decision must be made.

	Retrieved	Not Retrieved
Relevant	OK	Type II Error (False Negative)
Irrelevant	Type I Error (False Positive)	OK

Figure 6.1 Types of errors that can occur during document retrieval.

Retrieval begins by specifying a query Q. Documents are either relevant to the query Q or they are irrelevant to the query. The probability of relevance is Pr(Relevant) and the probability of irrelevance is Pr(Irrelevant) = 1 - Pr(Relevant). If one is considering a particular document D, then the probability of relevance is the conditional probability Pr(Relevant|D), and the probability of irrelevance is Pr(Irrelevant|D) = 1 - Pr(Relevant|D). The cost of retrieving this document is c_1Pr(Irrelevant|D) and the cost of not retrieving it is c_2Pr(Relevant|D). The ideal strategy is to retrieve the document when the cost of retrieval is less than the cost of nonretrieval or

$$Pr(Relevant|D)/Pr(Irrelevant|D) > c_1/c_2.$$

In practice, one does not explicitly specify either c_1 or c_2 or even their ratio. Rather, one attempts to arrange the documents in descending order by the ratio Pr(Relevant|D)/Pr(Irrelevant|D). The person requesting the query can then examine the documents in this order until it is found that the documents are no longer relevant. In other words, the ratio c_1/c_2 is implicitly determined by the researcher during examination of the document list.

The conditional probabilities Pr(Relevant|D) and Pr(Irrelevant|D) can be "reversed" by applying Bayes' law. Thus

$$Pr(Relevant \mid D) = Pr(D \mid Relevant)Pr(Relevant)/Pr(D)$$

and similarly for the probability of irrelevance. In the ratio of these two, the term Pr(D) cancels, and we obtain the following expression:

$$\frac{Pr(Relevant \mid D)}{Pr(Irrelevant \mid D)} = \frac{Pr(D \mid Relevant)}{Pr(D \mid Irrelevant)} \frac{Pr(Relevant)}{Pr(Irrelevant)}$$

The last factor in the equation above is a ratio that depends only on the query Q, not on the document D. Consequently, arranging the documents in descending order by the ratio Pr(D|Relevant)/Pr(D|Irrelevant) will produce exactly the same order as using the ratio Pr(Relevant|D)/Pr(Irrelevant|D). This is fortunate because the probabilities in the former ratio are much easier to compute.

To estimate the ratio Pr(D|Relevant)/Pr(D|Irrelevant), first consider the denominator. In a large corpus such as the web, with billions of pages, or Medline with over 12 million citations, one will rarely be interested in more than a very small fraction of all documents. Thus nearly all documents will be irrelevant. As a result, it is reasonable to assume that Pr(D|Irrelevant) is the same as Pr(D).

To estimate Pr(D|Relevant)/Pr(D) it is common to assume that the documents and queries can be decomposed into statistically independent terms. We will discuss how to deal with statistical dependencies later. Statistical independence implies that Pr(D|Relevant) is the product of Pr(T|Relevant) for all terms T in the document D, and Pr(D) is the product of the unconditional probabilities Pr(T). Because queries can also be decomposed into independent terms, there are two possibilities for a term T in a document D. It is either part of the query Q or it is not. If T is in the query Q, then by definition the term T is relevant, so Pr(T|Relevant) = 1. If T is not in the query Q, then the occurrence of T is independent of any relevance determination, so Pr(T|Relevant) = Pr(T). The ratio Pr(D|Relevant)/Pr(D) is then the product of two kinds of factor: 1/Pr(T) when T is in the query Q and Pr(T)/Pr(T) when T is not in the query Q. So all that matters are the terms in D that are also in Q.

If there are N documents in the entire collection, and if M of them contain a particular term T, then Pr(T) = M/N. The number M is the called the *document frequency* of the term. Since N is a constant for all documents, queries, and terms, it is not needed for determining the ranking of documents. The ratio 1/M is the *inverse document frequency* (IDF). Most term-weighting techniques make use of the IDF or some variation of it.

To see why IDF is important, consider a term such as "human," which occurs in 66.2% of the documents in Medline. Knowing that a citation contains this word gives one less than 1 bit of information. By comparison, the term "normetanephrines" only occurs in five Medline citations, so knowing that a citation contains this term gives one much more information. To be precise, since there are over 12 million Medline citations, this term gives approximately 21 bits of information.

Returning to the computation of Pr(D|Relevant)/Pr(D), we find that it is the product of 1/Pr(T) for all terms that occur both in D and in the query Q. Since the logarithm preserves order and converts products to sums, it is convenient to take the logarithm of this product. This implies that one should order documents by the sum of log(1/Pr(T)) = -log(Pr(T)), for all terms that occur both in each document and the query. This representation has the further advantage that the terms in the sum can be interpreted as measurements of information. It also explains why it makes sense to interpret documents and queries as high-dimensional vectors. The vectors have one entry for each possible term. Define the vector for a document D to have log(1/Pr(T)) as the entry for the term T, whenever T is in D, and to have 0 if T is not in D. Define the vector for a query Q to have 1 as the entry for the term T, whenever T is in Q, and to have 0 if T is not in Q. In general, if $v=(v_1,v_2,\ldots,v_n)$ and $w=(w_1,w_2,\ldots,w_n)$ are two vectors, then the *dot product* (also called the *inner product*) of v and w is given by this sum:

$$v \cdot w = \sum_{1}^{n} v_i w_i$$

Consequently, the logarithm of the ratio Pr(D|Relevant)/Pr(D) is the dot product of the vector for D and the vector for Q.

When the same term occurs more than once in a document, there is a question of how to incorporate this in the term-weighting scheme. One way is to treat each occurrence as being independent, so that one should add the IDF each time. The number of occurrences of a single term in one document is called its *term frequency*. The resulting term-weighting scheme is called the *term frequency, inverse document frequency* (TFIDF) weighting scheme. This scheme is by far the most commonly used term-weighting scheme information retrieval using the vector space model.

The interpretation of documents and queries as vectors gives a versatile and intuitive geometric approach for information retrieval. The analysis above made a number of assumptions which generally do not hold in practice. However, the vector space model can be adjusted to some extent when these assumptions do not hold.

1. *Documents can be decomposed into statistically independent terms.* Statistical dependencies among terms are manifest in the vector space model as axes in the vector space that are not orthogonal. For example, "integrin" occurs in about 28,000 PubMed citations, "primate" occurs in about 116,000 citations, and "fibronectin" occurs in about 23,000 citations. Given that PubMed has about 15 million citations, one would expect that about 200 citations would have both "integrin" and "primate" and that about 40 citations would have both "integrin" and "fibronectin." In fact, there are about 300 citations in the former case and almost 5000 citations in the latter. This suggests that "integrin" and "primate" are nearly independent, but that "integrin" and "fibronectin" are significantly correlated. One can incorporate correlations into the vector space model by using a nonorthogonal basis. In other words, the terms are no longer geometrically at right angles to one another.

2. *Queries can be decomposed into statistically independent terms.* Query terms may have dependencies just as document terms can be dependent. However, queries are usually much smaller, typically involving just a few terms, and any dependencies can be presumed to be the same as the ones for documents.

3. *Queries are highly specific.* In other words, the set of relevant documents is relatively small compared with the entire collection of documents. This holds when the queries are small (i.e., have very few terms), but it is less accurate when queries are large (e.g., when one compares documents with other documents). However, modern corpora (such as Medline or the World Wide Web) are becoming so immense, that even very large documents are small compared with the corpus.

The dot product has a nice geometric interpretation. If the two vectors have unit length, then the dot product is the cosine of the angle between the two vectors. For any nonzero vector v there is exactly one vector that has unit length and has the same direction as v. This vector is obtained by dividing v by its length: $\frac{v}{|v|}$. Thus the angle between vectors v and w is given by $\frac{v \cdot w}{|v||w|}$. The length $|v|$ of a vector is also called its *norm*, hence $\frac{v}{|v|}$ is called the *normalization* of v. Some systems normalize the vectors of documents so that all documents have the same "size" with respect to information retrieval, and so that the dot product is the cosine of the angle between vectors. Normalization does not have a probabilistic interpretation, so it is not appropriate for information retrieval using a query. However, it is useful when documents are compared with one another. In this case, the cosine of the angle between the document vectors is a measure of similarity that varies between 0 and 1. A value of 0 means that the documents are unrelated, while a value of 1 means that the documents use the same terms with the same relative frequencies. One can use similarity functions such as the cosine as a means of classifying documents by looking for clusters of documents that are near one another. All of the clustering algorithms mentioned in section 1.5 can be used to cluster documents either hierarchically or by using some other organizing principle. Clustering techniques based on similarity functions are still in use, but they have been superseded to some extent by citation-based techniques, to be discussed in section 6.4.

In spite of the logical elegance of the vector space model, it has several deficiencies.

1. In many languages, words are composed of letters which can be in more than one "case." In English, letters can be upper- or lower-case. Computers actually deal with *characters*, not letters, so the upper-case variant differs from the lower-case variant. To deal with this ambiguity, most search techniques ignore case distinctions when comparing words.

   Unfortunately, case distinctions are sometimes important. For example, acronyms are usually written using upper-case letters to prevent confusion with the ordinary word. Thus "COLD" (which is the acronym for chronic obstructive lung disease) can be distinguished from "cold" (which has several meanings) by the use of upper-case letters.

2. Many languages, including English, also vary the form of a word for grammatical purposes. This is known as *inflection*. For example, English words can be singular or plural. For example, while "normetanephrines" only occurs in five PubMed citations, the singular form "normetanephrine" occurs in 1207 citations. Although the singular and plural forms have different meanings, such distinctions are rarely important during a search.

It is difficult to map the inflected forms of an English word to a single concept because inflection is highly irregular and ambiguous.

3. The vector space model treats the document as just a collection of unconnected and unrelated terms. There is no meaning beyond the terms themselves.

4. It presumes that the terms are statistically independent, both in the collection as a whole and in each document. The vector space model in general allows for terms that are correlated, but it is computationally difficult even to find correlations between pairs of terms, let alone sets of three or more terms, so very few retrieval engines attempt to find or to make use of such correlations.

5. By focusing exclusively on terms, it cannot take advantage of document structure. webpages and XML documents have a hierarchical structure whose elements are tagged. XML document elements are especially meaningful, but none of this meaning is expressible in the vector space model.

6. By treating documents as independent entities, the vector space model cannot take advantage of interdocument links such as the citations that occur in scientific research papers and the hypertext links that occur in webpages.

Some systems attempt to alleviate these problems by adding dependencies between terms such as how close the terms are to each other in the document. However, these improvements do not address the fundamental weaknesses of this approach.

Ontologies can be useful tools for dealing with these deficiencies, and some of the techniques are introduced in the next section.

Summary

- Words have different degrees of selectivity.
- In the vector space model each document and query is represented by a vector where each component of the vector is the term weight for a word that can occur in a document.
- The most common term weight is the TFIDF weight which is the product of the number of times that the word occurs in the document times the logarithm of the inverse of the number of documents that have the word.

- A query is evaluated by computing the inner product of the query vector and each document vector and sorting. The documents are arranged (ranked) by the inner products.

- The vector space model is a geometric interpretation of the corpus which can be used to classify documents by looking for clusters of documents that are near one another.

- In spite of its elegance and geometric appeal, the vector space model depends on many assumptions and has a number of limitations.

6.3 Using Ontologies for Formulating Queries

Ontologies can address the shortcomings of traditional information retrieval in many ways. In this section, we look at how an ontology can be a context and a source of terminology which can be used to help formulate queries which are then given to an ordinary vector space retrieval engine.

The simplest way to browse an ontology is to use its hierarchical structure as a means of organizing the concepts. One first presents the top-level concepts, then the next level, and so on. This is the same approach used to organize directory structures (file system browsing) and XML documents. (See section 1.4, especially table 1.1.) Ontologies that can be browsed in this way are relatively small and simple, consisting of just a taxonomy.

However, biomedical ontologies can be very large repositories of terminology which require their own information retrieval systems. Consider, for example, the Unified Medical Language System (UMLS), which was introduced in subsection 5.1.1. With 4.5 million terms, the UMLS is much too large to be browsed in any casual manner, and a variety of tools have been introduced to assist one in this task. Some tools are designed for general ontologies, while others are specialized for specific biological or medical ontologies.

The Medical Subject Headings (MeSH) browser is possibly the best known biomedical terminology browser. It is available at `www.nlm.nih.gov/mesh`. The MeSH browser is a specialized browser for MeSH. MeSH is the controlled vocabulary thesaurus of the National Library of Medicine (NLM). MeSH consists of sets of terms naming descriptors in a hierarchical structure that permits searching at various levels of specificity.

The MeSH hierarchy has 11 levels. There are currently 22,568 descriptors in MeSH and more than 139,000 headings called supplementary concept records, available separately. There are 23,887 cross-references that assist in finding the most appropriate MeSH heading, and 106,651 other entry points. The MeSH thesaurus is used by the NLM for indexing articles from 4600 of the world's leading biomedical journals for the Medline repository and the PubMed retrieval system.

The MeSH browser is a vocabulary lookup aid for a specific ontology. It is designed to help quickly locate descriptors of possible interest and to show the hierarchy in which descriptors of interest appear. The MeSH is a stand-alone service that is not directly linked to any retrieval system.

The MeSH browser finds all matches to a query, and orders them alphabetically by primary subject heading. One can ask for exact matches, matches to all of the words in the query, or matches to any of the words in the query. The entries and descriptors contain a great deal of information about the concept. Searches can be restricted in a variety of ways.

The MeSH browser is a very useful browser for medical terminology, but the MeSH thesaurus is very small compared to the UMLS, which was described in more detail in subsection 5.1.1. The UMLS includes MeSH as well as terminology from over 100 other sources. The UMLS itself can be licensed from the NLM, and the distribution includes a simple browser. More sophisticated browsers for the UMLS are available from a number of companies, either commercially or for free on the web. Know-ME is freely available on the web at (Know-Me 2004). It is not known how much of the UMLS is covered by Know-ME. The Apelon DTS covers 8 of the over 100 source vocabularies of the UMLS. It is commercially available, but freely available to government employees. The SKIP Knowledge Browser from SemanTx Life Sciences (Jarg 2005) is available on the web at `www.semantxls.com` and covers all of the source vocabularies of the UMLS, subject to copyright restrictions.

The more sophisticated browsers described above allow one to use ontologies in a number of ways during information retrieval. The most important feature of an ontology is its terminology. The terms of an ontology are called the *controlled vocabulary*. They allow one to formulate a query using the keywords that have been used in a corpus of documents. However, an ontology consists of more than just a controlled vocabulary. It also organizes concepts hierarchically and has many relationships between concepts. When one is browsing an ontology, one can navigate from concept to concept in several ways:

1. One can use more general concepts, when more specific concepts do not find the desired information. This is known as "broadening" the query.

2. One can use more specific concepts, when more general concepts find too much information. This is known as "narrowing" the query.
3. One can use concepts that are related in ways that are nonhierarchical. For example, a nucleolus is a part of the nucleus of a cell. This is a *query modification* which neither broadens nor narrows the query.

Summary

- Ontologies are an important source of terminology that can be used to formulate queries.
- Biological and medical ontologies can be so large and complex that specialized browsing and retrieval tools are necessary.
- Several browsers are now available for the UMLS.
- One can use ontologies as a means of query modification when a query does not return satisfactory results.

6.4 Organizing by Citation

The popularity of the World Wide Web has led to many new search techniques that attempt to utilize its structure. One such technique was developed by Kleinberg in (Kleinberg 1998; Chakrabarti et al. 1998; Gibson et al. 1998). A variation on this technique has since been implemented with considerable success by Google (Page and Brin 2004).

The underlying structure that is utilized by the Kleinberg algorithm is the graph structure of documents in which one document refers to other documents. In the terminology of directed graphs, the documents are nodes and each reference from one document to another is represented by a directed edge from the node of the referring document to the node of the document being referenced.

Like any search algorithm, the Kleinberg algorithm begins with a query. This query is processed using a form of vector space retrieval to obtain a collection of candidate documents. Unlike search techniques based solely on the vector space model, there is no need to be very precise or careful about how term weighting is performed because the final ranking of the retrieved documents uses a very different technique from the one used by vector space retrieval engines. After obtaining the initial set of candidates, the set is expanded somewhat by including documents which refer to or are referenced by the initial candidates. Many other refinements can also be employed to improve the set of candidates.

The next step is unique to the Kleinberg algorithm. A matrix is computed in which the entries represent references of one document to another. For example, suppose that one has three documents called D, E, and F, such that D refers to E and F, while F refers to E. None of the documents refer to D and document E does not refer to any other document in this set. The matrix is given as follows:

	D	E	F
D	0	1	1
E	0	0	0
F	0	1	0

For example, the 1 in the first row and second column of the matrix indicates that document D refers to document E. Note that the rows and columns have been labeled for ease in understanding the meaning of the entries. This matrix is called the *adjacency matrix* of the graph. It is usually designated by the letter A. We now compute the matrix products $A^T A$ and AA^T, where the superscript T means that the matrix has been transposed. The following are these two matrices:

	D	E	F
D	0	0	0
E	0	2	1
F	0	1	1

	D	E	F
D	2	0	1
E	0	0	0
F	1	0	1

The original matrix A will not be symmetric in general, but both of the products will be. In fact, both matrices are positive semidefinite. In other words, the eigenvalues will be nonnegative. The largest eigenvalue is called the principal eigenvalue, and its eigenvectors are called the principal eigenvectors. While it is difficult in general to compute eigenvalues and eigenvectors of large matrices, it is relatively easy to find a principal eigenvector. The space of principal eigenvectors is called the principal component. Principal components analysis (PCA) is a commonly used statistical technique for accounting for the variance in data.

In the case of graphs, the entries in a principal eigenvector measure the relative importance of the corresponding node with respect to the links. Each of the two matrices has a different interpretation. The matrix A^TA is the *authority matrix*. The principal eigenvector ranks the documents according to how much they are referred to by other documents. In this case the principal eigenvector is (0, 1, 0.618). Document D is not referred to by any other document in this set, so it is no surprise that it is not an authority. Document E is referred to by two other documents, and document F is referred to by just one other document. Thus E is more of an authority than F.

It is interesting to compare the Kleinberg algorithm with what one would obtain using simple citation counts, as is often done in the research literature. Since E has twice as many citation counts, one would expect that E would be twice as authoritative as F. However, the principal eigenvector adjusts the authoritativeness of each citation so that the authority weights are consistent. In effect, the algorithm is implicitly assigning a level of "quality" to the citations. In other words, being cited by a more authoritative document counts more than being cited by a less authoritative source.

The matrix AA^T is the *hub matrix*. A *hub* or *central source* is a document that refers to a large number of other documents in the same set. In the research literature, a survey article in a field would be a hub, and it might not be an authority. This would be the case shortly after the survey article has been published and before it has been cited by other articles. The principal eigenvector ranks the documents according to how much of a hub it is for the particular query. In the example, the principal eigenvector is (1, 0, 0.618). Since document E does not refer to any other documents in this set, it is not a hub. Document D is the main hub, since it refers to two other documents, while document F is much less of a hub since it only refers to one other document.

One interesting and useful feature of the Kleinberg algorithm is that it is possible for a document to have considerable importance either as an authority or a hub even when the document does not match any of the terms in the query. As a result, the Kleinberg algorithm improves both selectivity and coverage of retrieval. It improves selectivity by improving the ordering of documents so that the most relevant documents are more likely to occur at the beginning of the list. It improves coverage by retrieving documents that do not match the query but which are cited by documents that do match.

However, the Kleinberg algorithm does have its weaknesses. Because it focuses on the principal eigenvector, the other eigenvectors are ignored even when they may represent the actual focus of interest of the researcher. This is the case when a relatively small community uses the same terminology as a much larger community. For this reason, commercial search engines like Google that are based on the Kleinberg algorithm do not implement it in its original form.

Google, for example, uses a formula which differs from the Kleinberg algorithm in several ways:

1. The rank of a document is normalized by dividing it by the total number of references made by that document. Thus a document with a large number of references will have its influence reduced a great deal, while documents with a small number of references will have more influence. Presumably this was done to prevent the algorithm from being easily subverted.

2. Instead of the normal eigenvector equation, an additional term was added that serves to "dampen" the process of computing the rank, but which adds some arbitrariness to the computed rank.

3. The original adjacency matrix is used rather than either the authority or hub matrix. Thus the algorithm is measuring a form of popularity rather than whether the document is authoritative or a central source.

Current search engines have another weakness. The original set of candidate documents is obtained using simple word-matching strategies that do not incorporate any of the meaning of the words. As a simple example, try running these two queries with Google: "spinal tap" and "spinal taps." From almost any point of view these two have essentially the same meaning. Yet, the documents displayed by Google have completely different rankings in these two cases. Among the first ten documents of each query there is only one document in common. Although the spinal tap query is problematic because there is a popular movie by that name, one can easily create many more such examples by just varying the inflection of the words in the query or by substituting synonymous words or phrases.

One obvious way to deal with this shortcoming of Google would be to index using concepts rather than character strings. This leads to the possibility of search based on the meaning of the documents. Many search engines, including Google, are starting to incorporate semantics in their algorithms. We discuss this in section 6.6.

Summary

- Citations (such as hypertext links) can be used to rank documents relevant to a query according to various criteria:
  1. Authoritativeness
  2. Central source
  3. Popularity
- Citation ranking improves both selectivity and coverage.
- However, citation ranking has a number of weaknesses:
  1. Current systems are based on matching words in the query with words in documents and do not consider the meaning of the words.
  2. Only the principal eigenvector is used, so smaller communities will be masked by larger ones.

6.5 Vector Space Retrieval of Knowledge Representations

One of the main assumptions of the vector space model is that documents are composed of collections of terms. While some systems attempt to take advantage of correlations between terms, such correlations are difficult to determine accurately, and the number of correlations that must be computed is huge. In any case, the terms are still disjoint from one another. Knowledge representations change this situation. Terms can now be complex concept combinations that are built from simpler terms. Thus a term like "flu vaccine" contains both "flu" and "vaccine" as well as the complex relationship between these two concepts which expresses the effect of the vaccine on the influenza virus as well as the the derivation of the vaccine from the virus and in response to it. In the UMLS, all three of these are concepts, and they are related to one another.

To see how natural, as well as how subtle, concept combinations can be, try juxtaposing two commonly used terms in different orders. For example, "test drug" and "drug test." Although these two have completely different meanings, most search engines give essentially the same answer for both. Indeed, "test drug" can be interpreted in two ways depending on whether "test" is a verb or adjective. The term "drug test" also has several meanings. As an exercise, try some other pairs of terms to see how many meanings you can extract from them. Concept combination could be a powerful information retrieval mechanism, provided it is properly interpreted. With a relatively small number of basic concepts along with a small number of conventional relationships, one can construct a very large number of concept combinations.

Concept combination, also called *conceptual blending* and *conceptual integration*, is an active area of research in linguistics. The meaning of a concept combination requires a deeper understanding of the relationship between words and the phenomena in the world that they signify. Based on the earlier work of Peirce, de Saussure, and others in the field of semiotics, Fauconnier and Turner (1998, 2002) have developed a theory of conceptual blending that explains how concepts can be blended. However, this theory is informal. Goguen has now developed a formal basis for conceptual blending (Goguen 1999; Goguen and Harrell 2004). Furthermore, Goguen and his students have developed software that automates the blending of concepts, and their system has been used to understand and even to create poetry and other narratives. Concept combination is closely connected with human categorization and metaphor. For an entertaining account of these topics, see Lakoff's book with the intriguing title *Women, Fire and Dangerous Things: What Categories Reveal about the Mind* (Lakoff 1987).

The tool developed by Goguen and his students, mentioned above, is capable of finding a wealth of concept combinations even when the concepts are relatively simple. For the words "house" and "boat," their tool finds 48 complete blends and 736 partial blends. Two of these have become so common that they are considered single words; namely, "houseboat" and "boathouse." Others are less obvious, but still make sense, such as a boat used for transporting houses, an amphibious recreational vehicle, or a boat used permanently on land as a house.

As one might imagine, the combinatorial possibilities for combinations become enormous when there are more than two words being combined. A typical title for a biomedical research article can have a dozen words. Understanding the meaning of such a title can be a formidable undertaking if one is not familiar with the subject matter of the article, as we pointed out in section 1.6. Goguen and Harrell (2004) pointed out that conceptual blending alone is not sufficient for understanding entire narratives that involve many such blends (Goguen and Harrell 2004). They introduced the notion of *structural blending*, also called *structural integration*, to account for the meaning of whole documents.

Having introduced concept combinations, one still has the problem of how such topics can be handled by an information retrieval system. Unlike ordinary terms, concept combinations are necessarily dependent on the concepts that were combined (both semantically and statistically). As we noted in section 6.2, terms are considered to be statistically independent by default. No such assumption will be valid for concept combinations. Fortunately, the vector space model can incorporate such dependencies.

To see how this is done, consider the term "flu vaccine." In the usual vector space model, this is just two independent terms, "flu" and "vaccine." These two terms represent two of the factors in the ratio

$$Pr(D \mid Relevant)/Pr(D \mid Irrelevant)$$

from section 6.2 which determines the degree of relevance of a document D to a query. If we presume that "flu vaccine" is a concept combination which has been indexed, then the two factors for "flu" and "vaccine" should be replaced by the single factor for "flu vaccine." In other words, the concept combination is a new term which supersedes the terms that were combined. However, this is done only when both the document and the query use the concept combination. If only one of them has the combination, then the individual terms must still be used to measure relevance.

Although the vector space model can be adapted to deal with concept combinations, it still suffers from the deficiencies already enumerated in section 6.2. Techniques that deal more directly with the meaning of the documents and queries are considered in section 6.6 and in chapter 8.

Summary

- Concepts can be combined in many ways which are much deeper than just the juxtaposition of the words used.
- The vector space model can be extended to deal with concept combinations, but it is still subject to deficiencies because it does not deal with the meaning of words.

6.6 Retrieval of Knowledge Representations

Information retrieval systems, including Google, generally do not make use of the meaning of the information in the document. As a result, searches will necessarily be hit-or-miss activities. Sometimes one will get lucky, other times the retrieval will be useless. These engines use a variety of mechanisms for overcoming this limitation, but they can never completely eliminate it.

This is in striking contrast with relational database queries which always return all of the items specified and no others. Using the terminology of information retrieval, relational queries always have 100% coverage and selectivity. It is natural to imagine that one could try to achieve the same coverage and selectivity with information retrieval. To do so one must overcome several difficult problems:

1. The meaning of natural language text is complex and difficult to represent in a manner necessary for retrieval using database query languages.

2. Even if one could develop a representation language for natural language, it is very difficult to translate from natural language into the representation language.

The first problem above is addressed by ontologies. While we do not yet have ontologies that are sufficiently deep and have enough coverage for the biomedical domain, the ontologies are improving steadily. Even the incomplete and relatively shallow ontologies that are available today (such as the UMLS) are recognized as important and useful resources for biomedical research.

The second problem is in many ways the more problematic one. Natural language text is not easily understood by computers. The process whereby natural language is translated from text to the representation language is called natural language processing (NLP). The result of applying NLP to a document is called the *knowledge representation* of the document. In a knowledge representation, all terms are expressed unambiguously as instances of classes in the ontology, that is, they refer to the corresponding concept in the ontology. Relationships between terms are also expressed unambiguously using the relationships in the ontology. To see what a knowledge representation looks like, see any of the XML documents shown in chapter 1, especially the ones in section 1.6.

Once natural language text has been converted to a knowledge representation, one can infer additional facts that were not explicitly stated. These inferences take advantage of the ontology which can have many rules that allow such inferencing. For example, if one speaks of a patient that has leukemia, one can infer that this individual is a human who has cancer.

NLP techniques are not yet sufficiently well developed to be able to produce knowledge representations that always exactly represent the meaning of every natural language statement. However, it is possible to extract knowledge representations that are good enough to be used for information retrieval. Some commercial systems are available already that extract knowledge representations from biomedical text and that index these knowledge representations for rapid retrieval.

Once the knowledge representations have been extracted, there are two approaches to querying the documents:

1. Use a precise and unambiguous query language. For relational databases one can use the SQL language, and such languages are now being developed for XML.

2. Query the documents with natural language. This means that NLP techniques must be used to extract the knowledge representation of the query. The query knowledge representation can then be matched against the knowledge representations of the documents.

Although both approaches support inferencing, they differ in many ways.

In the first approach the query uses a specialized query language that has little resemblance to natural language. There is now a standard query language for XML documents, and chapter 8 discusses it in some detail. The advantage is that one can be confident that the query will return exactly every item that is relevant to the query. However, if the knowledge representations of the documents are inaccurate, then the query results will also be inaccurate. Since NLP techniques are still not perfect, one cannot expect that query results will also be perfect. The disadvantage of specialized query languages is that one must learn how to program in the language, and this can require a significant amount of effort.

The second approach allows a person to query the document corpus by using natural language queries. The queries are expressed as a knowledge representation in the same way that documents are expressed. The retrieval system answers the query by looking for matching knowledge representations in the corpus. The match can be complete or partial. This approach is less brittle than the specialized query language approach, and so is better suited to knowledge representations that are somewhat inaccurate. However, the fact that it is less precise can be a disadvantage when the knowledge representations are known to be good and one would like to extract precise information from the corpus.

One example of the second approach is the Semantic Knowledge Indexing Platform (SKIP). A demonstration is available online at `www.semantxls.`

com. Figure 6.2 is an example of a query presented to SKIP. The query in this case is "What regulates the adhesiveness of integrins at the plasma membrane of lymphocytes, and is responsible for association of PSCDs with membranes?" The database used by the public demonstration is the NCBI Reference Sequences (RefSeq) database (NIH 2004a). One can further restrict the query by specifying that it must include some concepts and that it must exclude others.

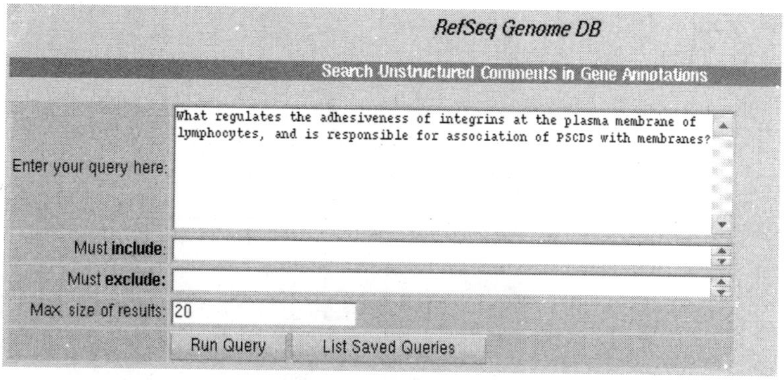

Figure 6.2 Query screen for the SKIP retrieval system.

After clicking on "Run Query", the query text is converted to a knowledge representation as shown in figure 6.3. In this figure the knowledge representation for the query is to the right of the query. At the bottom of the figure are references to the two documents that match the query best. Other documents that match in other ways are shown in figure 6.4. The boxes in the knowledge representation represent instances of the concepts shown, and the arrows between the boxes represent relationships between the concepts. Note that "plasma membrane" is a single concept in the UMLS so it is represented using a single box rather than two boxes joined by a relationship.

SKIP uses a high-performance indexing technology that was inspired by biological sequence matching. As discussed in chapter 7, one can find homologous sequences by using short sequences to index the sequence database. One then extracts short sequences from the query and matches them with the ones in the index. For nucleotide sequence matching this can actually be done chemically by synthesizing the query sequence as an oligonucleotide and hybridizing it with the target DNA (in single-stranded form). The SKIP index generalizes this technique to find knowledge representations that are homologous to a query (Baclawski 1997a).

The matching documents are arranged by how well they match the query.

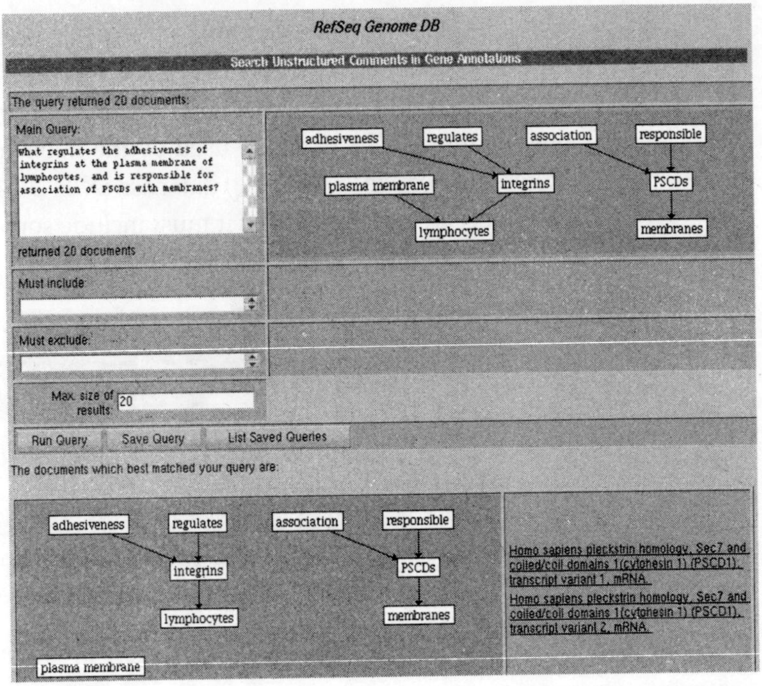

Figure 6.3 Top part of the result screen for the SKIP retrieval system showing the knowledge representation of the query and the document that matches the query the best.

The first two matching documents match the best. They contain all of the concepts in the query, and all but two of them are related as in the query. The only relationship that was not found in these documents was between "plasma membrane" and "lymphocytes." The knowledge representation to the left of the document link shows the part of the query that was found in the document.

Scrolling down the result screen gives figure 6.4 which shows a number of other documents that contain fewer concepts and relationships than the documents that match the best. Continuing to scroll the results screen will show many more matching documents, but these match less and less of the original query.

This approach to retrieval has some advantages that were already discussed above. Another advantage specific to SKIP is that the retrieved documents are arranged in groups and labeled by how they match using an intuitive and visually appealing graphical structure.

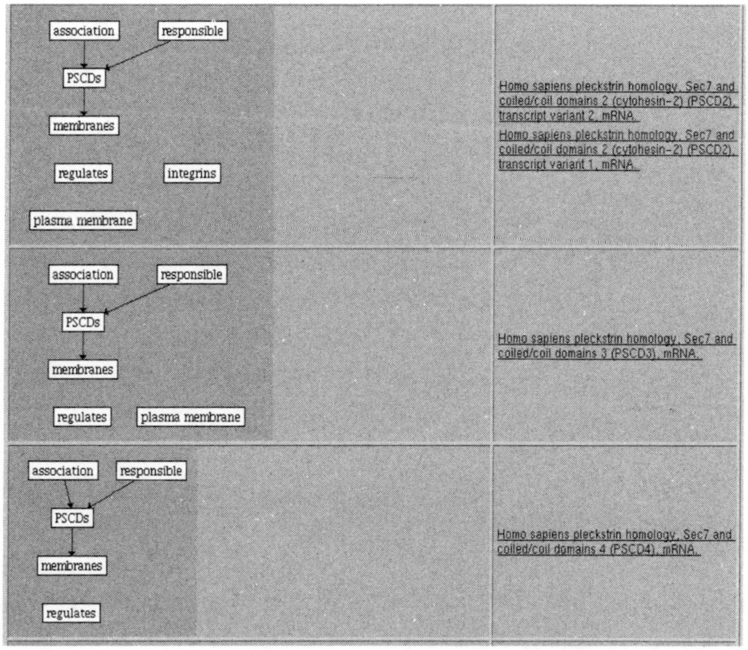

Figure 6.4 Other documents that match a query. The knowledge representations shown on the left show the part of the query that occurs in the documents on the right.

Summary

- Translating natural language text to a representation language that captures meaning remains an unsolved problem, but reasonably good knowledge representations are possible.
- Knowledge representations can be queried in two ways:
  1. Using a formal query language
  2. Matching against another knowledge representation
- Systems that use knowledge representation matching are now available. Such a system allows one to query a corpus using natural language.

7 *Sequence Similarity Searching Tools*

Information retrieval can take many forms, and does not have to be based on natural language. In bioinformatics, it is very common to base queries on biological sequences, the biochemical language of cells. Indeed, most predictions of biological function are obtained by comparing new sequence data (for which little is known) with existing data (for which there is prior knowledge). The comparison is performed by using the new sequence data as a query to retrieve similar sequence data in a corpus of such data. Such comparisons are of fundamental importance in computational biology. Similar sequences are referred to as being *homologous*.

In this chapter we present the basic concepts necessary for sequence similarity and the main approaches and tools for sequence similarity searching. The most commonly used sequence similarity searching tools in computational biology are FASTA, Basic Local Alignment Search Tool (BLAST), and the many variations of BLAST. All these algorithms search a sequence database for the closest matches to a query sequence. It should be noted that all three algorithms are database search heuristics, which may completely miss some significant matches and may produce nonoptimal matches. Of these three tools, BLAST is the most heavily used sequence analysis tool available in the public domain.

7.1 Basic Concepts

Like information retrieval, sequence similarity searching is a process whereby a relatively small "query" sequence is compared with a large genomic "corpus" of sequence information. In a perfect match the query sequence occurs as a subsequence of the corpus. In practice such perfect matches seldom occur so it is necessary to have a measure of similarity. Each potential match is a way of lining up the residues in the query sequence with part of a sequence in the corpus. Such a lining up is called an *alignment*. In an alignment, the match can fail to be an exact match in two ways: aligned residues can be different and there may be gaps in one sequence relative to the other. For each alignment one can compute a similarity measure or *score* based on the residues that match or fail to match and the sizes of the gaps. Matches generally contribute positively to the overall score while mismatches and gaps contribute negatively. The *scoring matrix* specifies the contribution to the overall score of each possible match and mismatch. This contribution can be dependent on the position of a residue in the query sequence, in which case the scoring matrix is called a *position-specific scoring matrix* (PSSM). Such matrices are also called "profiles" or "motifs." If the contributions do not depend on positions, then the scoring matrix specifies the score associated with a substitution of one type of residue for another. Such a scoring matrix is called a *substitution matrix*. The *gap penalties* specify the effect of gaps on the score. The objective of a sequence similarity matching tool is to find the alignments with the best overall score.

There are a number of ways to compute the alignment score. The primary distinction is between nucleotide sequences and amino acid sequences. The scoring for amino acid sequence similarity is more complicated because there are more kinds of amino acids and because amino acid properties are more complicated than nucleotide properties. For example, chemical structures and amino acid frequencies can both be taken into consideration. If two aligned residues have a very low probability of being homologous, a heavy penalty score is given for such a mismatch. Protein evolution is believed to be subject to stronger forces than DNA evolution, so that some amino acid substitutions (which result in Mendelian disorders) are much less functionally tolerant than others because natural selection processes select against them.

The two most commonly used substitution matrices for amino acids are the point accepted mutation (PAM) (Dayhoff et al. 1978) and the blocks substitution matrix (BLOSUM) (Henikoff and Henikoff 1992). BLOSUM is more popular than PAM. In both cases, the entries in the matrix have the form $s_{ij} = Clog_C(r_{ij})$, where C determines the units by which the entries are scaled (usually 2 for BLOSUM and 10 for PAM) and r_{ij} is the ratio of the estimated frequency with which the amino acids i and j are substituted due to evolutionary descent, to the frequency with which they would be substituted by chance. The numerator of this ratio is computed by using a sample of known alignments. This formula is known more succinctly as the *log-odds* formula. Logarithms are used so that total scores can be computed by adding the scores for individual residues in the alignment. Vector space retrieval for text databases uses the same technique. For convenience, s_{ij} is often rounded to the nearest integer.

BLOSUM matrices were based on data derived from the BLOCKS database (Henikoff and Henikoff 1991, 1994), which is a set of ungapped alignments of protein families (i.e., structurally and functionally related proteins). Using about 2000 blocks of such aligned sequence segments, the sequences of each block are sorted into closely related clusters, and the probability of a meaningful amino acid substitution is calculated based on the frequencies of substitutions among these clusters within a family. The number associated with a BLOSUM matrix (such as BLOSUM62) indicates the cut-off value for percentage sequence identity that defines the clusters. In particular, BLOSUM62 scores alignments with sequence identity at most 62%. Note that a lower cut-off value would allow for more diverse sequences into groups, and the corresponding matrices are therefore appropriate for examining more distant relationships.

The PAM matrices are based on taking sets of high-confidence alignments of many homologous proteins and assessing the frequencies of all substitutions. The PAM matrices were calculated based on a certain model of evolutionary distance from alignments of closely related sequences (about 85% identical) from 34 "superfamilies" grouped into 71 evolutionary trees and containing 1572 point mutations. Phylogenetic trees were reconstructed based on these sequences to determine the ancestral sequence for each alignment. Substitutions were tallied by type, normalized over usage frequencies, and then converted to log-odds scores. The value in a PAM1 matrix represents the probability that 1 out of 100 amino acids will undergo substitution. Multiplying PAM1 by itself generates PAM2, and more generally $(\text{PAM1})^n$ is a scoring matrix for amino acid sequences that have undergone n multiple and independent steps of mutations. Thus, the PAM250 matrix has undergone 130 more steps of mutations than the PAM120 matrix. Hence, for aligning closely related amino acid sequences, PAM120 matrix is a good choice; for aligning more distantly related amino acid sequences, PAM250 matrix is a more appropriate choice. It should be noted that errors can be amplified during the multiplication process, and thus higher-order PAM matrices are more error-prone. By comparison, in a BLOSUM62 matrix, each value is calculated by dividing the frequency of occurrence of the amino acid pair in the BLOCKS database, "clustered" at the 62% level, by the probability that the same amino acid pair aligns purely by chance. PAM matrices are scaled in

$10log_{10}$ units, which is roughly the same as third-bit units. BLOSUM matrices are usually scaled in half-bit units. In either type of scoring matrix, if the score is 0, then the alignment of the amino acid pair is equivalent to being coincidental; if the score is positive, the alignment of the amino acid pair is found to be more often than by chance; if the score is negative, the alignment of the amino acid pair is found to be even less often than by chance. The NCBI BLAST tool allows one to choose from a variety of scoring matrices, including PAM30, PAM70, BLOSUM45, BLOSUM62, and BLOSUM80. A more complete roster of scoring matrices (PAM10-PAM500, and BLOSUM30-BLOSUM100) is available at the following ftp site: `ftp://ftp.ncbi.nlm.nih.gov/blast/matrices`.

Mutational events include not only substitutions but also insertions and deletions. Consequently one must also consider the possibility of alignment gaps. However, gaps are a form of sequence mismatch, so they affect the score negatively. During the process of alignment, the initiation of a new gap adds a penalty called an *opening* gap penalty, while the widening of an existing gap adds an *extension* gap penalty. For amino acid sequences, it is common to set extension gap penalties to be lower than opening gap penalties because certain protein domains evolve as a unit, rather than as single residues.

Summary

- Sequence similarity search is a process whereby a query sequence is compared with sequences in a database to find the best matches.
- The score depends on the scoring matrix and the gap penalties.
- The scoring matrix can be position-independent (substitution matrix) or position-sensitive.
- The most commonly used substitution matrices are PAM and BLOSUM.

7.2 Dynamic Programming Algorithm

The first algorithm that was used for sequence matching was a dynamic programming algorithm, called the Needleman-Wunsch algorithm (Needleman and Wunsch 1970). A dynamic programming algorithm finds an optimal solution by breaking the original composite problem recursively into smaller and smaller problems until the smallest problems have trivial solutions. The smaller solutions are then used to construct the solutions for the larger and larger parts of the original problem until the original problem has been solved. In this case, the composite problem is to determine the optimal alignment of the two sequences at their full lengths. This alignment problem is split by breaking down the two sequences into smaller segments. The splitting continues recursively until the subproblem consists of comparing two residues. At this point the score is obtained from the scoring matrix. The resulting alignment is guaranteed to be globally optimal. Smith and Waterman (1981) modified the Needleman-Wunsch algorithm to make it run faster but only guaranteeing that the alignment is locally optimal.

Although the exact dynamic programming algorithm are guaranteed to find the optimal match (either global or local), they can be very slow. This is especially true for a full search of the very large sequence databases such as GenBank for nucleotide sequences and SWISS-PROT for amino acid sequences that are commonly used today. To deal with this problem, a number of heuristic techniques have been introduced, such as FASTA and BLAST, that give up the guarantee of optimality for the sake of improved speed. In practice, the effect on optimality is small, so the improvement in performance is worth the compromise. These new algorithms search for the best local alignment rather than the best global alignment.

Summary

- The earliest sequence similarity searching algorithms applied exact dynamic programming either globally or locally.
- Current algorithms are heuristic methods that still use dynamic programming but apply approximations to improve performance.

7.3 FASTA

FASTA is both a collection of programs and a widely used format (Pearson and Lipman 1988). The programs are available from `www.ebi.ac.uk/fasta33`. This site is also a web service for performing FASTA sequence searching.

The FASTA algorithm was the first widely used algorithm for amino acid and nucleotide similarity searches. The first step of the FASTA algorithm is to find exactly matching "words" of length $ktup$. The default value of *ktup* is 2 for amino acid and 6 for nucleotide sequences. The next step is to extend the matches of length *ktup* to obtain the highest scoring ungapped regions. In the third step, these ungapped regions are assessed to determine whether they could be joined together with gaps, taking into account the gap penalties. Finally the highest scoring candidates of the third step are realigned using the full Smith-Waterman algorithm, but confining the dynamic programming matrix to a subregion around the candidates. The trade-off between speed and sensitivity is determined by the value of the *ktup* parameter. Higher values of *ktup*, which represent higher "word" sizes, will give rise to a smaller number of exact hits and hence a lower sensitivity, but will result in a faster search. For the purpose of tuning, the *ktup* parameter will generally be either 1 or 2 for amino acid sequences and can range from 4 to 6 for nucleotide sequences.

A sequence file in FASTA format can contain several sequences. Each sequence in FASTA format begins with a single-line description, followed by lines of sequence data. The description line must begin with a greater-than symbol (>) in the first column. An example sequence in FASTA format is shown in figure 7.1.

```
>gi|11066424|gb|AF200505.1|AF200503S3 Pongo pygmaeus
GGCGCTGATGGACGAGACCATGAAGGAGTTGAAGGCCTACAAATCGGAAC
TGGAGGAACAACTGACCCCGGTGGCGGAGGAGACGCGGGCACGGCTGTCC
AAGGAGCTGCAGGCGGCGCAGGCCCGGCTGGGCGCGGACATGGAGGACGT
GCGCGGCCGCCTGGTGCAGTACCGCGGCGAGGTGCAGGCCATGCTCGGCC
AGAGCACCGAGGAGCTGCGGGCGCGCCTCGCCTCCCACCTGCGCAAGCTG
CGCAAGCGGCTCCTCCGCGATGCCGATGACCTGCAGAAGCGTCTGGCAGT
GTACCAGGCCGGGGCCCGCGAGGGCGCCGAGCGCGGCGTCAGCGCCATCC
GCGAGCGCCTGGGGCCCCTGGTGGAACAGGGCCGCGTGCGGGCCGCCACT
GTGGGCTCCGTGGCCGGCAAGCCGCTGCAGGAGCGGGCCCAGGCCTGGGG
CGAGCGGCTGCGCGCGCGGATGGAGGAGATGGGCAGCCGGACCCGCGACC
GCCTGGACGAGGTGAAGGAGCAGGTGGCGGAGGTGCGCGCCAAGCTGGAG
GAGCAGGCCCAGCAGATACGCCTGCAGGCCGAGGCCTTCCAGGCCCGCCT
CAAGAGCTGGTTCGAGCCCCTGGTGGAAGACATGCAGCGCCAGTGGGCCG
GGCTGGTGGAGAAGGTGCAGGCTGCCGTGGGCACCAGCGCCGCCCCTGTG
CCCAGCGACAATCACTGA
```

Figure 7.1 FASTA format of a 718-bp DNA sequence (GenBank accession number AF200505.1) encoding exon 4 of *Pongo pygmaeus* apolipoprotein E (ApoE) gene.

Summary

- FASTA is a set of sequence similarity search programs.
- FASTA is also a sequence format, and this is currently the main use for FASTA.

7.4 BLAST

The most widely used tool for sequence alignment is BLAST (McGinnis and Madden 2004), and it plays an important role in genome annotation (Muller et al. 1999). BLAST uses a heuristic approach to construct alignments based on optimizing a measure of local similarity (Altschul et al. 1990, 1997). Because of its heuristic nature, BLAST searches much faster than the main dynamic programming methods: the Needleman-Wunsch (Needleman and Wunsch 1970) and Smith-Waterman (Smith and Waterman 1981) algorithms.

In this section we begin by explaining the BLAST algorithm. The algorithm is then used for a number of types of search, as presented in subsection 7.4.2. The result of a BLAST search is a collection of matching sequences (or "hits"). Each hit is given a number of scores that attempt to measure how well the hit matches the query. These scores are explained in subsection 7.4.3. We end the section with some variations on the BLAST algorithm.

7.4.1 The BLAST Algorithm

BLAST has become the most popular tool used by biologists. There are two main versions of BLAST:

NCBI BLAST `www.ncbi.nlm.nih.gov/blast`
This is the version that is most commonly used (Altschul et al. 1990, 1997).

WU-BLAST `blast.wustl.edu`
Washington University BLAST (Altschul et al. 1990; Gish and States 1993; States and Gish 1994)

The central idea of the BLAST algorithm is that a statistically significant alignment is likely to contain a high-scoring matching "word." BLAST is a heuristic that attempts to optimize a specific measure of sequence similarity, based on a "threshold" parameter. In terms of time complexity, the BLAST algorithm requires time proportional to the product of the lengths of the query sequence and the target database.

Before discussing the algorithmic details of BLAST, it is helpful to mention some key concepts. A *segment* is a consecutive sequence of letters from the DNA or amino acid alphabet. Dashes are used to denote gaps in a sequence. A *segment pair* is a pair of segments having the same length.

The BLAST algorithm consists of the following steps:

1. **Preprocessing step.** BLAST masks (omits) simple, low-complexity regions in the query sequence because these regions are not biologically informative.

2. **Word generation step.** BLAST generates a list of *words* (i.e., short sequences) from the query sequence. The default word lengths are 3 (for amino acid sequences) and 11 (for nucleotide sequences). The words are then matched with the database to find the matches whose score exceeds a given threshold T. Such matches are called *hits*. BLAST uses BLOSUM62 as the default scoring matrix for amino acids to find the hits. No gaps are allowed during this step of the algorithm. A higher threshold T increases the speed but also increases the probability that biologically significant segment pairs will be missed. Thus there is a tradeoff between speed and sensitivity that can be adjusted according to individual needs and the resources available.

   A search across the entire target sequence database for exact matches of the hits is then performed. The search makes use of database indexes for efficiency. As a result, this part of the algorithm is very fast. Matches obtained in this part of the algorithm are the *seeds* for a potential alignment between the query and database sequences.

3. **Word extension step**. In this step some of the hits obtained in the word generation step are extended to find full alignment matches. The original BLAST algorithm (now called the *one-hit* method) uses all of the hits with a score above the threshold. It then attempts to extend the alignment from each matching word in both directions as long as the alignment score is no worse than an amount X below the maximum score attained so far. The resulting alignment of an extended word is called a high-scoring segment pair (HSP). The extension step is computationally expensive: if T and X are chosen to attain a reasonable sensitivity, the extension step will typically account for more than 90% of the execution time of BLAST. The original BLAST algorithm did not allow gaps in matching alignments.

   More recently the BLAST algorithm has used a different technique for selecting the hits that will be extended. In the *two-hit* method, an HSP is detected if it contains two nonoverlapping words of length W whose scores are at least the threshold T, with starting positions that differ by no more than A residues. The gapped BLAST algorithm uses the two-hit method (Altschul et al. 1997). The two hits are extended in both directions by means of dynamic programming. In the two-hit method, a smaller threshold T can be used because the requirement that two hits occur near each other limits the number of hits that qualify. As the name suggests, the gapped BLAST algorithm can introduce gaps to matching alignments.

4. **Evaluation step**. BLAST determines the statistical significance of each of the HSPs obtained in the word extension step and gives a report on the HSPs that have been found. This report is discussed in more detail in subsection 7.4.3 below.

   Sometimes two or more segment pairs can be merged into a single, longer segment. In such cases, a joint assessment of the statistical significance can be made using the Poisson method or the sum-of-scores method. The earliest BLAST versions used the Poisson method, while more recent BLAST versions (including WU-BLAST and gapped BLAST) use the sum-of-scores method.

FASTA differs from BLAST primarily in that FASTA strives to get exact "word" matches, whereas BLAST uses a scoring matrix (such as the default BLOSUM62 for amino acid sequences) to search for words that may not match exactly, but are high-scoring nevertheless. FASTA does not have a preprocessing step as in BLAST, and FASTA does not use the BLAST strategy of extending seeds using sophisticated dynamic programming. Both FASTA and BLAST have a word generation step which does not allow gaps, followed by a Smith-Waterman alignment step that can introduce gaps.

Summary

- BLAST uses a heuristic approach to find alignments quickly.
- The BLAST algorithm consists of these steps:
  1. Preprocessing: omit uninformative regions of the query.
  2. Word generation: generate small seed matches.
  3. Word extension: extend single seeds or pairs of seeds.
  4. Evaluation: compute measures of significance.

7.4.2 BLAST Search Types

BLAST can be used to perform the following types of sequence similarity searches.

blastn
The query is compared against a nucleotide sequence database, using parameters appropriate for nucleotides.

blastp
Query against an amino acid sequence database, using parameters appropriate for amino acid sequences.

blastx
The query is first translated into each of the six possible reading frames, then compared against amino acid sequence databases.

tblastn
The query is taken to be an amino acid sequence, but it is compared against a nucleotide sequence database after translating each database entry into an amino acid sequence using all six reading frames.

tblastx
The query is first translated into each of all six possible reading frames, then compared against nucleotide databases, with each database sequence translated into an amino acid sequence in each of its possible reading frames.

The advantage of using one of the `blastx`, `tblastn`, or `tblastx` search methods is that it allows one to find matches to distantly related sequences. The disadvantage is that the searches are computationally intensive and may take an inordinate length of time. An example of the use of `blastx` for a DNA sequence similarity search is shown in figure 7.2.

BLAST searches can be obtained either by using a publicly available web service (e.g., `www.ncbi.nlm.nih.gov/blast/`) or by downloading the BLAST program and running it locally. Both of these techniques require that the queries be in FASTA format. The web services are convenient but only accept a single query at a time. This can take a long time if one needs to run a large number of queries. It also has the disadvantage that there are only a limited number of customization options. For example, there are typically only a few choices for the substitution matrix used in an amino acid sequence query.

Because of the large computational requirements of BLAST, it is becoming increasingly common to run BLAST searches on a cluster of computers.

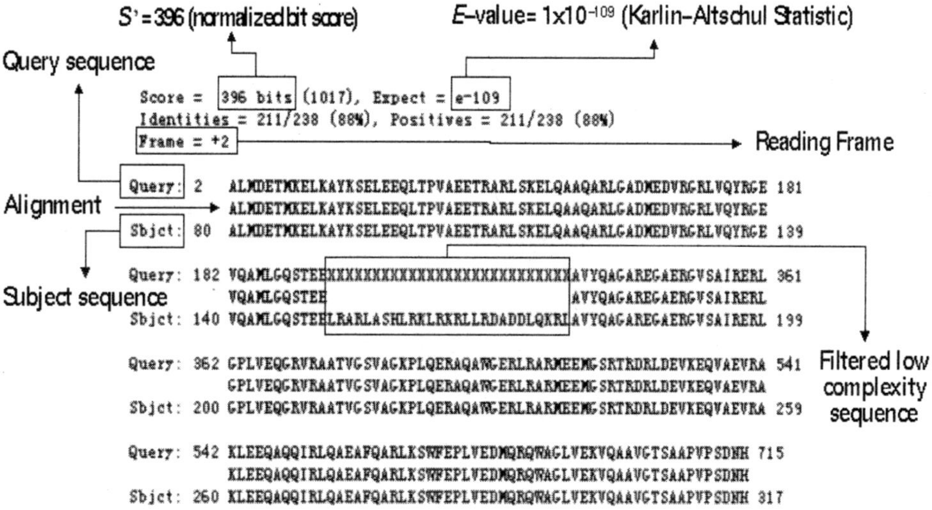

Figure 7.2 Illustration of blastx (version 2.2.10) output using a 718-bp DNA sequence (GenBank accession number AF200505.1) encoding exon 4 of *Pongo pygmaeus* ApoE gene.

This is necessary if one must process a large number of queries. Managing such a collection of queries becomes a problem in itself, and software has been developed for this task. For example, `BeoBLAST` is a Perl program that distributes individual BLAST jobs across the nodes of a cluster (Grant et al. 2002). Although `BeoBLAST` was designed for use on Linux Beowulf clusters, it can be used on any collection of computers that satisfy a few basic requirements such as having a BLAST program, a web server, and the GNU queue service. For more information about how one can configure a computer for BeoBLAST, download it from `bioinformatics.fccc.edu/software/OpenSource/beoblast/beoblast.shtml` and read the installation instructions.

As a general rule, if a query is an amino acid sequence, then it is better to search against an amino acid sequence database rather than against a nucleotide sequence database. There are several reasons for this. First, the genetic codes are degenerate (i.e., several different genetic codes encode the same amino acid). Direct amino acid sequence alignment eliminates the noise that results from the degeneracy. Second, amino acid databases tend to be more sparsely populated than nucleotide sequences because constraints during protein evolution are more severe than during DNA evolution. Unlike DNA, a protein must fold into a functionally competent 3D structure.

Summary

- In addition to BLAST searches for nucleotide and amino acid sequences, there are search types that take into account the translation from nucleotide to amino acid.

- There are publicly available BLAST web services for searches done with one sequence at a time.

- Clusters of computers are frequently used for performing large batches of BLAST searches.

7.4.3 Scores and Values

The output of a BLAST search consists of a set of HSPs annotated with various measures of their statistical significance. The score of each HSP is usually denoted by S and is called the *raw score*. The raw score depends on the various customization parameters of the search such as the scoring matrix. The *normalized score* adjusts the raw score so that alignment scores from different searches can be compared (Altschul et al. 1997). The normalized score is $S' = (\lambda S - \ln K)/\ln 2$, where λ and K are the Karlin-Altschul statistics (Karlin and Altschul 1990, 1993). The reason why one divides by $\ln 2$ is so that the units of the normalized score are in *bits*, a term borrowed from information theory (Altschul 1991). As a result, S' is also called the *bit score*.

The HSP with the largest score is called the *maximal-scoring segment pair* (MSP). Because the MSP is the best match of the query, it is the most important. One should be careful when using MSP scores from multiple queries. Since the MSP score is a maximum, its probability distribution is given by the *extreme value distribution*, also known as the *Fisher-Tippett* or *log-Weibull* distribution. This distribution is not the same as a normal distribution even when scores in general are normally distributed. This distribution is shown in figure 7.3 where it is compared with the normal distribution.

Sequence similarity searches are commonly used to determine the functionality of a sequence by comparing it with sequences whose functionality is known. Inferring functionality is reasonable only when the similarity is statistically significant. To determine statistical significance one compares the actual search result with what would be expected for a search using a random query sequence. The *expectation value* for a score is the number of

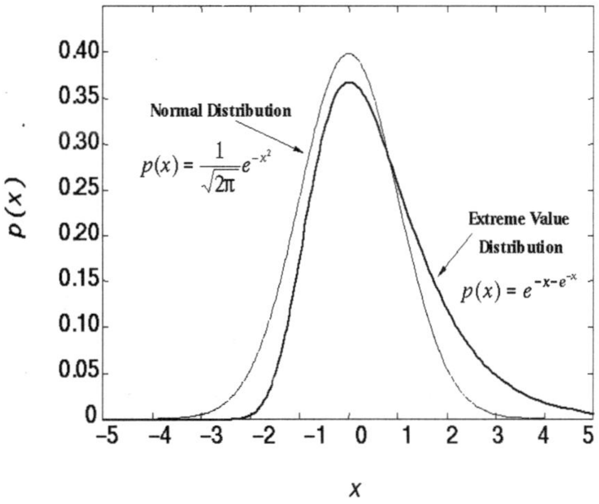

Figure 7.3 Comparison of the extreme value distribution with the normal distribution.

distinct HSPs that would have that score or higher entirely by chance. The expectation value is written E and is approximated by a Poisson distribution (Karlin and Altschul 1990; Altschul 1991). In terms of the normalized score S', the expectation value E is given by $mn2^{-S'}$, where m is the size of the query and n is the size of the database. The expectation value is probably the most useful in the BLAST output. The threshold for significance is usually set at either 10% or 5%. In other words, when E is less than 0.1 or E is less than 0.05, then the HSP is considered to be statistically significant (Altschul et al. 1997).

Strictly speaking, the E-value is not a probability, so it should not be used to determine statistical significance. However, it is easy to convert E to a probability by using the formula $P = 1 - e^{-E}$. The P-value is the probability that a search with a random query would produce at least one HSP at the same score or higher. Table 7.1 shows the relationship between E and P. For E-values below 0.01, there is essentially no difference between E and P. The reason for this is that the Taylor expansion of e^x is $1 + x + x^2/2! + x^3/3! + \ldots$ so that for x close to 0, we have e^x is approximately equal to $1 + x$ and thus, when E is close to 0, $P = 1 - e^{-E}$ is approximately equal to $1 - (1 - E) = E$.

The usual way to use BLAST is to find those sequences in a database that are homologous to a given query sequence. This process compares sequences in the database with the query sequence, but it does not compare the database sequences with each other. If one wishes to learn about the evolution of

E-value	P-value
10	0.9999546
5	0.9932621
1	0.6321206
0.1	0.0951626
0.05	0.0487706
0.001	0.0009995 (about 0.001)
0.0001	0.0001000

Table 7.1 Comparison between E-values and their corresponding P-values.

the homologous sequences, it is necessary to compare them with each other and to organize them to show important features such as highly conserved regions and other subtle similarities. This is known as *multiple sequence alignment* (MSA). We have already mentioned MSAs in section 5.4 where they are used in the PROSITE database. MSAs are also used in some BLAST variants.

Summary

- The raw score S for a search can be normalized so that the results of different searches can be compared.
- The expectation value E can be used to test whether the result of a search has statistical significance.

7.4.4 BLAST Variants

Since the launch of the original version of BLAST, many BLAST variants have been developed. The major variants are presented in the following.

`bl2seq` `www.ncbi.nlm.nih.gov/blast/bl2seq/bl2.html`
In many cases, biologists only want to compare two sequences, rather than embarking on a time-consuming journey of a full database search. The BLAST 2 SEQUENCES program is specifically designed for pairwise comparisons of DNA or amino acid sequences (Tatusova and Madden 1999).

PSI-BLAST `bioinfo.bgu.ac.il/blast/psiblast_cs.html`
Queries using PSSMs differ from queries using substitution matrices in some important ways. Unlike substitution matrices, there are no default or standard PSSMs. In fact, the PSSM is an important part of the query itself.

Another difference is that the result of a PSSM search can be expressed in terms of the probability that a type of residue occurs in each position. In other words, the output of a PSSM search is another PSSM. PSI-BLAST, the position-specific iterated BLAST, algorithm takes advantage of these features of PSSM searches to improve sensitivity by iterating the BLAST algorithm. In other words, the output of a PSSM search is expressed as another PSSM and used for another PSSM search. This process is then repeated. PSI-BLAST is often much better at detecting relatively weak relationships than noniterated sequence similarity queries (Taylor 1986; Dodd and Egan 1990). Another advantage of PSI-BLAST is that motif boundaries can be more precisely defined. Ordinary BLAST relies on cumbersome extension and trimming processes to determine the optimal boundary.

The first step in the PSI-BLAST algorithm is to find all database segments that match the query sequence with an E-value below a user-defined threshold (say 0.01). The matching database segments are then organized as an MSA. The next step following the construction of the MSA is to construct a PSSM. Closely related sequences in the MSA are given relatively smaller weights to avoid biasing the probability distributions. The BLAST algorithm is then applied with this PSSM, and the whole process is iterated a large number of times.

One disadvantage of PSI-BLAST is that false positives (with a low E-value) could kick in and cause corrupted PSSMs that eventually lead to spurious results in subsequent iterations. To deal with this problem a modified version of PSI-BLAST has been developed that incorporates composition-based statistics (Schaffer et al. 2001). This technique significantly improves the accuracy of PSI-BLAST by suppressing the corruption of constructed PSSMs.

PHI-BLAST `bioinfo.bgu.ac.il/blast/psiblast_cs.html`
PHI-BLAST, the pattern-hit initiated BLAST program, is a hybrid strategy that addresses a question frequently asked by researchers; namely, whether a particular pattern seen in a protein of interest is likely to be functionally relevant or occurs simply by chance (Zhang et al. 1998). This question is addressed by combining a pattern search with a search for statistically significant sequence similarity. The input to PHI-BLAST consists of an amino acid or DNA sequence, along with a specific *pattern* occurring at least once within the sequence. The pattern consists of a sequence of residues or sets of residues, with "wild cards" and variable spacing allowed. PHI-BLAST helps to ascertain the biological relevance of patterns detected within sequences, and in some cases to detect subtle similarities that escape a regular BLAST

search. The disadvantage of PHI-BLAST is that it is designed to combine pattern search with the search for statistically significant sequence similarity, rather than to maximize search sensitivity.

Note that PHI-BLAST uses the same website as PSI-BLAST.

WU-BLAST2 `www.ebi.ac.uk/blast2`

This is a new version of WU-BLAST that uses sum statistics for gapped alignments (Altschul and Gish 1996).

MegaBLAST `www.ncbi.nlm.nih.gov/blast/mmtrace.shtml`

This algorithm introduces a "greedy" alignment algorithm that can perform much faster than the traditional dynamic programming algorithm for sequence alignment (Zhang et al. 2000). See also `www.ncbi.nlm.nih.gov/blast/tracemb.shtml`.

RPS-BLAST

`genopole.toulouse.inra.fr/blast/rpsblast.html`

Reverse PSI-BLAST searches a query sequence against a database of profiles (Marchler-Bauer et al. 2002). This is the opposite of PSI-BLAST, which searches a profile against a database of sequences. RPS-BLAST uses a BLAST-like algorithm, finding single- or double-word hits and then performing an ungapped extension on these candidate matches. If a sufficiently high-scoring ungapped alignment is produced, a gapped extension is performed and the gapped alignments with sufficiently low expectation value are reported. RPS-BLAST uses a BLAST database, but also has some other files that contain a precomputed lookup table for the profiles to allow the search to proceed faster. RPS-BLAST is available at both the NCBI web server and the BLOCKS web server.

MPBLAST `blast.wustl.edu`

This is a program that increases the throughput of batch `blastn` searches by multiplexing (concatenating) query sequences to reduce the number of actual database searches performed (Korf and Gish 2000). Throughput was observed to increase in inverse proportion to the component sequence length. In other words, shorter component queries benefit more from MPBLAST than longer component queries. For component DNA queries of length 500, an order of magnitude speedup has been observed.

PathBLAST

`www.pathblast.org/bioc/pathblast/blastpathway.jsp`

This is a network alignment and search tool for comparing protein interaction networks across species to identify protein pathways and complexes that have been conserved by evolution (Kelley et al. 2004). The basic method searches for high-scoring alignments between pairs of protein interaction paths, for which proteins of the first path are paired with putative orthologs occurring in the same order in the second path.

BLAT genome.ucsc.edu/cgi-bin/hgBlat

The BLAST-Like Alignment Tool is a very fast DNA/amino acid sequence alignment tool written by Jim Kent at the University of California, Santa Cruz (Kent 2002). It is designed to quickly find sequences of 95% and greater similarity of length 40 bases or more. It will find perfect sequence matches of 33 bases, and sometimes find them down to 22 bases. BLAT on proteins finds sequences of 80% and greater similarity of length 20 amino acids or more. In practice DNA BLAT works well on primates, and protein BLAT on land vertebrates. It is noted that BLAT may miss more divergent or shorter sequence alignments.

BLAT is similar in many ways to BLAST. The program rapidly scans for relatively short matches (hits), and extends these into HSPs. However, BLAT differs from BLAST in some significant ways. For instance, where BLAST returns each area of homology between two sequences as separate alignments, BLAT stitches them together into a larger alignment. BLAT has a special code to handle introns in RNA/DNA alignments. Therefore, whereas BLAST delivers a list of exons sorted by exon size, with alignments extending slightly beyond the edge of each exon, BLAT effectively "unsplices" mRNA onto the genome giving a single alignment that uses each base of the mRNA only once, and which correctly positions splice sites. BLAT is more accurate and 500 times faster than popular existing tools for mRNA/DNA alignments and 50 times faster for amino acid alignments at sensitivity settings typically used when comparing vertebrate sequences.

BLAT's speed stems from an index of all nonoverlapping sequences of fixed length in the sequence database. DNA BLAT maintains an index of all nonoverlapping sequences of length 11 in the genome, except for those heavily involved in repeats. The index takes up a bit less than a gigabyte of RAM. The genome itself is not kept in memory, allowing BLAT to deliver high performance on a reasonably priced computer. The index is used to find areas of probable homology, which are then loaded into memory for a detailed alignment analysis. Protein BLAT works in a similar manner, except with sequences of length 4. The protein index takes a little more than 2 gigabytes.

BLAT has several major stages. It uses the index to find regions in the genome likely to be homologous to the query sequence. It performs an alignment between homologous regions. It stitches together these aligned regions (often exons) into larger alignments (typically genes). Finally, BLAT revisits small internal exons possibly missed at the first stage and adjusts large gap boundaries that have canonical splice sites where feasible.

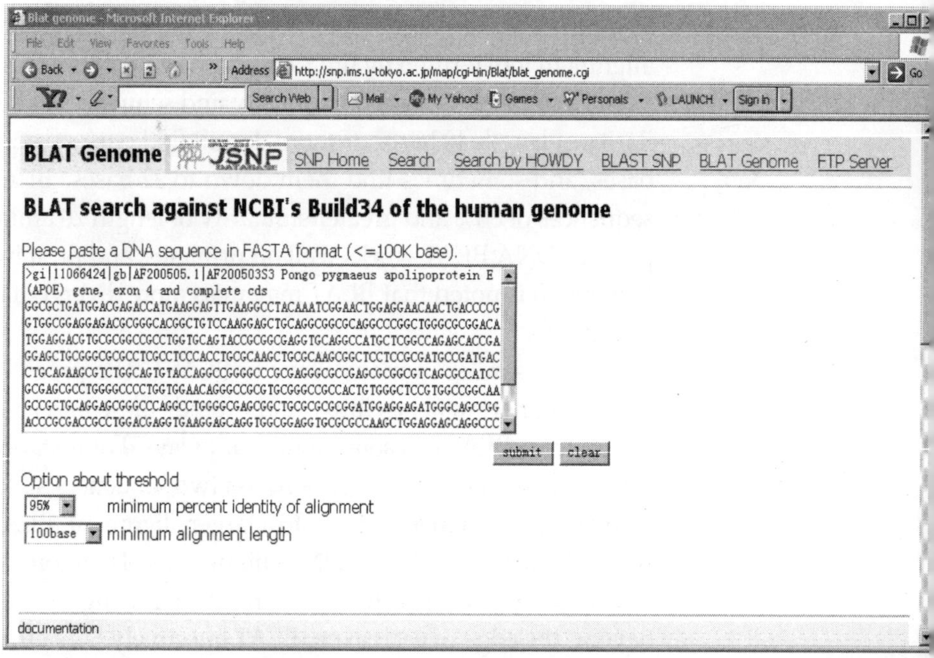

Figure 7.4 BLAT input window.

We illustrate the use of BLAT with a search of a 718-bp DNA sequence (GenBank accession number AF200505.1) encoding exon 4 of the *Pongo pygmaeus* ApoE gene using the BLAT server at `snp.ims.u-tokyo.ac.jp/map/cgi-bin/Blat/blat_genome.cgi`. Figure 7.4 shows the input window, and figure 7.5 shows the output window, where the human ApoE gene was found to be similar to the query DNA sequence.

Summary

Many variations and enhancements of BLAST have been introduced:

1. Performance improvements in special cases or in general:

   - `bl2seq`: 2 sequence comparison;

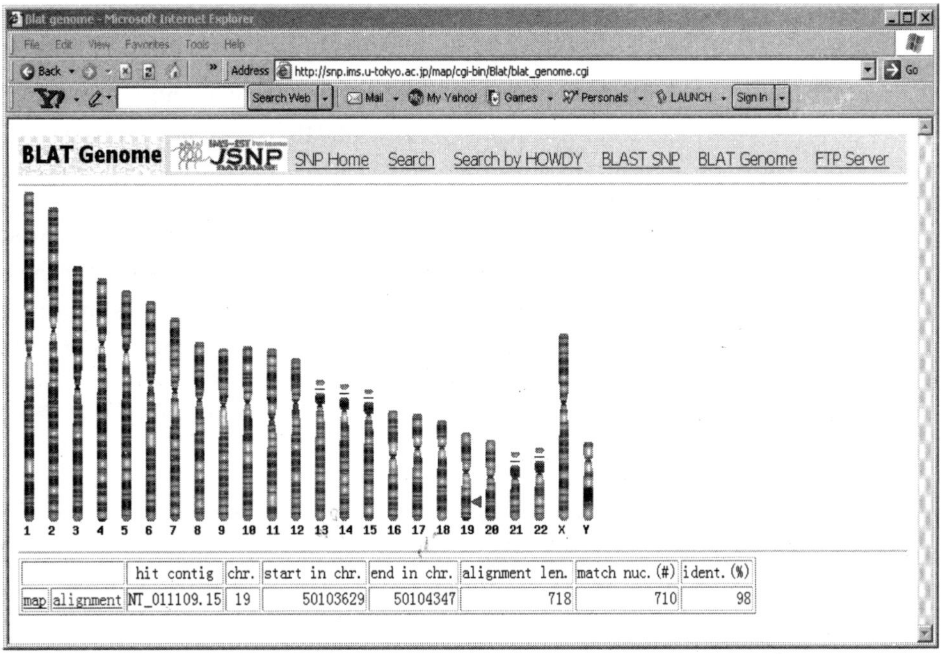

Figure 7.5 BLAT output window.

- MPBLAST: sets of small queries;
- MegaBLAST: greedy algorithm;
- WU-BLAST2: new version of WU-BLAST;
- BLAT: high-performance BLAST.

2. Iterated BLAST:

- PSI-BLAST: iterate to construct new motifs;
- RPS-BLAST: reverse iterate to find a known motifs.

3. Integrate BLAST with other search criteria:

- PHI-BLAST: combine with a pattern search;
- PathBLAST: compare protein interaction networks.

7.5 Exercises

1. If $m = 100$ and $n = 120,000,000$, what normalized bit score S' is necessary to achieve an E-value of 0.01? If the E-value threshold is lowered by 200 times (i.e., lowered to 5×10^{-5}), what normalized bit score is necessary?

2. The probability of finding exactly k HSPs with a raw score S that is at least S_0 follows a Poisson distribution. Suppose that the expected number of HSPs with a raw score $S \geq S_0$ is 0.01. What is the probability of finding no HSPs with score at least S_0? What is the probability of finding at least 2 HSPs with score at least S_0?

8 *Query Languages*

For relational databases the standard query language is SQL. The main purpose of SQL is to select records from one (or more) tables according to selection criteria. Having selected the relevant records, one can then extract the required information from the fields of the relevant records. Because of the success and popularity of SQL, it was natural to imitate SQL when a language was developed for querying XML. However, XML documents have a hierarchical structure that relational databases do not possess. Consequently, XML querying involves three kinds of operation:

1. *Navigation.* This is the process of locating an element or attribute within the hierarchical structure of an XML document.

2. *Selection.* Having located desirable elements and attributes, one selects the relevant ones.

3. *Extraction.* The last operation is to extract required information from the relevant elements and attributes.

The first kind of operation is unique to XML querying, while the other two are similar to what one does in SQL.

Navigation is so important to XML that a separate language was developed to deal with it, called *XPath* (W3C 1999). This language is introduced in section 8.1. XPath has been incorporated into many other languages and tools, and so it is widely available. One such language is *XQuery* (W3C 2004c) which is the standard query language for XML, covered in section 8.2. If one has some experience with relational databases and SQL, it will look familiar, although there are a few differences. The main difference is that XQuery supports navigation using XPath. Indeed, a query using XQuery can consist of nothing more than an XPath expression, and in many cases that is all one needs.

8.1 XML Navigation Using XPath

XPath is a language for selecting parts of an XML document (W3C 1999). If one has used computer file systems, then XPath navigation should be familiar. For example, in the health study database, one can obtain all interviews with this query:

```
HealthStudy/Interview
```

One specifies locations in an XML document by using the same notation that is used for locating directories (folders) and files in most operating systems (except that in Windows, a backward slash is used where a forward slash would be used in XPath). Queries in XPath are called *paths* because they describe the path to be followed to obtain the desired information.

In a document that has a much deeper hierarchical structure, one can use a double slash to mean "skip any number of intermediate levels." For example, in the Medline database, one can obtain all substances by using this path:

```
//NameOfSubstance
```

To obtain this set of elements without the double slash, one would have to specify this path:

```
MedlineCitation/ChemicalList/Chemical/NameOfSubstance
```

Attributes are specified by using the at-sign character (@). To get a list of all of the body mass index (BMI) values in the health study, use this path:

```
//Interview/@BMI
```

The format of the result of a path will vary with the specific tool being used. A typical result of the path above would look like this:

```
BMI = 18.66 BMI = 26.93 BMI = 33.95 BMI = 17.38
-> 4 item(s)
```

A path consists of a sequence of *steps*. Each step selects one or more desired *nodes*. There are many kinds of node. The following are the most important:

1. **element**. To select an element, simply give its name. The name can include a namespace prefix as in section 1.7. To select every element at one level, use an asterisk (*). The asterisk is also known as the "star" or the "wild card." For example, in a MedlineCitation, one can obtain every child element of every Chemical node by using this path:

```
//Chemical/*
```

2. **attribute**. To select an attribute, use an `@` followed by the name of the attribute. Wild cards can also be used for attributes. To obtain every attribute of every Interview element, use this path:

```
//Interview/@*
```

3. **text**. Text contained in an element can be explicitly selected by using the `text()` function. This is not usually necessary. If an element only contains text, then selecting the element will also select its text content.

One can navigate from one step to another in several ways. XPath calls this the *axis* of navigation. The most common are the following:

1. **child element**. This is the normal way to navigate from one node to another. If no other axis of navigation is explicitly specified, then the path navigates to the child element.

2. **descendant element**. A double slash will navigate any number of levels to a matching node. This is handy when one has a deep hierarchy.

3. **parent element**. One can go up one level by using a double dot. To obtain the PubMed identifier (PMID) of every Chemical node, use this path:

```
//Chemical/../../PMID
```

4. **ancestor element**. One can go up any number of levels by using the ancestor:: axis. For example, to obtain the PMID of every Chemical node, even when it is located several levels above, use this path:

```
//Chemical/ancestor::PMID
```

5. **root element**. A slash at the start of a path means to start at the highest level of the document. If a slash is not specified at the start of a path, then the path starts at the current element. This depends on the context in which XPath is used.

While directory paths and XML paths are very similar, there is a distinction mentioned in table 1.1 which is important for navigation; namely, there can be many child elements with the same name. A molecule, for example, can have many atoms (see figure 1.6). To select a particular atom, such as the first one, use this path:

```
/molecule/atomArray/atom[position()=1]
```

This path will select the first atom of every molecule. One can abbreviate the path above to the following:

```
/molecule/atomArray/atom[1]
```

which makes it look like the array notation used in programming languages, except that child numbering begins with 1, while programming languages usually start numbering at 0. However, this notation is an abbreviation that can be used in this case only, and it should not be used in more complicated selection expressions.

XPath brackets are a versatile mechanism for selecting nodes. In addition to selection by position, one can also select by attribute and node values. For example, to select the nitrogen atoms in nitrous oxide, use this path:

```
/molecule[@name='nitrous oxide']//atom[@elementType='N']
```

XPath has many numerical and string operations. Some of these are shown in table 8.1. Selection criteria can be combined, using the Boolean operators. For example, if one would like the carbon and oxygen atoms in hydroquinone, then use this path:

```
/molecule[@name='hydroquinone']
  //atom[@elementType='C' or @elementType='O']
```

The XPath query above should have been on a single line, but it was shown on two lines for typographical purposes. Other XPath queries in this chapter have also been split to fit in the space available.

When using more complicated expressions, one cannot use abbreviations. For example, if one would like the last atom of each molecule, but only if it is a carbon atom, then use this path:

```
/molecule
  //atom[@elementType='C' and position()=last()]
```

sum(set)	Sum of a set of matching elements or attributes
count(set)	Number of matching elements or attributes in a set
position()	The position in the current context
last()	The last position in the current context
and	The logical and operator
or	The logical or operator
not	The logical negation operator
+	Addition of two values
-	Subtraction. Because names can have hyphens always put spaces before and after the minus sign.
div	Division
mod	The remainder after division
*	Multiplication
round(number)	Round a number to the nearest integer
floor(number)	Round down to an integer
ceiling(number)	Round up to an integer
.	Navigate to the current element
..	Navigate to the parent element
/	Navigate to a child element
//	Navigate to any contained (descendant) element
|	Match either of two choices
starts-with(text)	Test that text begins in a specified way
substring(start,end,text)	A part of some text
string-length(text)	The number of characters in some text
=	Test for equality
!=	Test for not equal to
<	Test for less than. Note that < has a meaning in XML, so one must use the "escaped" notation.
>	Test for greater than
<=	Test for less than or equal to
>=	Test for greater than or equal to

Table 8.1 Some of the XPath operators

The XPath language is already a versatile query language. However, it is not a general query or transformation language. XQuery is a general query language which extends XPath. XQuery is introduced in the next section. XSLT is a general transformation language which uses XPath for navigation and computation. XSLT is discussed in chapter 11.

Summary

- XPath is a language for navigating the hierarchical structure of an XML document.
- Navigation uses paths that are similar to the ones used to find files in a directory hierarchy.
- Navigation consists of steps, each of which specifies how to go from one node to the next. One can specify the direction in which to go (axis), the type of node desired (node test), and the particular node or nodes when there are several of the same type (selection).
- An axis can specify directions such as: down one level (child), down any number of levels (descendant), up one level (parent), up any number of levels (ancestor), and the top of the hierarchy (root).
- Node tests include: elements, attributes (distinguished using an at-sign) and text.
- One can select nodes using a variety of criteria which can be combined using Boolean operators.

8.2 Querying XML Using XQuery

XQuery is a powerful and convenient language designed for processing hierarchically structured data such as XML documents (W3C 2004c). To run an XQuery query, one must have an XQuery engine. There are many XQuery engines available, both open source and commercial. Furthermore, XQuery has been incorporated into other tools.

The first step in any query is to specify which XML document is to be queried. One can do this by using the document function. For example, the query

```
document("healthstudy.xml")//Interview
```

will return all of the interview records in the health study database.

Alternatively, one can specify a collection of XML documents for which one will perform a series of queries. Such a collection is called the "database" or *corpus*. The specification of the corpus will vary from one XQuery engine to another. Once the corpus is ready, the queries do not have to mention any documents.

So far we have only considered XPath expressions. Queries can be far more elaborate. A general XQuery query may have four kinds of clause, as follows:

1. `for` clause. This specifies a loop or iteration over a collection. It says that a variable is to take on each value in the collection, one value at a time. For example,

   ```
   for $bmi in
     document("healthstudy.xml")//Interview/@BMI
   ```

   will set the `$bmi` variable to each `BMI` attribute. All variables in XQuery start with a dollar sign. This clause corresponds to the FROM clause in SQL queries, except that in SQL one has only one FROM clause, while an XQuery expression can have any number of `for` clauses. Most programming languages (including Perl, C, C++, and Java) use "for" to indicate an iteration process, and the meaning is the same.

2. `where` clause. This restricts which values are to be included in the result of the query. This clause corresponds to the `where` clause in SQL queries. For example, if one were only interested in `BMI` values larger than 30, then the query would be

   ```
   for $bmi in
     document("healthstudy.xml")//Interview/@BMI
   where $bmi > 30
   ```

   Programming languages like Perl prefer to use the word `if` to indicate a restriction.

3. `return` clause. A query can have any number of variables. The ones that should be printed are specified in the `return` clause. All of the queries given so far are actually incomplete, since they did not indicate what should be printed. To print the `BMI` values, the query above would look like this:

```
for $bmi in
    document("healthstudy.xml")//Interview/@BMI
where $bmi > 30
return $bmi
```

The return clause corresponds to the SELECT clause in an SQL query. Programming languages like Perl also use the word "return," but the meaning is completely different.

4. let clause. It is often convenient to do computations in a query. To simplify the computations, it is helpful to be able to introduce variables that are set to expressions using the other variables. This is done using let clauses. There is no analogous capability in SQL. For example, the following query has the same result as the one above:

```
let $bmilist :=
    document("healthstudy.xml")//Interview/@BMI
for $bmi in $bmilist
where $bmi > 30
return $bmi
```

The $bmilist variable is set to the whole collection of BMI values. The for clause then sets $bmi to each of the values in this collection, one at a time.

XQuery uses variables in ways that are different from how they are used in programming languages such as Perl. In Perl, the dollar sign is used to indicate that a variable is a scalar. A different symbol, the at-sign (@), is used to indicate variables that can have an array of values. In XQuery, there is no distinction between scalars and arrays: a variable can be either one. More significantly, Perl variables can be assigned to a value any number of times. XQuery will only assign a variable to different values in a for clause. A variable can only be given a value once by a let clause. Subsequent lets for the same variable are not allowed.

One can build up more complicated XQuery expressions by combining a series of for and let clauses. These are followed by an optional where clause. The return clause is always the last clause. For example, suppose that one wants to obtain the major topics of the Medline articles in a corpus of Medline citations. One would use a query like this:

```
for $desc in
  document("medline.xml")//MeshHeading/DescriptorName
let $cite := ../../MedlineCitation
where $desc/@MajorTopicYN = "Y"
return $cite, $desc/text()
```

Summary

- XQuery is the standard query language for processing XML documents.
- Every XPath expression is a valid XQuery query.
- A general query is made of four kinds of clause:
  1. A for clause scans the result of an XPath expression, one node at a time.
  2. A where clause selects which of the nodes scanned by the for clauses are to be used.
  3. A return clause specifies the output of the query.
  4. A let clause sets a variable to an intermediate result. This is an optional convenience so that a complicated expression does not have to be written more than once.

8.3 Semantic Web Queries

Unlike XML which has standard navigation and query languages, the Semantic Web languages do not yet have a standard query language. Some suggestions have been made, but it is still unclear what the standard language will eventually look like.

There are several contenders for a Semantic Web query language:

1. A language similar to XQuery, which itself was inspired by SQL. Such a language would treat the knowledge base as a database of triples. From the point of view of the relational model, RDF has a very simple data model: a table with three columns, the subject, predicate, and object. The OWL languages are based on RDF triples so the same data model would apply. However, this data model ignores the semantic differences between relational databases and the Semantic Web languages.

2. A language based on rule engines. As we noted in section 3.1, a query is just a rule with no conclusion, just a condition. Thus any rule language provides, as a special case, a query language. However, the same problem arises here as in the possibility above. Rules engines generally do not support the logic used in the Semantic Web languages.

3. A language based on formal logic. Processing a language such as OWL requires a theorem prover in general. As discussed in section 3.3, one queries a theorem prover by presenting it with a conjecture. This is not the same as a query, because a theorem prover will only report whether the query can be satisfied or not.

None of these choices is completely satisfactory. In two of the cases, one would be ignoring the semantics of RDF and OWL, while in the third case, the query language would be unsatisfactory for nearly all uses. To understand the problem, suppose that one would like to know the number of atoms in the nitrous oxide molecule as in exercise 8.2. This is such a simple and obvious kind of query that it seems amazing that any query language would have any trouble with it. However, in the Semantic Web languages, the only statement that is entailed by the available information is that there are *at least* two atoms. The fact that there are *exactly* two atoms is not entailed. The reason for this has to do with the monotonicity of the Semantic Web languages, as explained in section 4.4. A statement is entailed when it is true in every model of the theory. Unfortunately, there are models for the nitrous oxide theory that have more than two atoms. One could certainly add additional statements to the nitrous oxide theory which would allow one to entail the fact that there are only two atoms in the molecule. However, that would only resolve this particular query. It is not feasible to add all of the additional statements to a knowledge base that would be needed to resolve all possible queries. In fact, one can prove that it is logically impossible to do this. As a result, there can never be a fully satisfactory query language that will preserve the logic and semantics of the OWL languages as they are currently defined. The same is true for the transformation tasks which are discussed in Chapters 10 and 11.

8.4 Exercises

1. Using the health study database in section 1.2, find all interviews in the year 2000 for which the study subject had a BMI greater than 30.

2. Given a BioML document as in figure 1.3, find all literature references for the insulin gene.

3. In the PubMed database, find all citations dealing with the therapeutic use of glutethimide. More precisely, find the citations that have "glutethimide" as a major topic descriptor, qualified by "therapeutic use."

4. Perform the same task as in exercise 8.3, but further restrict the citations to be within the last 6 months.

5. For the health study database in section 1.2, the subject identifier is a field named `SID`. Find all subjects in the database for which the BMI of the subject increased by more than 4.5 during a period of less than 2 years.

6. How many associations does GO term 0003673 have?

9 The Transformation Process

9.1 Experimental and Statistical Methods as Transformations

Biology experiments consist of performing complex recipes by which input materials are ultimately transformed into output measurements. Research papers specify the experiment in the Materials and Methods section. There are two different types of information presented in a Materials and Methods section. One is the list of initial materials used in the experiment. In other words, these are the ingredients. They are the source materials of the experiment. The second type of information in the Materials and Methods section is the description of the processes. Processes consist of a sequence of steps that transform input substances into output substances and measurements.

Consider the following excerpt from a Materials and Methods section in a biology paper (Stock and Stock 1987): "**Immunoaffinity chromatography**. IgG was purified from mouse ascites fluid by DEAE-Affi-Gel Blue (Bio-Rad) chromatography (5) followed by precipitation in 50% ammonium sulfate at 0°C. Purified IgG (5 mg/ml) was dialized against 0.1M sodium bicarbonate, pH 8.5, mixed with Affi-Gel 10 (Bio-Rad) at a ratio of 10 mg of IgG per ml of Affi-Gel 10, and incubated for 12 h at 40°C.. . . " The corresponding series of steps performed in this procedure is shown in figure 9.1. For simplicity, not all of the attributes of the actual procedure are shown. Note that materials can be introduced at steps other than just the first one. Sometimes measurements can occur at more than just the last step (although that does not occur in this example).

The information transformation process is similar to the processing performed during an experimental procedure. The ingredients are called the *source* or *content*. This is the material that is processed and transformed into the desired output information. As in experimental procedures, information

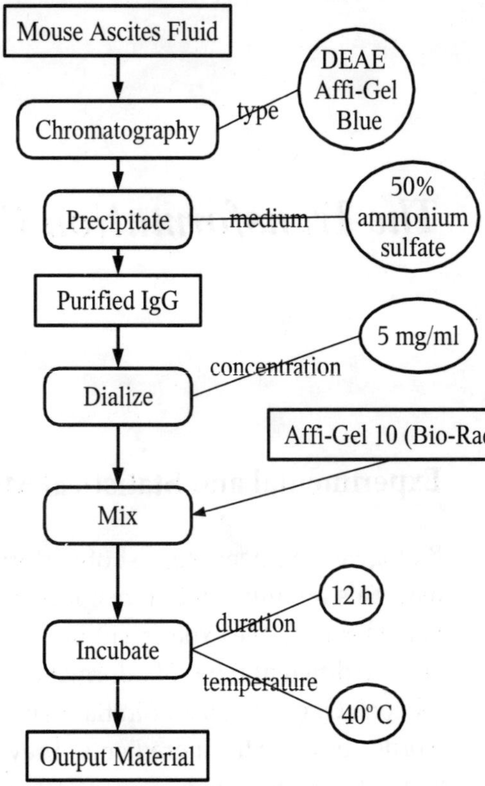

Figure 9.1 Immunoaffinity chromatography procedure for IgG shown as a series of processing steps. The materials are shown in rectangular boxes, the processing steps are shown in rounded boxes, and the attributes of a processing step are shown in ellipses.

transformation is commonly performed in a series of discrete steps. Each step transforms input information into output information according to the type of step and the attributes that have been chosen. An example of an information transformation process is shown in figure 9.2.

The myGrid project, `www.mygrid.org.uk`, has developed a toolkit for much more elaborate transformation processes. The Taverna workbench supports the scientific process for in silico experiments, including management, sharing and reusing the results, recording their provenance and the methods used to generate them. The myGrid project has developed a comprehensive loosely-coupled suite of middleware components specifically to support data intensive in silico experiments in biology. Workflows and query

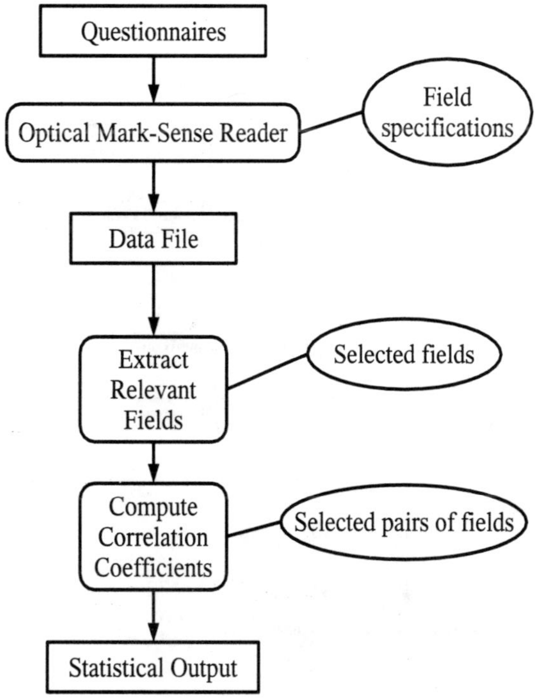

Figure 9.2 Information transformation process for statistical analysis of a health study.

specifications link together third party and local resources using web service protocols. Taverna is a GUI used for assembling, adapting and running workflows. Workflows that execute remote or local web services and Java applications are the chief mechanism for forming experiments. Legacy applications are incorporated using myGrid wrapper tools. In addition to services and applications, databases may be integrated using a query processor developed jointly with the UK OGSA-DAI project. An example of a myGrid workflow is shown in figure 9.3. The software can be freely downloaded.

Summary

- Biology experiments and statistical analyses are transformation processes.

  1. A biology experiment transforms biological and chemical materials into quantitative measurements.

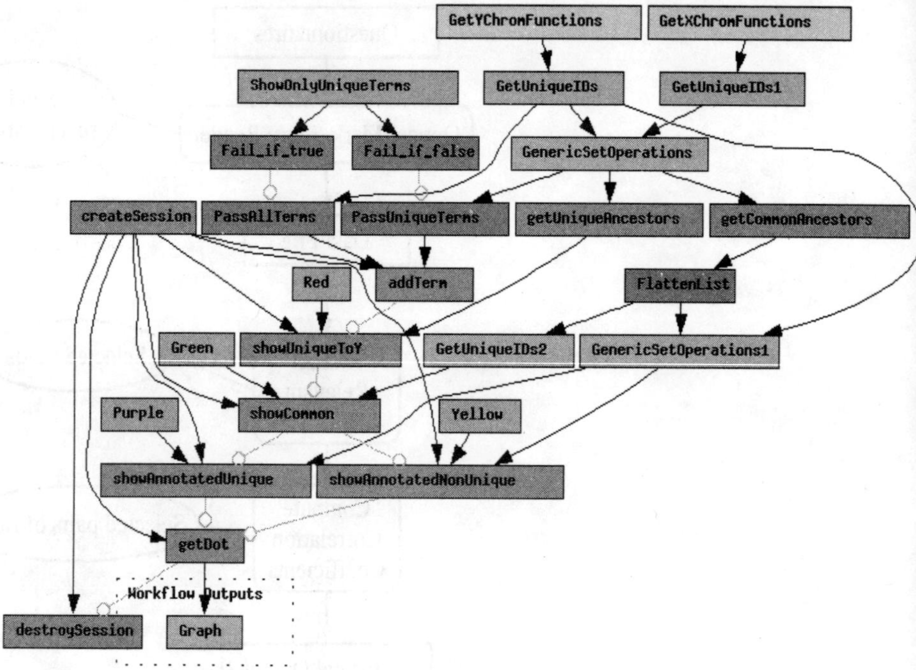

Figure 9.3 An example of a myGrid workflow. The type of action performed by each node in the workflow is indicated by its color (not shown).

2. A statistical analysis transforms survey information into statistical measurements.

- The myGrid project supports in silico experiments by providing tools for managing, sharing and reusing the results, recording their provenance and the methods used to generate the results.

9.2 Presentation of Information

The dissemination of biological knowledge is steadily being transferred from traditional print media to the web. Research papers are increasingly being used to document data that has been produced by a laboratory and that are available on the Internet. Many research contracts explicitly require that data be disseminated on the Internet, and data made available this way are heavily used by other laboratories. Accordingly, it is important that laboratory websites be properly maintained and up to date.

websites consist mainly of a collection of files written in the Hypertext Markup Language (HTML) format. There exist many tools for creating these files. However, constructing the files directly using such a tool is not a very good way to construct a website. It has a number of serious disadvantages:

1. Most of the effort is directed at the visual appearance (presentation) rather than the actual information (content) on each webpage.

2. The same information can appear on several pages, so updating information requires one to find the information and update it in all its locations.

3. Small changes in the content can result in a large number of changes to the website. For example, adding a new person to a department requires that a series of pages be constructed for that person along with references to those pages in numerous other pages. Similarly, when a person leaves, one must delete many pages and remove references on other pages to the deleted pages. The effect is reminiscent of the "butterfly effect" from chaos theory, but on a smaller scale.

4. It is very difficult to maintain a uniform style throughout the website. Small changes to the style require that a large number of webpages be updated. The effort involved can be substantial.

The most effective way to deal with these problems is to divide the overall process of constructing a website into a series of transformation steps. This is called "separating concerns." A typical example of how this is done is shown in figure 9.4. The concerns in this example are:

1. *Source content.* For a biology laboratory this would include the current projects, the people working in the laboratory, published papers, data and software, scheduled events, and so on.

2. *Logical structure.* The source content is subdivided into overlapping views. For a laboratory, each project has its own view that includes just those people working on the project and the papers, data, and software for that project. There are many ways in which these views overlap.

3. *Presentation.* The logical views of the previous step must be converted into a format such as HTML, a spreadsheet format, or PDF so that it can be presented to the person or program that is requesting it.

The process begins with source information that could be stored in files or databases in a large variety of formats. The data are extracted and converted

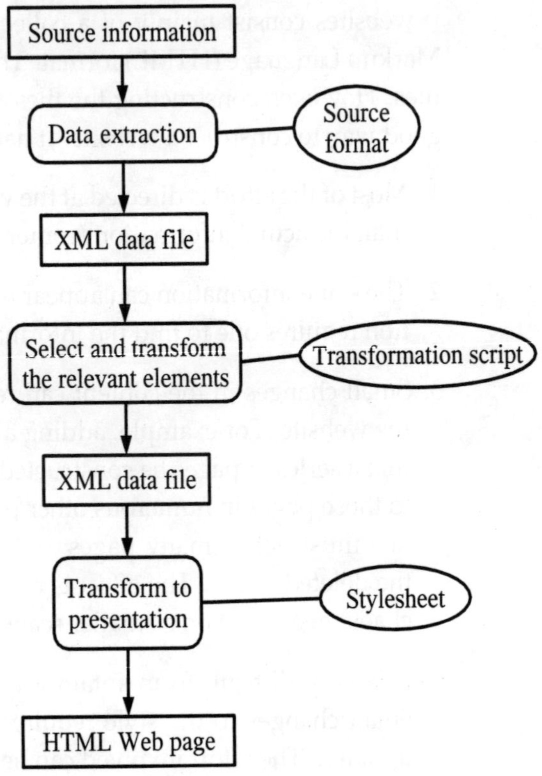

Figure 9.4 The transformation process for constructing a website. The process separates the content, logical structure, and presentation. Each of the steps can be developed and maintained by different individuals or groups.

to a single format: XML. If the original source data are already in XML, then this step simply reads the file. Here is what it might look like:

```
<?xml version="1.0" encoding="UTF-8"?>
<!DOCTYPE Source SYSTEM "source.dtd">
<Source>
  <Person id="K.Baclawski">
    <Name>Kenneth Baclawski</Name>
    <PubName>K. Baclawski</PubName>
    <Email>Ken@Baclawski.com</Email>
    <PersonalPage>http://baclawski.com/ken</PersonalPage>
  </Person>
```

```
</Source>
```

The next step in the process selects and transforms the relevant source information. For example, if one is producing a webpage for a project, then only information relevant to the project is extracted, such as its title, description, personnel, and so on. Here is what a project might look like after the source information for one project has been extracted from the source:

```
<?xml version="1.0" encoding="UTF-8"?>
<!DOCTYPE Project SYSTEM "project.dtd">
<Project>
    <Title>Genetic basis of non-insulin dependent
diabetes mellitus
    </Title>
    <Description>This study examines the literature
dealing with the genetic basis for type II or non-insulin
dependent diabetes mellitus (NIDDM).  Substantial progress
has been made toward understanding the etiology of NIDDM.
By organizing the literature of NIDDM, it is expected that
one will be able to identify new therapeutic targets for
treatment of this common disease more effectively.
    </Description>
...
```

This step is mainly concerned with selecting relevant source information and rearranging it appropriately. For the most part it will not modify the source information significantly.

The last step is to transform the selected information to a presentation format such as HTML or PDF. Unlike the previous step, this can result in a substantially different format. The PDF format is completely different from XML. Here is what the example above might look like in HTML:

```
<HTML>
  <HEAD>
    <META http-equiv="Content-Type"
      content="text/html; charset=UTF-8">
    <TITLE>
      Harvard Medical School Bioinformatics Web Site
    </TITLE>
  </HEAD>
  <BODY BGCOLOR="#FFFFFF">
```

```
<TABLE CELLPADDING="2" CELLSPACING="0" BORDER="0">
  <TR>
    <TD VALIGN="TOP" ALIGN="LEFT">
      <TABLE WIDTH="120">
      ...
      </TABLE>
    </TD>
    <TD WIDTH="10"> </TD>
    <TD WIDTH="100%" VALIGN="TOP" ALIGN="LEFT">
      <TABLE BGCOLOR="#336699" WIDTH="100%"
        CELLPADDING="3" CELLSPACING="0" BORDER="0">
        <TR>
          <TD ALIGN="CENTER">
            <B>
              <FONT SIZE="+1" COLOR="#FFFFFF">
                Genetic basis of non-insulin dependent
                diabetes mellitus
              </FONT>
            </B>
          </TD>
        </TR>
        <TR>
          <TD>This study examines the literature
dealing with the genetic basis for type II or non-insulin
dependent diabetes mellitus (NIDDM).  Substantial progress
has been made toward understanding the etiology of NIDDM.
By organizing the literature of NIDDM, it is expected that
one will be able to identify new therapeutic targets for
treatment of this common disease more effectively.
          </TD>
          ...
```

Needless to say, the presentation information in this HTML file overwhelms the content, and this particular style is relatively simple compared with what one commonly encounters on the web.

Dividing the production of a website into steps does more than just reduce the effort involved. It allows the presentation style to be developed independently by completely different individuals. Thus a large organization can much more easily enforce a single style for all webpages on its website. Instead of sending a memo explaining the style to everyone in the company, which will be interpreted differently by different people, the company can specify its style using a stylesheet, and it will be uniformly and accurately enforced. Furthermore, if the company chooses to change its style, it can do so by simply changing the stylesheets that define the style. Neither the content nor the logical structure needs to be changed, and no employee needs to be involved.

Summary

- Transformation is an effective means for controlling how data are presented.
- Information transformation is performed in a series of steps to reduce the overall effort and to separate concerns.
- Different individuals and groups of individuals are concerned with each step of the transformation process.

9.3 Changing the Point of View

An XML document is an organized repository of information. Information is stored in a series of nested elements that act in some ways like the directories and files of a file system, as discussed in section 1.4. However, hierarchical structures are inflexible: the organization requires a commitment to a particular point of view. In this section we discuss what this involves and how to deal with it using transformations.

Consider microarray information. This information consists of binding potentials (which can be expressed in several ways) of proteins to a collection of small molecules. One could choose to represent these data from the protein point of view in which case the XML document would look something this:

```
...
<Protein id="Mas375">
 <Substrate id="Sub89032">
  <BindingStrength>5.67</BindingStrength>
  <Concentration unit="nm">43</Concentration>
 </Substrate>
 <Substrate id="Sub8933">
  <BindingStrength>4.37</BindingStrength>
  <Concentration unit="nm">75</Concentration>
 </Substrate>
 ...
```

```
</Protein>
<Protein id="Mtr245">
 <Substrate id="Sub89032">
  <BindingStrength>0.65</BindingStrength>
  <Concentration unit="um">0.53</Concentration>
 </Substrate>
 <Substrate id="Sub8933">
  <BindingStrength>8.87</BindingStrength>
  <Concentration unit="nm">8.4</Concentration>
 </Substrate>
 ...
</Protein>
...
```

This certainly represents the data well, but one could equally well have chosen to take the point of view of the substrates instead of the proteins, as follows:

```
...
<Substrate id="Sub89032">
 <Protein id="Mas375">
  <BindingStrength>5.67</BindingStrength>
  <Concentration unit="nm">43</Concentration>
 </Protein>
 <Protein id="Mtr245">
  <BindingStrength>0.65</BindingStrength>
  <Concentration unit="um">0.53</Concentration>
 </Protein>
 ...
</Substrate>
<Substrate id="Sub8933">
 <Protein id="Mas375">
  <BindingStrength>4.37</BindingStrength>
  <Concentration unit="nm">75</Concentration>
 </Protein>
 <Protein id="Mtr245">
  <BindingStrength>8.87</BindingStrength>
  <Concentration unit="nm">8.4</Concentration>
 </Protein>
 ...
```

```
</Substrate>
...
```

The data are the same, but the point of view has changed. The point of view depends strongly on the purpose for which the data were collected. Changing the purpose generally requires that one also change the point of view. The point of view is especially important when information is being displayed. There are many examples of this. One can present a list of research papers organized in many different ways: by topic, by author, by publication date.

The point of view is also important when data are being processed. The processing program expects to receive the data in a particular way. Even when the source document has all of the necessary data, the data can easily be in the wrong form. Indeed, unless the source document and program were developed together (or they conform to the same standard), it is very unlikely that they will be compatible.

The process of changing the point of view of an XML document is an example of a transformation which is also called "repackaging" or "repurposing." Transformations in general are sometimes called "stylesheets" because they were first used as a means of specifying the style (visual appearance) of a document. Separating display characteristics from the content of a document was one of the original motivations for the development of XML and its predecessors.

Summary

- Transformation is the means by which information in one format and for one purpose is adapted to another format for another purpose.
- Information transformation is also called repackaging or repurposing.

9.4 Transformation Techniques

One of the most powerful features of XML is that it is especially well suited to being transformed. There are many reasons for transforming data:

1. The structure must be rearranged as in the microarray example in section 9.3 above.
2. Some element and attribute names need to be changed.

3. Attributes need to be changed into elements, or vice versa.
4. One must infer new information.
5. Several documents should be merged into a single document or a single document should be split into several.
6. Information must be selected from one or more documents. This is essentially the same as querying.
7. An entirely different kind of document is required, such as an HTML document suitable for a web browser, a comma-separated values (CSV) file suitable for a spreadsheet, even a LaTeX file suitable for typesetting.
8. Element information has to be combined. Processing can range from relatively simple operations such as computing totals and averages to using sophisticated algorithms.

Transformation is performed by means of a program. There are many programming languages that can be used for transformation, and there are many variations on how the transformation process can be carried out. The one that is best will depend not only on the nature of the transformation but also on one's background and experience.

Traditional programming languages such as Perl and Java can be used for XML transformation. If one is already familiar with one of these languages, then it might be best to stay with it. Even so there are two distinctly different approaches to transformation using traditional programming languages. A third possibility that is becoming increasingly popular is to use a rule-based language specifically designed for XML transformation. We now discuss each of these three approaches.

The first approach is called *event-based parsing* or, more succinctly, *parsing*. The document is read as input, and it identifies the interesting events, such as the beginning of an element, the end of an element, the content of an element, and so on. The events occur in exactly the same order as they appear in the document. When each event occurs, a corresponding procedure is called, and the features of the event are available as parameters. The procedures that are called form the application programming interface (API). Event-based parsing for XML most commonly uses the simple API for XML (SAX). For example, when the beginning of an element is encountered, the `startElement` procedure is called. The parameters include the name of the element and its attributes. This approach is covered in detail for Perl in subsections 10.2.2 and 10.2.5.

Event-based parsing has an intuitive appeal. Most programs in bioinformatics act upon files that have a flat structure where each line of the file represents one record or event. The program consists of an operation that is performed on each input record. This works well for many problems, especially those that involve computation of statistics. However, event-based parsing can be very difficult to use for any nontrivial transformation task, such as the microarray example in section 9.3. The difficulty is that the transformation may require information that is not immediately available at the time the event occurs. Thus one must save information for later use. Creating data structures that serve this function requires a great deal of time and experience.

The second approach is called *tree-based processing*. In this approach the entire document is read into memory using a standard data structure. The data structure is known as a "tree" to computer scientists, which is why this form of processing is called tree-based. The most commonly used standard for the data structure is called the document object model (DOM). The advantage of this approach is that all information in the document is available at all times. No additional data structures need to be developed just for the sake of ensuring that information is always available when needed. However, the DOM model is complicated and takes some time to understand.

Although traditional programming languages are an effective means of processing documents, most transformation tasks can be accomplished much more easily by using languages designed specifically for this task. The disadvantage is that one must learn yet another language. This can be a very serious disadvantage if one is not going to be using the language very often. However, if one is performing relatively simple tasks, then one does not need to know very much of the language.

Specialized transformation languages have the advantage that they emphasize the meaning of the document (its *semantics*) rather than its appearance (its *syntax*). This is done by using rule-based (declarative) programming rather than the more traditional procedural (imperative) programming style. By focusing on the content rather than low-level details, one can develop transformations much more effectively.

This approach has a long history going back to the style files of LaTeX that are still in use today. The idea was to allow the writer of a document to focus on its meaning rather than typesetting details. The typesetting details were specified in a separate *style file*. In a LaTeX file one can specify the overall style of the document as well as the style to be used for more specialized purposes such as for the bibliography. One can change the style of a document by simply changing one line of the document. In a similar way, one can specify the style of an XML document using a *stylesheet* as follows:

```
<?xml version="1.0"?>
<?xsl-stylesheet type="text/xml" href="transform.xsl"?>
...
```

In the example above, the XML document would be transformed by the stylesheet file named "transform.xsl." The stylesheet is a separate file just as in LaTeX. There are many stylesheet languages for XML. Because of this history, XML transformation programs are often called "stylesheets" even when the transformation has nothing to do with presentation style.

The three approaches to transformation are covered in the next two chapters. Event-based parsing and tree-based processing are covered using the Perl programming language in subsections 10.2.2 and 10.2.5. Rule-based transformation is covered by using the XML Transformation Language in chapter 11.

Summary

A transformation step is performed using one of three main approaches:

1. Event-based parsing

2. Tree-based processing

3. Rule-based transformation

9.5 Automating Transformations

So far we have been assuming that transformations between ontologies will be specified by people, usually domain experts. It is natural to suppose that when two ontologies refer to the same concepts, it ought to be possible to transform from one ontology to the other using some automated process without the need for human effort. The problem of automating the process of finding semantic correspondences between different ontologies has been studied for many years. Most of the work has been for relational database schemas, but there has also been some recent work on this problem for XML DTDs and even for the more sophisticated ontologies of the Semantic Web.

There are many names for the process of discovering transformations. Reconciling differing terminology in various ontologies is called *ontology mediation*. For relational databases, the problem is called *schema integration* for which there is a large literature. See, for example, (Rahm and Bernstein 2001) for a survey of schema integration tools. Similar structures and concepts that appear in multiple schemas are called "integration points" (Bergamaschi et al. 1999). When the data from a variety of sources are transformed to a single target database, then the process is called *data warehousing*. Data warehousing for relational databases is an entire industry, and many data warehousing companies now also support XML. If a query using one vocabulary is rewritten so as to retrieve data from various sources, each of which uses its own vocabulary, then it is called *virtual data integration*. Another name for this process is *query discovery* (Embley et al. 2001; Li and Clifton 2000; Miller et al. 2000).

Ontology mediation and transformation depend on identifying semantically corresponding elements in a set of schemas. (Do and Rahm 2002; Madhavan et al. 2001; Rahm and Bernstein 2001) This is a difficult problem to solve because terminology for the same entities from different sources may use very different structural and naming conventions. The same name can be used for elements having totally different meanings, such as different units, precision, resolution, measurement protocol, and so on. It is usually necessary to annotate an ontology with auxiliary information to assist one in determining the meaning of elements, but the ontology mediation and transformation is difficult to automate even with this additional information.

For example, in ecology, the species density is the ratio of the number of species by the area. In one schema one might have a species density element, while in another, there might be elements for both the species count and area. As another example, in the health study example in section 9.1, the BMI attribute is a ratio of the weight by the square of the height. Another database might have only the weight and height, and these attributes might use different units than in the first database. Consequently, a single element in one schema may correspond to multiple elements in another. In general, the correspondence between elements is many-to-many: many elements correspond to many elements.

Many tools for automating ontology mediation have been proposed and some research prototypes exist. There are also some commercial products for relational schema integration in the data warehousing industry. However, these tools mainly help discover simple one-to-one matches, and they do not consider the meaning of the data or how the transformation will be used.

Using such a tool requires significant manual effort to correct wrong matches and add missing matches. In practice, schema matching is done manually by domain experts, and is very time-consuming when there are many data sources or when schemas are large or complex.

Automated ontology mediation systems are designed to reduce manual effort. However, such a system requires a substantial amount of time to prepare input to the system as well as to guide the matching process. This amount of time can be substantial, and may easily swamp the amount of time saved by using the system. Unfortunately, existing schema-matching systems focus on measuring accuracy and completeness rather than on whether they provide a net gain. Schema-matching systems have now been proposed (Wang et al. 2004) that address this issue. However, such systems are not yet available. The best that one can hope for from current systems is that they can help one to record and to manage the schema matches that have been detected, by whatever means.

One example of a schema integration tool is COMA, developed at the University of Leipzig (Do and Rahm 2002; Do et al. 2002), but there are many others. See (Rahm and Bernstein 2001) for a survey of these tools. Some of these tools also deal with XML DTDs (Nam et al. 2002). Unfortunately, they are only research prototypes and do not seem to be available for downloading.

There are many ontology mediation projects, and some have developed prototypes, such as PROMPT (Noy and Musen 2000) from the Stanford Medical Informatics laboratory and the Semantic Knowledge Articulation Tool (SKAT), also from Stanford (Mitra et al. 1999), but as with schema integration, none seem to be available for public use, either via open source software or commercial software.

Summary

- Reconciling differing terminology has many names depending on the particular context where it is done, such as: ontology mediation, schema integration, data warehousing, virtual data integration, query discovery, and schema matching.

- Automated ontology mediation systems attempt to reduce manual effort, but they rarely provide a net gain.

- Most automated ontology mediation systems are still research prototypes.

10 *Transforming with Traditional Programming Languages*

There are many programming languages, but the one that has been especially popular in bioinformatics is Perl. It is designed to "make easy jobs easy without making the hard jobs impossible" (Wall et al. 1996). On the other hand, this does not mean that hard jobs are best done with Perl. It is still the case that programming languages such as Java and C++ are better suited for major system development than Perl. It is likely that there will always be a need for a variety of programming languages. Indeed, this is perfectly compatible with the Perl slogan, "There's More Than One Way to Do It" (TMTOWTDI).

Perl is especially well suited to data transformation tasks, and this is what will be emphasized here. Perl is much too large a language for complete coverage in even several books, let alone a chapter in just one book. However, the coverage should be adequate for most transformation tasks.

In keeping with the TMTOWTDI philosophy of Perl, there are many ways to approach any given transformation task. There are also many kinds of transformation tasks. This chapter is organized first around the kind of transformation task, and then for each kind of transformation, a number of approaches are given, arranged from the simpler to the more complex. When one is facing a particular task, whether you think of it as transformation, conversion, or reformatting, first look to the main classification to choose the section for your task. Within a section, all of the techniques accomplish the same basic task. The earlier ones are simple and work well in easy cases, but get tedious for the harder tasks of this kind. The later ones require a more careful design, but the result is a smaller program that is easier to maintain. Accordingly, just scan through the possibilities until you reach one that is sufficient for your needs.

Another aspect of TMTOWTDI is that one can omit punctuation and variables if Perl can understand what is being said. This increases the possibilities for what a program can look like enormously. It can also make it difficult for a person to read some Perl programs even if the Perl compiler has no difficulty with it. Except for some common Perl motifs and an example in the section 10.1 below, the examples in this chapter will try to use a programming style that emphasizes readability over cleverness as much as possible.

Some of the most common programming tasks can be classified as being transformations. Even statistical computations are a form of data transformation. To organize the transformation tasks, the world of data will be divided into XML and text files. The text file category includes flat files as well as the text produced by many bioinformatics tools. This lumps together a lot of very different formats, but it is convenient for classification purposes. The many file formats (such as PDF, Word, spreadsheet formats, etc.) that require specialized software for their interpretation will not be considered unless the format can be converted to either an XML file or a text file.

The first section of the chapter deals with non-XML text processing, and the second section of the chapter deals with XML processing. Many techniques from the first part reappear in the second, but some new notions are also required.

10.1 Text Transformations

The subsections of this part of the chapter deal with increasingly complex data and transformations of the data. The first two subsections consider data having a structure that is uniform, as in flat files and database tables. The first subsection shows how to process such information one line or record at a time; the second introduces arrays which allow one to process the information in some other fashion than as it is being received. The third subsection acts as an interlude between the first two and the last two subsections. It covers procedures which are important for organizing programs as they get larger. The last two subsections consider data with more complicated structures. The fourth subsection shows how to extract information from complicated text, which is processed as it is extracted. The fifth introduces data structures which allow one to process complicated data in some other fashion than as it is being extracted.

Perl can be invoked in many ways, but one of the most common is to use a command such as this:

```
perl program.perl file.txt > result.txt
```

where `program.perl` is the name of the file containing the Perl program to perform the transformation, `file.txt` is the name of the file to be transformed (also called the "input" file), and `result.txt` is the transformed file that is produced by your program. One can specify more than one file to be transformed, but there is only one file produced as a result of the transformation. It is as if there were a single input file made up of the data in all of the files put together in order.

The programs in this part of the chapter consider input file formats and transformation tasks that get progressively more complex. The early ones use simple flat files with fixed-width format, and the later ones use more complex formats. Early tasks make no changes to the data in the input files; the task is just to change the format. Later tasks perform statistical computations.

10.1.1 Line-Oriented Transformation

The simplest approach to transformation is just to read the file one line at a time, transforming each line as it is read. The program for this looks a lot like a book or paper: it has an introduction, a main body, and a conclusion. The introduction takes care of tasks that precede the transformation such as printing a report title, and the conclusion performs tasks such as printing summary information. Sometimes the introduction or the conclusion will be omitted, but there will always be a body, as that is where the transformation takes place.

Consider the task in which the health study mentioned in section 1.1 is to be transformed from the fixed-width format to a variable-width format that is more readable for people. The input file has lines that start like this:

```
011500    18.66  0   0 62    46.27102
011500    26.93  0   1 63    68.95152
020100    33.95  1   0 65    92.53204
020100    17.38  0   0 67    50.35111
```

The output should look like this:

```
Health Study Data

1/15/2000 18.66 normal 62 cm 46.27 kg 102 lb
1/15/2000 26.93 overweight 63 cm 68.95 kg 152 lb
2/1/2000 33.95 obese 65 cm 92.53 kg 204 lb
2/1/2000 17.38 normal 67 cm 50.35 kg 111 lb
```

```
print("Health Study Data\n\n");

while (<>) {
  $month = substr($_, 0, 2) + 0;
  $day = substr($_, 2, 2) + 0;
  $yr = substr($_, 4, 2) + 0;
  $year = 1900 + $yr;
  $year = 2000 + $yr if $yr < 20;
  $bmi = substr($_, 6, 8) + 0;
  $status = "normal";
  $status = "obese"
    if substr($_, 14, 3) + 0 > 0;
  $status = "overweight"
    if substr($_, 17, 3) + 0 > 0;
  $height = substr($_, 20, 3) + 0;
  $wtkgs = substr($_, 23, 8) + 0;
  $wtlbs = substr($_, 31, 3) + 0;
  print("$month/$day/$year $bmi $status");
  print(" $height cm $wtkgs kg $wtlbs lb\n");
}
```

Program 10.1 Convert fixed-width fields to variable-width fields

One of the ways that this transformation can be accomplished is shown in program 10.1. This program has only two parts: an introduction and a body. It does not have any concluding part. Examples of programs that have a concluding part are given later.

The first or introductory part of the program prints the title of the report. The "\n" is called a *newline;* it ends the line at that point. Two newlines in a row will insert an extra blank line between the title and the rest of the report.

The body of this program performs the same operation on all of the lines of the file to be transformed. The *while* statement means that everything in the braces is to be repeated as long as the condition (in parentheses) is true. The <> angle brackets are used to get the next line of the input file. It succeeds as long as there is another line, and it fails (i.e., it tells the while statement that it should stop) when there is nothing left in the file.

The statements in the transformation block use *variables*. The variables have names such as `month`, `day`. The dollar signs are not part of the name but rather indicate that the variables are *scalars*, that is, numbers or strings (ordinary text). The line that was just read is available in the variable whose name is an underscore character. One extracts parts of a string by using the `substr` function (short for "substring"). Scalars have a kind of "split personality" since they can be either numbers or strings. The `substr` function produces a string, but all of the substrings being extracted in this program are supposed to be numbers. One can change a scalar to a number by adding 0 to it. If the scalar is already a number this does nothing. If the scalar is a string, then this will find some way to interpret the string as being a number. Perl is very flexible in how it interprets strings as numbers. For example, if there is some text in the string that could not be part of a number, it (and everything after it) is just ignored. For example "123 kgs" will be interpreted as the number 123, and "Hello 123" will be interpreted as the number 0.

The computation of the year is somewhat problematic because there are only two digits in the original file, but the full year number is expected in the report. This is handled by adding conditions after the statements that compute the full four-digit year. The assumption is that all years are between 1921 and 2020.

Program 10.1 is certainly not the only way to perform this task in Perl. The style of programming was chosen to make it as easy as possible to read this program. The use of angle brackets for obtaining the next line of the file which is then represented using an underscore is a bit obscure, but it is a commonly used motif in Perl. It is relatively easy to get used to it.

To illustrate some of the variations that are possible in Perl, program 10.1 could also have been written as in program 10.2. This program avoids the use parentheses as much as possible, and when it does use them, it does so differently than the first program. In general, one can omit parentheses in functions, and it is only necessary to include them when Perl would misinterpret your intentions. If the parentheses were omitted in the tests for obesity and overweight, then Perl would have compared 0 with 3 rather than with the number extracted from the original file. Notice also that the semicolon after the last statement can be omitted because it occurs immediately before a right brace.

The next task to consider is the computation of summary information. One common use of data from a study is to compute the mean and variance. Program 10.3 computes the mean, variance, and standard deviation of the BMI column of the health study. Running this program on the four records at the beginning of this section produces this report:

```
print "Health Study Data\n\n";

while (<>) {
  $month = 0 + substr $_, 0, 2;
  $day = 0 + substr $_, 2, 2;
  $yr = 0 + substr $_, 4, 2;
  $year = $yr < 20 ? 2000 + $yr : 1900 + $yr;
  $bmi = 0 + substr $_, 6, 8;
  $status = "normal";
  $status = "obese" if (substr $_, 14, 3) > 0;
  $status = "overweight" if (substr $_, 17, 3) > 0;
  $height = 0 + substr $_, 20, 3;
  $wtkgs = 0 + substr $_, 23, 8;
  $wtlbs = 0 + substr $_, 31, 3;
  print "$month/$day/$year $bmi $status";
  print " $height cm $wtkgs kg $wtlbs lb\n"
}
```

Program 10.2 Alternative version of program 10.1

```
Health Study Data

1/15/2000 18.66 normal 62 cm 46.27 kg 102 lb
1/15/2000 26.93 overweight 63 cm 68.95 kg 152 lb
2/1/2000 33.95 obese 65 cm 92.53 kg 204 lb
2/1/2000 17.38 normal 67 cm 50.35 kg 111 lb

Number of records: 4
Average BMI: 24.23
BMI Variance: 59.9052666666668
BMI Standard Deviation: 7.73984926640479
```

This program uses all three parts of a typical program. The first part prints the title of the report as before, and the body processes the records in the health study file, but now there is also a concluding part that prints the statistics. The processing of the records has some additional computations. The `count` variable has the number of records, the `bmisum` variable has the sum of the BMI values for all records, and the `bmisumsq` has the sum of the squares of the BMI values. These are set to 0 in the introductory part of the

```
print("Health Study Data\n\n");
$count = 0;
$bmisum = 0;
$bmisumsq = 0;

while (<>) {
  $month = substr($_, 0, 2) + 0;
  $day = substr($_, 2, 2) + 0;
  $yr = substr($_, 4, 2) + 0;
  $year = 1900 + $yr;
  $year = 2000 + $yr if $yr < 20;
  $bmi = substr($_, 6, 8) + 0;
  $status = "normal";
  $status = "obese" if substr($_, 14, 3) + 0 > 0;
  $status = "overweight" if substr($_, 17, 3) + 0 > 0;
  $height = substr($_, 20, 3) + 0;
  $wtkgs = substr($_, 23, 8) + 0;
  $wtlbs = substr($_, 31, 3) + 0;
  print("$month/$day/$year $bmi $status");
  print(" $height cm $wtkgs kg $wtlbs lb\n");
  $count = $count + 1;
  $bmisum = $bmisum + $bmi;
  $bmisumsq = $bmisumsq + $bmi ** 2;
}

print("\n");
print("Number of records: $count\n");
$bmimean = $bmisum / $count;
print("Average BMI: $bmimean\n");
$bmivar =
  ($bmisumsq - $count * $bmimean ** 2) / ($count - 1);
print("BMI Variance: $bmivar\n");
$bmisd = $bmivar ** 0.5;
print("BMI Standard Deviation: $bmisd\n");
```

Program 10.3 Computation of statistical information

program, and modified for each record in the health study file. The mean, variance, and standard deviation are computed from these three values by well-known formulas.

It is not actually necessary to initialize the three variables to 0. In other words these three lines in the first part of the program could have been omitted:

```
$count = 0;
$bmisum = 0;
$bmisumsq = 0;
```

Perl will automatically set any variable to a standard default value the first time it is used. For numbers the default initial value is 0. For strings it is the empty string.

Many commonly occurring statements can be abbreviated. For example, the statement

```
$bmisum = $bmisum + $bmi;
```

can be abbreviated to the more succinct statement

```
$bmisum += $bmi;
```

Similar abbreviations are available for all the arithmetic operations such as subtraction, multiplication, and division, as well as for other kinds of operation. Incrementing a variable (i.e., increasing it by 1) is so common that it has its own special operator. As a result, one can abbreviate

```
$count = $count + 1;
```

to the more succinct

```
$count++;
```

So far the transformation tasks have been on data files in a fixed-width format. The transformed data have used fields that vary in length. The next task is to transform a data file with variable-length fields separated from one another by spaces. This requires a new kind of variable as well as operations for them. An *array* or *list* is a sequence of values. Perl indicates an array variable with an initial `@` character. To transform a data file it must first be *split* into fields. The result of the splitting operation is an array.

```
print "Health Study Data\n\n";

while (<>) {
  chomp;
  @record = split(" ", $_);
  @date = split("/", $record[0]);
  $month = $date[0] + 0;
  $day = $date[1] + 0;
  $yr = $date[2] + 0;
  if ($yr < 20) { $year = 2000 + $yr; }
  elsif ($yr < 1000) { $year = 1900 + $yr; }
  else { $year = $yr; }
  $bmi = $record[1];
  if ($record[2] + 0 > 0) { $status = "obese"; }
  elsif ($record[3] + 0 > 0) { $status = "overweight"; }
  else { $status = "normal"; }
  $height = $record[4] + 0;
  $wtkgs = $record[5] + 0;
  $wtlbs = $record[6] + 0;
  print("$month/$day/$year $bmi $status");
  print(" $height cm $wtkgs kg $wtlbs lb\n");
}
```

Program 10.4 Reformatted health study data

Program 10.4 does the same transformation as program 10.1, except that it assumes that the data file uses variable-length fields separated by spaces, and that the month, day, and year in a date are separated from one another by using the forward slash ("/") character. This program will transform the following data file:

```
1/15/2000 18.66 0 0 62 46.27 102
1/15/2000 26.93 0 1 63 68.95 152
2/1/2000 33.95 1 0 65   92.53 204
  2/1/2000 17.38 0 0    67 50.35 111
```

to the following:

```
Health Study Data
```

```
1/15/2000 18.66 normal 62 cm 46.27 kg 102 lb
1/15/2000 26.93 overweight 63 cm 68.95 kg 152 lb
2/1/2000 33.95 obese 65 cm 92.53 kg 204 lb
2/1/2000 17.38 normal 67 cm 50.35 kg 111 lb
```

The first step is to remove any extra space at the beginning and end of the line. This is done with the `chomp` function. This was not necessary in previous programs because the fields were in fixed locations in the line. It is a good idea to use `chomp` whenever the input has variable-length fields. The next step is to split the record into fields using the `split` operator. This produces the `@record` array. The values in this array are denoted by `$record[0]`, `$record[1]`, `$record[2]`, `...` The number in brackets is called the *index* or *position* of the value in the array. If one uses a negative index, then it specifies a position starting from the last value (i.e., starting from the other end of the array). Notice that the array as a whole uses `@` but the individual values use `$`. Remember that in Perl, the initial character on a variable (and there are more than just `$` and `@`) denotes the kind of value. It is not part of the name of the variable.

The next step is to split the first field (i.e., the date) into its parts. The parts are separated by slashes. The rest of the program is the same as the first program except that array values are used instead of substrings. Another difference is that the conditional statements are written with the if-conditions first rather than after the statement. The first condition is indicated by `if`. Subsequent conditions (except the last one) are indicated using `elsif` which is short for "else if". The last case is indicated by `else` which is used for those cases not handled any other case. Putting the conditional after the statement as in the first program is best if you think of the statement as being subject to a condition. Putting a conditional before the statement is best if you are thinking in terms of a series of cases, such that only one of them applies to each record.

The split statement

```
@record = split(" ", $_);
```

could have been abbreviated to

```
@record = split(" ");
```

The default for splitting is to split up the value of `$_`. One can even abbreviate further to

```
@record = split;
```

This is actually better than the previous form because it will treat all forms of "white space" (such as tab characters) as being the same as spaces. Finally, one can abbreviate all the way to `split;` except that now the array containing the fields of the line is `@_` instead of `@record`.

The opposite of `split` is `join`. One can put the split array back together after splitting by using

```
join(" ", @record);
```

This can be handy if one would like to separate the fields with a character other than a space. For example,

```
join(",", @record);
```

would use a comma to separate the fields.

Statistics for an entire population as were just computed are generally of limited interest. It is far more interesting to look for correlations between various characteristics of the population. When characteristics, such as ages, have a temporal significance, it is also interesting to look for trends. Consider the task of computing the BMI as a function of the month and year. In statistical terminology one is interested in the conditional probability distribution of the BMI given the month and year. The month and year are specified by the first and third parts of the first field of the health study record (the second being the day of the month). The task is to compute the mean and variance of the BMI grouped by the month and year.

Using scalar variables or arrays is not enough to accomplish this task. We need a way to group information by month. The information belonging to a month is said to be *associated* with the month and year. Each month is called a *key* for the associated information, and each key is *mapped* to its associated information. A mapping from keys to associated information is called an *associative array, hash table,* or just *hash,* for short. Perl uses the character `%` to distinguish hashes from scalars and arrays. In the following program, statistics are computed for each month separately using three hashes:

```
print "Health Study Data\n\n";
%count = ();
%bmisum = ();
%bmisumsq = ();
```

```
while (<>) {
  chomp;
  @record = split(" ", $_);
  @date = split("/", $record[0]);
  $month = $date[0] + 0;
  $day = $date[1] + 0;
  $yr = $date[2] + 0;
  if ($yr < 20) { $year = 2000 + $yr; }
  elsif ($yr < 1000) { $year = 1900 + $yr; }
  else { $year = $yr; }
  $bmi = $record[1];
  if ($record[2] + 0 > 0) { $status = "obese"; }
  elsif ($record[3] + 0 > 0) { $status = "overweight"; }
  else { $status = "normal"; }
  $height = $record[4] + 0;
  $wtkgs = $record[5] + 0;
  $wtlbs = $record[6] + 0;
  print("$month/$day/$year $bmi $status");
  print(" $height cm $wtkgs kg $wtlbs lb\n");
  $m = "$month/$year";
  $count{$m}++;
  $bmisum{$m} += $bmi;
  $bmisumsq{$m} += $bmi ** 2;
}
foreach $m (sort(keys(%count))) {
  print("\nStatistics for $m\n");
  print("Number of records: $count{$m}\n");
  $bmimean = $bmisum{$m} / $count{$m};
  print("Average BMI: $bmimean\n");
  $bmivar = ($bmisumsq{$m} - $count{$m} * $bmimean ** 2)
            / ($count{$m} - 1);
  print("BMI Variance: $bmivar\n");
  $bmisd = $bmivar ** 0.5;
  print("BMI Standard Deviation: $bmisd\n");
}
```

The hashes are used in each of the three parts of the program. In the introductory part, the three hashes are declared and initialized to empty hashes. As in the case of scalars, it is not necessary to declare and initialize hashes. If they are not declared and initialized, then they will be set to empty hashes. Arrays also do not have to be initialized. By default, arrays are initially empty.

In the main body of the program, the hashes are used at the end to compute the three statistics: count, sum, and sum of squares. First, the month and year are combined into a single string. This string is then called the *key* for the hash value. It is analogous to the index of an array, but the index for array values can only be an integer, while a hash key can be any scalar. This includes integers, other numbers, and strings. The value corresponding to a hash key is specified by using braces as, for example, in the expression `$bmisum{$m}`. By contrast, arrays use brackets to specify a value of an array.

The final and concluding part of the program introduces a new kind of statement: the *foreach* statement. This statement is used for performing some action on every element of a list. This is called *iteration* or *looping* because the same action is done repeatedly, differing each time only by which element is being processed. In this case the iteration is to be over all of the month-year combinations that are in the hashes. The body of the iteration should be performed once for each month-year combination. Each time it is performed `$m` will be a different month-year combination. The month-year combinations are the keys of any one of the three hashes. The program uses `%count`, but any one of the three hashes could have been used. The `keys` function gets the list of all keys of a hash. This list can be in any order, so one usually sorts the keys to get output that looks better. If the order does not matter, then one can omit using the sort function. The rest of the computation is nearly the same as before except that the values in the hashes are used instead of simple scalar statistics. Applying this program to the simple four-record example data file will print the following:

```
Health Study Data

1/15/2000 18.66 normal 62 cm 46.27 kg 102 lb
1/15/2000 26.93 overweight 63 cm 68.95 kg 152 lb
2/1/2000 33.95 obese 65 cm 92.53 kg 204 lb
2/1/2000 17.38 normal 67 cm 50.35 kg 111 lb

Statistics for 1/2000
Number of records: 2
Average BMI: 22.795
BMI Variance: 34.1964499999997
BMI Standard Deviation: 5.84777308041272

Statistics for 2/2000
```

```
Number of records: 2
Average BMI: 25.665
BMI Variance: 137.28245
BMI Standard Deviation: 11.7167593642611
```

As usual, there are many ways to abbreviate statements in this program. One common abbreviation is to omit some of the parentheses in the foreach statement. Thus

```
foreach $m (sort(keys(%count))) {
```

can be written

```
foreach $m (sort keys %count) {
```

One could also abbreviate it to

```
foreach (sort keys %count) {
```

in which case the scalar holding the key is `$_` instead of `$m`.

Although hashes use a different notation than arrays, one can use hashes to implement arrays. All one needs to do is write the index using braces instead of brackets. Thus one would write `$x{2}` instead of `$x[2]`. Using hashes instead of arrays is convenient when the indexes are not consecutive or do not start at 0. The one place where one must be careful when using hashes is when one is iterating. By default the `sort` function sorts the keys as strings. If the keys are actually numbers, the order will look rather strange. For example, "10" precedes "2" as strings. To sort in numerical order specify `{ $a <=> $b }` after the sort function. For example, if `%x` is a hash whose keys are numerical, then the following will print the values in numerical order:

```
foreach $i (sort { $a <=> $b } keys %x) {
  print("$i: $x{$i}\n");
}
```

Many other sorting orders can be used by varying the sort order used in the braces after the sort function. The default order is dictionary ordering which uses the `cmp` operator.

Summary

- Programs are commonly organized into three parts: introduction, body, and conclusion.
- While programs run, they store data in variables. As the program runs the data stored in each variable will change. The simplest kind of variable in Perl is a scalar, which holds a single string or number.
- The simplest way to transform data is to transform one line at a time. In such a program:
    1. The introduction prints the title and sets variables to initial values.
    2. The body reads each line, extracts data from it and prints the data in the required format.
    3. The conclusion computes summary information and prints it.
- Perl has many ways to abbreviate commonly used operations.
- Perl has two kinds of variables for holding collections of data items:
    1. An array holds a sequence of data items. Arrays are also called lists.
    2. A hash maps keys to associated information.
- The `foreach` statement is used for performing operations on each data item in a collection.

10.1.2 Multidimensional Arrays

The transformation technique we have used so far has been to transform one line of a file at a time. This works well for files that have one record on each line and each record has nearly the same structure as any other record. When the input file is not so well arranged, it may be better to process the file in some other order. In Perl it is very easy to read an entire file all at once as in program 10.5.

The `while` statement of the programs in subsection 10.1.1 has been replaced by an assignment of the input operator <>. When the first statement is finished, the `lines` array will have all of the lines of the input file.

The `scalar` function converts a variable of any kind to a scalar. This is a general-purpose function whose meaning depends on the particular variable being converted. In the case of arrays, `scalar` gives the size of the array. As

```
@lines = <>;
$size = scalar(@lines);
print("The input file has $size lines.\n");
```

Program 10.5 Reading an entire file into an array

one might expect, this is used frequently in Perl programs, although it does not necessary appear explicitly. In fact, in this case it can be omitted because assigning an array to a scalar tells Perl that the array is to be converted to a scalar. In the case of hashes, `scalar` gives information about the structure of the hash that is usually not very useful. To get the size of a hash use `scalar(keys(%h))`. This gives the number of keys in the hash.

One might think that one can simplify the last two lines of program 10.5 to the one statement

```
print("Table size is scalar(@table)\n");
```

but this does not work. In a quoted string, the special meaning of `scalar` is lost. The string "scalar" will be printed verbatim, and the `scalar` function will not be invoked. Quoted strings know how to deal with variables, but they do not understand computations in general. One can get around this restriction in two ways:

1. Compute the information in separate statements using a number of variables, and then combine them into the string to be printed. This is the technique that has been used so far.

2. Use string concatenation. Once a computation is outside a string, then it will be performed as expected. Strings are concatenated using the period character. In a print statement one can also use commas. The print statement above can be done using this statement:

   ```
   print("Table size is " . scalar(@table) . "\n");
   ```

   or by using commas instead of periods.

Arrays are a versatile technique that can be used for lists of values as well as for representing the mathematical notion of a vector. It is natural to consider how to represent other mathematical structures such as matrices and

```
while (<>) {
  chomp;
  push(@table, [split]);
}
$size = @table;
print("Table size is $size\n");
```

Program 10.6 Reading a file of records into a 2D array

tables. Once one has a concept of an array, it is easy to represent these other mathematical structures. A matrix, for example, is just a vector of vectors, so to represent it in Perl, one simply constructs a array whose items are themselves arrays.

Consider the task of reading all of the fields of all the records in an input file. The array will have one item for each record of the data file. Each item, in turn, will be an array that has one item for each field of the record. In other words, the data will be represented as an array of arrays, also called a *two-dimensional array* or *database table*. It is very easy to create such a table in Perl. In program 10.6, the array is constructed, and its size is printed.

The `push` procedure adds new items to the end of a list. In this case, it adds a new record to the table array. Each record is obtained by splitting the current line. Recall that `split` by itself splits the current line into fields that were separated by spaces.

The opposite of `push` is `pop`. It removes one item from the end of a list. There are also procedures for adding and removing items from the beginning of a list. The `shift` procedure removes the first item from a list. Unlike `push` and `pop`, the `shift` procedure changes the positions of all the items in the list (e.g., the one in position 1 now has position 0). The opposite of `shift` is `unshift`, which adds items to the beginning of a list.

The brackets around `split` tell Perl to maintain the integrity of the record. Without the brackets, the fields of the record would be pushed individually onto the array resulting in a very large one-dimensional array with all of the fields of all of the records.

Brackets around an array tell Perl that the array is to be considered a single unit rather than a collection of values. The term for this in Perl is *reference*. It is similar to the distinction between a company and the employees of a company. The company is a legal entity by itself, with its own tax identification number and legal obligations, almost as if it were a person. There are similar situations in biology as well. Multicellular organs and organisms are living entities that are made up of cells but which act as if they were single units.

Perl arrays are made into single entities (scalars) by using brackets. For example, the array `@lines` could be made into a scalar by writing `[@lines]`, and one can assign such an entity to a scalar variable as in

```
$var = [@lines];
```

One can put any number of arrays and scalars in brackets, and the result is (reference to) a single array. Hashes are made into single entities by using braces. One can combine hashes by putting more than one in braces, and one can add additional keys as in

```
$var = {
  name => "George",
  id => "123456",
  %otherData,
};
```

In this case `$var` refers to a hash that maps "name" to "George" and "id" to "123456," in addition to all of the other mappings in `%otherData`.

Now consider the same statistical task as in program 10.3; namely, compute the mean and standard deviation of the BMI. However, now use a database table rather than computing it as the file is being read to obtain program 10.7.

The statistical computation is done in the `for` statement. This is similar to the `foreach` statement. Whereas the `foreach` statement iterates over the items of a list, the `for` statement iterates over the numbers in a sequence. The statement specifies the three parts of any such sequence: where to start it, how to end it, and how to go from one number in the sequence to the next one. The three parts in this case specify:

1. Where to start: `$i = 0` means start the sequence at 0.
2. How to end: `$i < $count` means to stop just before the number of records in the database. The reason for ending just before the number of records rather than at the number of records is that numbering starts at 0.
3. How to go from one number to the next: `$i++` means increment the number to get the next one. In other words, the sequence is consecutive.

```
print "Health Study Statistics\n\n";

while (<>) {
  chomp;
  push(@table, [split]);
  $bmi = $record[1];
}

$count = @table;
for ($i = 0; $i < $count; $i++) {
  $bmisum = $bmisum + $table[$i][1];
  $bmisumsq = $bmisumsq + $table[$i][1] ** 2;
}

print("Number of records: $count\n");
$bmimean = $bmisum / $count;
print("Average BMI: $bmimean\n");
$bmivar = ($bmisumsq - $count * $bmimean ** 2)
          / ($count - 1);
print("BMI Variance: $bmivar\n");
$bmisd = $bmivar ** 0.5;
print("BMI Standard Deviation: $bmisd\n");
```

Program 10.7 Computing statistics using a 2D array

This way of specifying sequences is very common. It is used in all of the major programming languages, including C++ and Java.

The expression `$table[$i][1]` gets the second field of one record from the table. This is how one obtains one item in a 2D array in Perl (and, for that matter, also in C++ and Java). In the language of matrices, the first index is the row and the second index is the column.

Summary

- An array of arrays is a two-dimensional array, also called a table. One can construct arrays having any number of dimensions.
- Brackets are used for selecting one item from an array.
- Braces are used for selecting the value associated with a key in a hash.

- Brackets enclosing an array variable or braces enclosing a hash variable are used to refer to the array or hash as a single unit rather than as a collection of items.
- The `for` statement is used to perform some action for each number in a sequence.

10.1.3 Perl Procedures

While the programs above are very good for accomplishing their tasks, it should be clear that they would get rather tedious if one were computing statistics for more than a couple of fields. A real health study database will have hundreds of fields, and one would like to perform a large number of statistical computations. So some other way to deal with the transformations will be needed. In the case of statistics, the same kind of computation is required over and over again. Rather than program this same computation endlessly, one should just program it once, and then use that same program whenever it is needed.

In Perl, a collection of statements that can be performed as a unit is called a *procedure*. They are also called *subroutines* or *functions*. Performing a procedure is called *invocation*. One also says that the procedure is being *executed* or *called*. One invokes a procedure by using the name of the procedure and a list of *parameters*. Procedures have already been used in the programs presented so far. For example, `split` is a procedure that splits apart a string into an array of fields.

Returning to the health study, suppose that one would like to compute the mean and variance for many of the fields. The computation for each field is the same, so it is convenient to program it just once. Assuming that the database is in a table (2D array) as in program 10.7, it is easy to use this table to compute statistics with the following procedure named `stats`:

```
sub stats {
  my $count = @table;
  my $column = $_[0];
  my $sum = 0;
  my $sumsq = 0;
  for ($i = 0; $i < $count; $i++) {
    my $field = $table[$i][$column];
    $sum += $field;
    $sumsq += $field * $field;
```

```
  }
  my $mean = $sum / $count;
  my $var = ($sumsq - $count * $mean ** 2)
            / ($count - 1);
  return ($mean, $var);
}
```

This procedure introduces some new notation. The most noticeable change from the previous Perl programs is the use of `my` at the beginning of most lines. The variables that have been used so far are known as *global* variables. They are accessible everywhere in the program. In particular, they can be used in any of the procedures. The `@table` variable, for example, is used in the first line of this procedure. A *my* variable, on the other hand, belongs only to that part of the program in which it was declared. All but one of the `my` variables in this procedure belong to the procedure. The two exceptions are `$i` and `$field`. Both of these belong to the `for` statement. It is not necessary to declare that such variables are `my` variables, but it is okay to do so.

The advantage of `my` variables is that they prevent any confusion in case the same variable name is used in more than one procedure. While one can certainly use ordinary global variables for computation done in procedures, it is risky. To be safe, it is best for all variables that are only used within a procedure to be `my` variables.

The `stats` procedure is invoked by specifying the position of the field that is to be computed. For example, `stats(4)` computes the mean and variance of the column with index 4. This is actually the fifth column because Perl array indexes start at 0. The number 4 in `stats(4)` is called a *parameter* of the procedure. The parameters given to a procedure are available within the procedure as the `@_` array. In particular, the first parameter is `$_[0]`, and this explains the second line of the procedure:

```
my $column = $_[0];
```

which sets `$column` to the first parameter given to the procedure when it is invoked.

The `return` statement has two purposes. It tells Perl that the procedure is finished with its computation. In addition, it specifies the end result of the computation. This is what the program that invoked the procedure will receive. Note that a list of two values is produced by this procedure. One can use this list like any other. For example, the following program will print the statistics for two of the columns of the database:

```
while (<>) {
  chomp;
  push(@table, [split]);
}
($mean, $var) = stats(1);
print("Statistics for column 1:");
print(" mean $mean variance $var\n");
($mean, $var) = stats(4);
print("Statistics for column 4:");
print(" mean $mean variance $var\n");
```

Of course, the program above has yet another opportunity for a procedure; namely, one that prints the statistics:

```
sub printstats {
  my $column = $_[0];
  my ($mean, $var) = stats($column);
  print("Statistics for column $column:");
  print(" mean $mean variance $var\n");
}
```

One cannot help but notice that the scalar `$_` and the array `@_` are used frequently in Perl. Because of this it is a good idea to assign the parameters of a procedure to various `my` variables belonging to the procedure as soon as possible. It also makes it much easier for a person to understand what a procedure is supposed to do. In this case, `$column` is a lot more understandable than `$_[0]`.

Putting all of this together, one obtains the solution to the task in program 10.8.

The procedure definitions can go either before the main program or after it.

Summary

- A procedure is a collection of statements that can be performed as a unit.
- A procedure is invoked by using its name and giving a list of parameters.
- A `my` variable is limited to the part of the program in which it is declared.

```
while (<>) {
  chomp;
  push(@table, [split]);
}

printstats(1);
printstats(4);

sub stats {
    my $count = @table;
    my $column = $_[0];
    my $sum = 0;
    my $sumsq = 0;
    for (my $i = 0; $i < $count; $i++) {
        my $field = $table[$i][$column];
        $sum += $field;
        $sumsq += $field * $field;
    }
    my $mean = $sum / $count;
    my $var = ($sumsq - $count * $mean ** 2)
              / ($count - 1);
    return ($mean, $var);
}
sub printstats {
  my $column = $_[0];
  my ($mean, $var) = stats($column);
  print("Statistics for column $column:");
  print(" mean $mean variance $var\n");
}
```

Program 10.8 Computing statistics with procedures

- The `return` statement marks the end of the computation and specifies the value produced by the procedure.

10.1.4 Pattern Matching

So far the input files we have considered are record-oriented, that is, each line has one record and all records have the same structure. Unfortunately, a large amount of data does not have this simple structure. Consider the following file that was produced by BioProspector (Liu et al. 2001). Some parts of the file were omitted to save space.

```
****************************************
*                                      *
*      BioProspector Search Result     *
*                                      *
****************************************


Read input sequences.
Use following data to represent motif score distribution.
1.950
1.982
[26 similar lines omitted]
2.027
1.943
Null motif score distribution mean: 2.005, standard deviation: 0.052
Look for motifs from the original sequences.
Try #1  2.462   CGTTCCGGAGACCG CGGTCTCCGGAACG         36
Try #2  2.295   CTCGAGGAGCTTGG CCAAGCTCCTCGAG         32
[36 similar lines omitted]
Try #39 2.274   CGCTTCCAGCCCTC GAGGGCTGGAAGCG         32
Try #40 2.516   GAAGTTTCCCGACC GGTCGGGAAACTTC         40
The highest scoring 3 motifs are:

Motif #1:
******************************
[1 line omitted]

Blk1    A     G     C     T      Con  rCon Deg  rDeg
1      0.00  0.21  0.21  0.59     T    A    T    A
2      0.00  0.44  0.50  0.06     C    G    S    S
[10 similar lines omitted]
13     0.44  0.00  0.56  0.00     C    G    M    K
14     0.21  0.59  0.18  0.03     G    C    G    C

Seq #1  seg 1  r998    TCATCCAATCAGAG
Seq #2  seg 1  f91     TCAACCGAACAGAA
[30 similar lines omitted]
Seq #27 seg 1  r343    GGAACCAATCAGCG
Seq #27 seg 2  r261    TCAGCCAATGACCG
******************************

[The other motifs were omitted]
```

The information produced by BioProspector is not only complex, but the format is also complex. Furthermore, it is unique to BioProspector. There are many other motif-finding programs available such as AlignACE (Hughes

```
while (<>) {
  chomp;
  if (/Motif #1:/) {
    print "The first motif has been found!\n";
  }
}
```

Program 10.9 Using pattern matching to find one piece of data in a file

et al. 2000; Roth et al. 1998), CONSENSUS (Stormo and Hartzell III 1989; Hertz et al. 1990; Hertz and Stormo 1999), and Gibbs sampler (Lawrence et al. 1993; Liu et al. 1995), and all of them use their own output formats. No doubt many more formats already exist for motifs, and many more will be used in the future. A similar situation exists for virtually every other kind of bioinformatics information. Many tools are available for similar tasks, and each one uses its own input and output formats.

To process information such as the BioProspector file above, we make use of the pattern-matching features of Perl. Pattern matching is one of the most powerful features of Perl, and it is one of the reasons why Perl has become so popular.

Consider the task of extracting just the information about the first motif. A motif is defined as a sequence of probability distributions on the four DNA bases. We will do this in a series of steps. First we need to read the BioProspector file and find where the information about the desired motif is located, as shown in program 10.9.

Each motif description begins with a title containing "Motif #" followed by a number and ending with a colon. The condition `/Motif #1:/` is responsible for detecting such a title. The text between the forward slashes is the *pattern* to be matched. A pattern can be as simple as just some text that is to be matched, as in this case.

If one wanted the line that contained exactly this text, one would use the condition `$_ eq "Motif #1:\n"`. Note that string comparison uses `eq`, not the equal-to sign. Also note that every line ends with the newline character. In practice it is usually easier to use a pattern match condition than a test for equality. The pattern match will handle more cases, and one does not have to worry about whether or not the newline character might be in the line.

```
while (<>) {
  chomp;
  if (/Motif #[0-9]+:/) {
    print "A motif has been found!\n";
  }
}
```

Program 10.10 Using pattern matching to find all data of one kind in a file

The next task is to find where every motif begins, not just the first one. This is done by modifying the pattern so that it matches any number rather than just the number 1 as in program 10.10.

The pattern now has `[0-9]+` where it used to have the number `1`. Bracketed expressions in a pattern define *character classes*. This character class will match any character between `0` and `9`. The plus sign after the character class means that the line must have one or more characters in this class. Any character or character class can be followed by a *quantifier*:

+	One or more (i.e., at least one)
*	Zero or more (i.e., any number of them)
?	Zero or one (i.e., optional character)

Quantifiers can also be used to specify exactly how many times a character must occur as well as a range of occurrences. This is done by placing the number of times or the range in braces after the character or character class.

Character classes and quantifiers are specified using characters (such as brackets, braces, etc.) just as everything is specified in Perl. However, this means that these characters are special within a pattern. They are called *metacharacters*. When used within a pattern they do not match themselves. The metacharacters are backward slash, vertical bar, parentheses, brackets, braces, circumflex, dollar sign, asterisk, plus sign, question mark, and period. If a pattern should match one of the metacharacters, then use a backward slash. For example, `\?` means match the question mark character rather than quantify the preceding character or character class.

The next task is to obtain the motif number. In principle, one could get this number by using the `split` and `substr` functions, but there is a much easier way. When Perl matches a pattern, it keeps track of what succeeded in matching those parts of the pattern that are in parentheses. In this case,

```
while (<>) {
  chomp;
  if (/Motif #([0-9]+):/) {
    print "The motif $1 has been found!\n";
  }
}
```

Program 10.11 Extracting information from a file using pattern matching

```
while (<>) {
  chomp;
  if (/Motif #([0-9]+):/) {
    print "Probability distributions for motif $1\n";
  } elsif (/^[0-9]+ /) {
    split;
    print "A $_[1] G $_[2] C $_[3] T $_[4]\n";
  }
}
```

Program 10.12 Extracting an array of data from a file using pattern matching

we want the motif number so the number pattern is parenthesized as in program 10.11.

One can have any number of parenthesized subpatterns. The part that matched the first parenthesized subpattern is `$1`, the second is `$2`, and so on.

The next step in processing the BioProspector file is to find where the motif probability distributions are located. Looking at the file, one can see that the probability distributions are located on lines that begin with a number. Program 10.12 extracts the array. The ^ character means the "beginning of the line." The end of the line is denoted by `$`.

Summary

- Patterns are a powerful mechanism for extracting desired information.

- A pattern specifies the text that a string must have in order to match the pattern.

- When a pattern matches, Perl extracts the text that matches the whole pattern as well as text that matches each subpattern.

10.1.5 Perl Data Structures

While pattern matching is a powerful feature for finding information in an input file, it is not enough by itself when the information is arranged in a different order than is needed by the transformation task. Consider the following excerpt from the output produced by the CONSENSUS (Stormo and Hartzell III 1989; Hertz et al. 1990; Hertz and Stormo 1999) motif-finding program:

```
MATRIX 2
number of sequences = 19
unadjusted information = 13.2069
sample size adjusted information = 12.0373
ln(p-value) = -198.594     p-value = 5.64573E-87
ln(expected frequency) = -57.9937 expected frequency = 6.51143E-26
A |   0   0   0  18  16   0   7   0   0   0   4   0  19
T |  18   0   0   0   1   8   3  15  19   3   2   0   0   0
C |   1   0   2   1   0   0   6   4   0  16  12   7   0   0
G |   0  19  17   0   2  11   3   0   0   0   5   8  19   0
```

This excerpt shows the probability distributions for one motif (labeled "MATRIX 2"). There are two ways in which this file differs from what is necessary for the task. First, the distributions are given in terms of frequencies rather than probabilities. Second, the frequencies are listed by DNA base rather than by position in the motif. The first difference is easy to fix: one can just divide by the total number of sequences. The second difference is not so easily handled because the information has the wrong arrangement.

To rearrange information obtained from an input file, it is necessary to store information from several lines before printing it. This would be easy if the information consisted of a few scalars, but it gets much more complicated when substantial amounts of data must be organized. The technique for doing this in programming languages is called a *data structure*. Some data structures have already been used; namely, arrays and hashes. These are the simplest data structures. One constructs more complex data structures by using a technique called *nesting*. A *nested data structure* is a data structure whose items are themselves data structures. For example, one can have an array of hashes, or a hash of arrays, or a hash of hashes of arrays, and so on. There is no limit to how deeply nested a data structure can be. The special case of an array of arrays was already developed in subsection 10.1.2. Data structures extend the concept of multidimensional array to allow for dimensions that can be hashes as well as arrays. Furthermore, data structures in general can mix arrays, hashes, and scalars in a single "dimension." Thus it is possible to have a data structure consisting of an array some of whose items are hashes, some are arrays, and the rest are scalars. This kind of mixing is necessary for representing XML documents as Perl data structures. This is developed in subsection 10.2.3. However, one should avoid mixing arrays and hashes in a completely arbitrary fashion, as this can get very confusing. One technique that helps keep the program simple is to use only hashes and scalars. In other words, avoid arrays. As we saw in subsection 10.1.1, one can use a hash instead of an array.

Consider the task of representing a DNA motif. A motif is a sequence of probability distributions, so it should be represented as an array. Each item of this array is a probability distribution. This probability distribution assigns a number to each of the DNA bases. Such an assignment is most naturally represented using a hash. Thus a motif is an array of hashes. A motif-finding program produces several motifs, each with a label. The most natural way to label the motifs is to use a hash. So the result of a motif-finding program is a hash of arrays of hashes. However, to avoid mixing hashes and arrays, motifs will be represented using a 3D hash. Program 10.13 extracts the probability distributions from the output produced by CONSENSUS.

The program extracts information by using Perl patterns. The label of the motif is indicated by a line that starts with `MATRIX` and followed by a number. Note the use of the dollar sign to specify that the line has nothing else on it. The motif number is obtained from the pattern by putting parentheses around the subpattern for the number. The number of sequences is obtained in a similar fashion. Adding 0 to the number of sequences tells Perl that this is a number. The motif label, by contrast, may look like a number but it is being treated as being just text.

The most complicated part of the program is the part that extracts the probability distributions. The frequencies for one DNA base are on a line that begins with the name of the base, followed by a vertical bar. The rest of the line consists of frequencies. The frequencies are obtained by splitting the line and looping over the fields, starting with the third field (i.e., starting with index 2 because arrays always start with 0).

The data structure being constructed is called `motifs`. It is a 3D hash. An item in the the first dimension is a single motif and is determined by the motif label. An item in the second dimension is one position in the motif. One advantage of using a hash instead of an array is that the DNA positions need not start at 0, and they need not be contiguous. In this case, the frequency of

```
while (<>) {
  chomp;
  if (/^MATRIX ([0-9]+)$/) {
    $label = $1;
  } elsif (/^number of sequences = ([0-9]+)$/) {
    $numberOfSequences = $1 + 0;
  } elsif (/^[ACGT] [|]/) {
    @record = split;
    for ($i = 2; $i < scalar(@record); $i++) {
      $motifs{$label}{$i-2}{$record[0]} =
        $record[$i] / $numberOfSequences;
    }
  }
}
foreach $label (sort(keys(%motifs))) {
  print "Probability distributions for motif $label\n";
  %motif = %{ $motifs{$label} };
  foreach $position (sort(keys(%motif))) {
    foreach $base (A, C, T, G) {
      print("$base $motif{$position}{$base} ");
    }
    print("\n");
  }
}
```

Program 10.13 Extracting data structures from a file using pattern matching

the first DNA base is the third field on the line, the second frequency is the fourth field, and so on. So it is necessary to subtract 2 from the field position to get the DNA base position. Finally, an item in the third dimension is the probability for one of the four DNA bases. This is obtained by dividing the frequency by the number of sequences.

Having extracted the motifs, the next step is to print them. Since the motifs are in a 3D data structure, the most natural way to use the structure is with three nested loops. The first loop processes the motifs. The labels are the keys of the `motifs` hash, and it is customary to sort the keys of a hash so that they are printed in a reasonable order.

Given the label for a motif, one can obtain the motif by using the label as the key: `$motifs{$label}`. However, this is a scalar, not the hash of DNA positions. This is the trickiest part of the program. To get the hash of DNA positions, one must use the expression `%{$motifs{$label}}`. This may seem mysterious at first, but it all makes sense when one finds out that *every* use of the prefixes `$`, `%`, and `@` are actually supposed to look like this. Omitting the braces is an abbreviation that one can use for simple variable names.

Once the hash for one motif has been obtained, one just loops over the positions and then over the four bases. The program explicitly writes out the DNA bases, because it is printing them in an order that is not alphabetical. After printing the probability distribution, a newline is printed to end the line. The output of the program will look something like this:

```
Probability distributions for motif 1
A 0.037037037037037 C 0.111111111111111 T 0.851851851851852 G 0
A 0.037037037037037 C 0.037037037037037 T 0 G 0.925925925925926
A 0 C 0.62962962962963 T 0.185185185185185 G 0.185185185185185
...
```

Perl will always print everything that it knows about a number. In many cases the numbers will have far too many decimal places than are merited by the data. To specify the exact number of decimal places that should be printed one should use the `printf` statement. It would look like this:

```
printf('%s %5.3f ', $base, $motif{$position}{$base});
```

The first parameter of the `printf` statement is called the *format*. Its purpose is to specify what kinds of data are to be printed as well as the precise format to use for each one. Each format specification begins with a percent sign. This use of the percent sign has no connection with the notion of a Perl hash. The `%s` format means that the variable is to be printed verbatim. The `s` stands for "string." The `%5.3f` format means that the variable is to be printed as a number with three digits after the decimal point and five characters in all (including the decimal point). The `f` stands for "floating-point number." Using this format, the output of the program would look like this:

```
Probability distributions for motif 1
A 0.037 C 0.111 T 0.852 G 0.000
A 0.037 C 0.037 T 0.000 G 0.926
A 0.000 C 0.630 T 0.185 G 0.185
```

```
A 0.222 C 0.407 T 0.074 G 0.296
...
```

Summary

- One can represent complex data structures by nesting arrays and hashes, for example, by constructing an array of hashes,
- To keep the data structure simple, it is convenient to use only hashes.
- Individual elements of a nested data structure are obtained by using it as a multidimensional data structure.
- To process all of the elements of a nested data structure use a series of loops nested within each other.
- When printing numbers, one can specify how much precision will be used by using the formatted print statement, `printf`.

10.2 Transforming XML

This section introduces techniques for transforming XML. It builds on the techniques of the first section, but new concepts are also required. The first subsection introduces the notion of a Perl module which allows one to extend the basic Perl language with new features. Several such modules are then used to process XML files and to produce them. The processing of XML is covered first and producing XML is covered second. As in section 10.1, processing can be performed either as it is encountered, one XML element at a time, or in some other order, by means of a data structure. XML can be produced starting from text or from XML. Transforming XML to XML is especially important and it will be considered again in chapter 11.

10.2.1 Using Perl Modules and Objects

Perl has a mechanism for grouping together variables and procedures into separate units. In fact, it has several mechanisms for doing this, but they are very similar to one another. This idea is closely related to the notion of a *reference* that was mentioned in subsection 10.1.2. There is a distinction between a collection of entities and a collection regarded as a single unit. The examples in that section consisted of arrays and hashes so these were rather more like *tissues* than organs or organisms, because all of the entities that make up the collection are the same kind of entity. With modules and objects, the grouping includes entities that are dissimilar. Thus one can group together scalars, arrays, hashes, procedures, and so on, all in a single unit.

Modules are mainly used for publishing programs. One person or group of persons constructs a module for a specialized purpose. The module is then published, usually at the Comprehensive Perl Archive Network (CPAN) located at `cpan.org`. The modules can then be downloaded and installed by other people. If you have installed your own personal Perl library, then you can look for and install modules by running the `cpan` command. If you have Perl, but do not have `cpan`, then try the following command:

```
perl -MCPAN -e shell
```

If you don't have Perl, then you will need to install it.

The `cpan` command (or its equivalent) presumes that you know which modules you want to install. If you do not know which ones you would like, then use one of the CPAN search engines, such as `search.cpan.org`. There are over 100 packages that mention bioinformatics, plus there are many others related to biology and medicine.

Once a module has been installed, the most common way for it to be used is to construct a *module object*. Programs that use modules typically look something like this:

```
use moduleName;

$p = new moduleName;
...
```

The `use` statement tells Perl that the program will be using a module. One can use any number of modules. The `new` statement constructs an *object*. An object is a reference to a collection of scalars, arrays, hashes, procedures, and other objects, all of which have been grouped together in a single unit. The parts of an object are obtained by using a special operator, written `->`. For example, if one of the parts of the module object `$p` is a procedure named `computeAverage`, then the procedure is invoked by using the statement

```
$p->computeAverage;
```

Procedures that are in the context of an object are called *methods*.

Once bioinformatics data have been represented in an XML format, it can be transformed using a wide variety of tools. In keeping with the TMTOWTDI philosophy of Perl, there are a great number of ways to transform XML using Perl. Here are some of the Perl modules that can be used to process XML documents:

1. *XML::Parser* provides one with the ability to process XML one element at a time. It is analogous to reading a file one line at a time. However, because elements can contain other elements, it is important to know not only when one starts reading an element but also when an element is finished. This process is similar to the pattern-matching programs in subsection 10.1.4 such as program 10.12. The XML parser looks for the patterns that indicate when an element begins and when an element ends.

2. *XML::DOM* is analogous to program 10.5 in subsection 10.1.2. Instead of processing the document one line at a time, the entire document is read into a single data structure, and one is free to examine the parts in whatever order is convenient. Of course, XML has a hierarchical document structure, so the Perl data structure will also be hierarchical.

3. *XML::XPath* organizes the document like a directory of files, exactly as in section 8.1.

Summary

- A Perl module groups together scalars, arrays, hashes and procedures as a single unit.
- The cpan command, or its equivalent, can be used to install Perl modules that have been published on the CPAN website.
- The -> operator refers to one of the items in a module.
- Perl modules are available for processing and querying XML documents.

10.2.2 Processing XML Elements

The simplest way to process XML is to read the document one element at a time. This is analogous to reading a file one line at a time, as in program 10.1 of subsection 10.1.1. Processing an XML document is called *parsing*, which is the term that computer scientists use for processing any computer language. There is a Perl module that will parse XML documents

```
use XML::Parser;

$p = new XML::Parser(Handlers => { Start => \&start });
$p->parsefile($ARGV[0]);

sub start {
  $tag = $_[1];
  %attributes = @_;
  if ($tag eq "Interview") {
    print("Weight $attributes{Weight}\n");
  }
}
```

Program 10.14 Parsing XML attributes

called XML::Parser. Suppose that we would like to obtain the `Weight` attribute of every `Interview` in an XML document that looks like this:

```
<HealthStudy>
  <Interview Date='2000-1-15' Weight='46.27'.../>
  <Interview Date='2000-1-15' Weight='68.95'.../>
  <Interview Date='2000-2-1' Weight='92.53'.../>
  <Interview Date='2000-2-1' Weight='50.35'.../>
</HealthStudy>
```

This task can be accomplished by using program 10.14. The `use` statement imports the XML::Parser module. If this statement fails, then this module has not yet been installed. You can install it by using the `cpan` command or its equivalent as described in subsection 10.2.1.

The next two statements of the program construct the XML parser and parse the document. There are several styles for parsing. The style for processing the XML document one element at a time is called the "handlers" style. *Handlers* are Perl procedures that are invoked as various kinds of data are encountered in the XML document. A *Start* handler is invoked whenever an XML element is first encountered. There are many kinds of handler that will be discussed later. In this case the Start handler is a procedure called `start`.

The initial `\&` in front of `start` is telling Perl that one is passing the `start` procedure to the parser as a parameter. Without this Perl would simply invoke the `start` procedure at this place in the program. In this case we never explicitly invoke `start` in our own program. We want the parser to do this instead. It will be invoked five times for the sample document: once for the `HealthStudy` element and once each for the four `Interview` elements.

The parsing is actually done when the `parsefile` procedure is invoked. This procedure belongs to the parser and is not one of your own procedures (such as the `start` procedure), so it is invoked by using the -> operator on the module object. Procedures that belong to a module are called *methods*. The parameter is the name of the file to be parsed. In this case, the name of the XML file will be specified on the command line. The file names on the command line are in the `ARGV` array. Program 10.14 will be run by typing this line to the computer:

```
perl printweights.perl healthstudy.xml
```

The `start` procedure will be invoked with two kinds of information: the name of the element and the attributes of the element. The name of the element (also called its "tag") is the second parameter. The first statement of `start` sets the `tag` variable to the element name for later use. The rest of the parameters are the attributes of the element. The simplest way to use these parameters is to convert them to a hash and then look up the attributes that are needed. The second statement converts the parameters to a hash named `attributes`. The program prints the `Weight` attribute of every `Interview` element. The output of the program is

```
Weight 46.27
Weight 68.95
Weight 92.53
Weight 50.35
```

The XML::Parser handlers all have a first parameter that is a reference to an internal parsing procedure. This is used only if one wishes to get access to low-level parsing information.

One might be curious about what those => symbols mean in this program. As it happens, they are just another way of writing a comma. In other words, one could equally well have constructed the parser using this statement:

```
$p = new XML::Parser(Handlers, { Start, \&start });
```

The purpose of the => symbols is to make the program easier to understand. It is very common to specify parameters in pairs, where each pair consists of the name of the parameter and the value of the parameter. The => symbols are suggestive of this way of using parameters. This style for designing procedures is analogous to the attributes in an XML element. One first gives the attribute name and then the attribute value. In XML the attribute name and attribute value are separated by an equal-to sign. In Perl they are separated by => symbols.

Program 10.14 can only process information that is in XML attributes. XML content requires additional handlers. Consider the task of parsing the output of program 10.19 of subsection 10.2.4. The XML document in this case has no XML attributes at all, and all of the data are in XML content. Program 10.15 will accomplish the task. Just as a story has a beginning, a middle, and an end, there are now three handlers, one for when an element starts, one the content, and the last one for when an element ends. The `weightElement` variable is nonzero exactly when one is parsing a `Weight` element. This ensures that the `char` procedure will print the content only for `Weight` elements. In general, the `char` procedure will be invoked several times within a single element. It will usually be called once for each line of the content.

One of the most useful resources for general biomedical information is PubMed. This is a repository of citations to biomedical publications. More than half of the citations include abstracts. There are over 15 million citations available online using PubMed. These citations are available as XML documents. The following is what part of a typical PubMed citation looks like. The actual citation is over 130 lines long.

```
<MedlineCitation Owner="NLM" Status="Completed">
<MedlineID>99405456</MedlineID>
<PMID>10476541</PMID>
<DateCreated>
<Year>1999</Year>
<Month>10</Month>
<Day>21</Day>
</DateCreated>
<DateCompleted>
<Year>1999</Year>
<Month>10</Month>
<Day>21</Day>
</DateCompleted>
<DateRevised>
<Year>2001</Year>
<Month>11</Month>
<Day>02</Day>
</DateRevised>
<Article>
```

```
use XML::Parser;

$p = new XML::Parser(Handlers => { Start => \&start,
                                   End => \&end,
                                   Char => \&char });

$p->parsefile($ARGV[0]);

sub start {
  $tag = $_[1];
  if ($tag eq "Weight") {
    print("Weight ");
    $weightElement = 1;
  }
}
sub char {
  if ($weightElement) {
    print($_[1]);
  }
}
sub end {
  if ($weightElement) {
    print("\n");
    $weightElement = 0;
  }
}
```

Program 10.15 Parsing XML content

```
<Journal>
<ISSN>1083-7159</ISSN>
<JournalIssue>
<Volume>4</Volume>
<Issue>4</Issue>
<PubDate>
<Year>1999</Year>
</PubDate>
</JournalIssue>
</Journal>
<ArticleTitle>Breast cancer highlights.</ArticleTitle>
```

```
<Pagination>
<MedlinePgn>299-308</MedlinePgn>
</Pagination>
<Affiliation>Massachusetts General Hospital,
  Boston, Massachusetts 02114-2617, USA.
  Kuter.Irene@MGH.Harvard.edu</Affiliation>
<AuthorList CompleteYN="Y">
<Author>
<LastName>Kuter</LastName>
<ForeName>I</ForeName>
<Initials>I</Initials>
</Author>
</AuthorList>
<Language>eng</Language>
<PublicationTypeList>
<PublicationType>Congresses</PublicationType>
</PublicationTypeList>
</Article>
<MedlineJournalInfo>
<Country>UNITED STATES</Country>
<MedlineTA>Oncologist</MedlineTA>
<NlmUniqueID>9607837</NlmUniqueID>
</MedlineJournalInfo>
<ChemicalList>
<Chemical>
<RegistryNumber>0</RegistryNumber>
<NameOfSubstance>Antineoplastic Agents, Hormonal</NameOfSubstance>
</Chemical>
...
</ChemicalList>
<CitationSubset>IM</CitationSubset>
<MeshHeadingList>
...
<MeshHeading>
<DescriptorName MajorTopicYN="N">Piperidines</DescriptorName>
<QualifierName MajorTopicYN="N">therapeutic use</QualifierName>
</MeshHeading>
...
</MeshHeadingList>
</MedlineCitation>
```

An XML document would contain this citation as one of its elements. Consider the task of extracting the title of the article together with the list of all MeSH descriptors. The program for parsing an XML document to extract this information is shown in program 10.16. The result should look like this:

```
PubMed ID: 10476541
Title: Breast cancer highlights.
```

```
use XML::Parser;

$p = new XML::Parser(Handlers => { Start => \&start,
                                   Char => \&char });

$p->parsefile('pubmed.xml');

sub clear {
  $pmidElement = 0;
  $titleElement = 0;
  $descElement = 0;
}
sub start {
  if ($_[1] eq "PMID") {
    $pmidElement = 1;
  } elsif ($_[1] eq "ArticleTitle") {
    $titleElement = 1;
  } elsif ($_[1] eq "DescriptorName") {
    $descElement = 1;
  }
}
sub char {
  if ($pmidElement) {
    print("PubMed ID: $_[1]\n");
  } elsif ($titleElement) {
    print("Title: $_[1]\n");
  } elsif ($descElement) {
    print("Descriptor: $_[1]\n");
  }
  clear;
}
```

Program 10.16 Parsing PubMed citations

```
Descriptor: Antineoplastic Agents, Hormonal
Descriptor: Antineoplastic Combined Chemotherapy Protocols
Descriptor: Breast Neoplasms
Descriptor: Chemotherapy, Adjuvant
Descriptor: Estrogen Antagonists
Descriptor: Female
Descriptor: Hematopoietic Stem Cell Transplantation
Descriptor: Human
Descriptor: Lymphatic Metastasis
Descriptor: Neoplasm Staging
Descriptor: Piperidines
Descriptor: Raloxifene
Descriptor: Randomized Controlled Trials
Descriptor: Risk Factors
Descriptor: Tamoxifen
```

The handlers style of processing XML documents is the most efficient way to process XML. In fact, all other styles are based on the handlers style. However, the handlers style is difficult to use when one needs to do more complicated processing of the document. Subsection 10.2.3 presents another style that is better suited to more complex tasks.

Summary

- One way to process XML documents is to parse the document one element at a time. This is called the handlers style.

- In the handlers style, one specifies procedures that are invoked by the parser. Most commonly one specifies procedures to be invoked at the start of each element, for the text content of the element, and at the end of the element.

- A common way to design procedures is for the parameters to be in pairs: a parameter name and a parameter value. To make this easier to read, one should separate the parameter name from the parameter value with the => symbol.

- The handlers style for parsing XML documents is efficient and fast but is only appropriate when the processing to be done is relatively simple.

```
use XML::DOM;

$p = new XML::DOM::Parser;
$doc = $p->parsefile($ARGV[0]);

$weights = $doc->getElementsByTagName(Weight);
for ($i = 0; $i < $weights->getLength; $i++) {
  $weight =
    $weights->item($i)->getFirstChild->getNodeValue;
  print("Weight $weight\n");
}
```

Program 10.17 Converting an entire XML document using a Perl data structure

10.2.3 The Document Object Model

Although transforming XML element by element is capable of accomplishing any transformation task, it gets very complicated very quickly. The problem is that the information one needs at a given time might not be in the element being processed. It may, for example, be in the parent element. To deal with this, one can parse the entire document and then extract the parts that are needed. For example, suppose that one would like to extract the date, height, and weight of each interview. Program 10.17 uses the "whole document" approach to this task. This program uses the XML::DOM package. DOM stands for document object model. It reads an entire XML document into a single module object. Just as in the XML::Parser package, one constructs a parser, but no handlers need to be defined. The parser is then invoked to parse a document, and the module object containing the document is assigned to the `doc` variable. After that, one extracts information about the document by using DOM methods.

There are many DOM methods. One of the most popular methods is `getElementsByTagName` which extracts all of the elements within the current element which have a particular tag. Most DOM methods return an object, so one uses the -> operator to extract information from it. The `item` method gets one of the elements extracted by `getElementsByTagName`. The `item` method is yet another way to extract one of the items in a list. One uses brackets to get one item in an array, and braces to get one item in a hash.

For DOM lists one uses the `item` method.

DOM uses the single word "node" for anything that can occur in an XML document. Elements, attributes, and text content are all examples of DOM nodes. There are many methods of DOM nodes. The following are the most important methods:

- **getAttribute**. The value of one attribute of an element.
- **getChildNodes**. A DOM list consisting of all the nodes contained in an element.
- **getFirstChild**. The first child node in an element.
- **getParentNode**. The containing element.

Summary

- The whole document style of XML processing reads the entire document into a single Perl data structure.
- DOM methods are used to extract information from an XML document.
- The entities that occur in an XML document are represented by DOM nodes.
- DOM lists are used for holding a collection of DOM nodes.

10.2.4 Producing XML

While it is becoming increasingly common for bioinformatics data to be represented in XML, it is still the case that large amounts of data are still (and will continue to be) in various text formats. As a result, one common task will be to convert from text formats to XML format. This section is concerned with transformations from text files to XML. This can be done in Perl by the same kind of command that was used for transforming text files to text files; namely, a command such as this:

```
perl program.perl file.txt ... > result.xml
```

except that the result file is an XML file.

The programs in this section consider input file formats and XML structures that get progressively more complex. The early ones use simple flat (record) structures, and the later ones consider more complex structures.

Consider that health study again:

```
print("<HealthStudy>\n");
while (<>) {
  $month = substr($_, 0, 2) + 0;
  $day = substr($_, 2, 2) + 0;
  $yr = substr($_, 4, 2) + 0;
  $year = 1900 + $yr;
  $year = 2000 + $yr if $yr < 20;
  $bmi = substr($_, 6, 8) + 0;
  $status = "normal";
  $status = "obese" if substr($_, 14, 3) + 0 > 0;
  $status = "overweight" if substr($_, 17, 3) + 0 > 0;
  $height = substr($_, 20, 3) + 0;
  $weight = substr($_, 23, 8) + 0;
  print("<Interview Date='$year-$month-$day'");
  print(" BMI='$bmi' Status='$status'");
  print(" Height='$height' Weight='$weight'/>\n");
}
print("</HealthStudy>\n");
```

Program 10.18 Converting flat file information to XML attributes

```
011500   18.66  0  0 62   46.27102
011500   26.93  0  1 63   68.95152
020100   33.95  1  0 65   92.53204
020100   17.38  0  0 67   50.35111
```

The task is to convert this file to an XML document like this:

```
<HealthStudy>
<Interview Date='2000-1-15' BMI='18.66' .../>
<Interview Date='2000-1-15' BMI='26.93' .../>
<Interview Date='2000-2-1' BMI='33.95' .../>
<Interview Date='2000-2-1' BMI='17.38' .../>
</HealthStudy>
```

The solution is shown in program 10.18. This program stores all information in XML attributes. Another way to store information is to use XML content instead. For the health study example, the XML document would then look like this:

```
<HealthStudy>
<Interview>
  <Date>2000-1-15</Date>
  <BMI>18.66</BMI>
  <Status>normal</Status>
  <Height>62</Height>
  <Weight>46.27</Weight>
</Interview>
<Interview>
  <Date>2000-1-15</Date>
  <BMI>26.93</BMI>
  <Status>overweight</Status>
  <Height>63</Height>
  <Weight>68.95</Weight>
</Interview>
<Interview>
  <Date>2000-2-1</Date>
  <BMI>33.95</BMI>
  <Status>obese</Status>
  <Height>65</Height>
  <Weight>92.53</Weight>
</Interview>
<Interview>
  <Date>2000-2-1</Date>
  <BMI>17.38</BMI>
  <Status>normal</Status>
  <Height>67</Height>
  <Weight>50.35</Weight>
</Interview>
</HealthStudy>
```

To produce the XML document above use program 10.19.

It should be clear at this point that as the complexity of the XML file increases, producing it starts becoming tedious, even with procedures. What is especially bad is that it is very hard to see the structure of the XML file in the program. Templates are a technique that remedies this to some degree. Instead of burying the output text in strings scattered throughout the program, templates bury the program inside the output text. To use an old term for this, templates are closer to being WYSIWIG ("What You See Is What You Get"). There are many examples of the template style in use today. Active Server Pages (ASP) and Java Server Pages (JSP), are examples of this style. In both cases, the programs are HTML pages in which program code has been inserted in appropriate places.

When using the Perl Template Toolkit, one will need both a Perl program and a template file. Consider the task of producing an XML file that has just

```
print("<HealthStudy>\n");
while (<>) {
  $month = substr($_, 0, 2) + 0;
  $day = substr($_, 2, 2) + 0;
  $yr = substr($_, 4, 2) + 0;
  $year = 1900 + $yr;
  $year = 2000 + $yr if $yr < 20;
  $bmi = substr($_, 6, 8) + 0;
  $status = "normal";
  $status = "obese" if substr($_, 14, 3) + 0 > 0;
  $status = "overweight" if substr($_, 17, 3) + 0 > 0;
  $height = substr($_, 20, 3) + 0;
  $weight = substr($_, 23, 8) + 0;
  print("<Interview>\n");
  print("  <Date>$year-$month-$day</Date>\n");
  print("  <BMI>$bmi</BMI>\n");
  print("  <Status>$status</Status>\n");
  print("  <Height>$height</Height>\n");
  print("  <Weight>$weight</Weight>\n");
  print("</Interview>\n");
}
print("</HealthStudy>\n");
```

Program 10.19 Converting flat file information to XML element content

two elements, one inside the other, that looks like this:

```
<Main>
  <Part id='p1'>XML Example</Part>
</Main>
```

except that the Part id and the content of the Part element are obtained from an input file:

```
p1:XML Example
```

One can do this in Perl without templates by using program 10.20.

This program accomplishes the task, but as the transformation task gets more complicated, it is difficult to understand what is being done by this

```
print("<Main>\n");
while (<>) {
  chomp;
  split(/:/);
  print("  <Part id='$_[0]'>$_[1]</Part>\n");
}
print("</Main>\n");
```

Program 10.20 Converting text data to XML

```
<Main>
  <Part id='[% name %]'>[% content %]</Part>
</Main>
```

Template 10.1 Perl template for converting text data to XML

program because the XML text is spread throughout the Perl code. The Perl template is shown in template 10.1.

Notice how the Perl template looks much more like the output that is to be produced than the Perl program. The parts of the template in bracketed percent signs are the *variable* parts of the template. The rest of the template is the *constant* part. The constant part is printed exactly as shown. The variable parts are *instantiated* with the values of what look like variables. However, the names `id` and `content` are actually hash keys, not variables. The template is used from program 10.21.

The first line of the program tells Perl that the Template Toolkit package is being used. The data are obtained by reading the first line of the input file and extracting the data to be used in the template. The last part of the program invokes the template package. The first statement constructs the template processor using the Template Toolkit package. The second statement constructs a hash that tells the template processor the data that should be used for instantiating the template. As noted earlier, what look like variables in the template are actually hash keys. The third statement actually does the processing. The template processor needs two parameters: the name of the template file and the hash containing the data to be used for instantiation of the template.

```
use Template;

while (<>) {
  chomp;
  split(/:/);
  $name = $_[0];
  $content = $_[1];
}
$tt = new Template;
$vars = {
    name => $name,
    content => $content,
};
$tt->process('part.tt', $vars);
```

Program 10.21 Using Perl templates

Now consider a more interesting transformation task: the first task of this chapter. To use a Perl template, the data extracted from the input file must be organized into a data structure to be used by the template processor for instantiating the template as in program 10.22. The `while` statement constructs an array of hashes. Each hash gives the information about one interview of the health study. In other words, each hash represents one record of the health study database. The template processor is given this array in the same way as in the earlier program, except that now there is just one hash key: `HealthStudyInterviews`. This will be the name of the array within the template. The template is shown in template 10.2. Notice that one iterates over the elements of the array in almost the same way as in Perl. The Template Toolkit, however, uses a more simplified notation than Perl:

1. Variables usually have no initial character such as `%` or `@`. The Template Toolkit does use the `$` but only when one is substituting a value in an expression. For example, if one has a variable named `status` whose value is "obese," then the expression `i.status` would have two different meanings. Should it mean `$i{status}` or should it mean `$i{obese}`? In the Template Toolkit one specifies the second one by writing `i.$status`. The $ prefix in the Template Toolkit means "substitute the value of the variable here."

```
use Template;

while (<>) {
  $interviews[$i]{month} = substr($_, 0, 2) + 0;
  $interviews[$i]{day} = substr($_, 2, 2) + 0;
  $interviews[$i]{year} = 2000 + substr($_, 4, 2);
  $interviews[$i]{bmi} = substr($_, 6, 8) + 0;
  $status = 'normal';
  if (substr($_, 14, 3) + 0 > 0) { $status = 'obese'; }
  if (substr($_, 17, 3) + 0 > 0)
    { $status = 'overweight'; }
  $interviews[$i]{status} = $status;
  $interviews[$i]{height} = substr($_, 20, 3) + 0;
  $interviews[$i]{weight} = substr($_, 23, 8) + 0;
  $i++;
}
$tt = new Template;
$vars = {
  HealthStudyInterviews => [@interviews],
};
$tt->process("health.tt", $vars);
```

Program 10.22 Perl program that uses a template

```
<HealthStudy>
[% FOREACH i IN HealthStudyInterviews %]
<Interview Date='[% i.year %]-[% i.month %]-[% i.day %]'
  BMI='[% i.bmi %]' Status='[% i.status %]'
  Height='[% i.height %]' Weight='[% i.weight %]'/>
[% END %]
</HealthStudy>
```

Template 10.2 Perl template to convert fixed-width fields to variable-width fields

2. Selection of a part of a data structure is specified in Perl using brackets for array indices (e.g., `$x[1]`) and braces for hash keys (e.g., `$h{name}`. The Template Toolkit uses the dot notation for both of these (e.g., `x.1` and `h.name`, respectively).

3. Keywords such as `FOREACH` are written using all capital letters in the Template Toolkit, but using lower-case letters in Perl.

The Template Toolkit can simplify its notation because it supports a very limited range of features compared with Perl.

Next consider a more difficult transformation such as transforming the output produced by BioProspector as in subsection 10.1.4. The Perl program for extracting the motifs must be modified so that the information is kept in a Perl data structure which is given to the template in the usual way, as shown in program 10.23. The corresponding template is shown in template 10.3. Running this program on the BioProspector file produces output that begins like this:

```
<MotifData>

    <Motif id='1'>

        <DNA>
            <A>0.00</A>
            <C>0.21</C>
            <T>0.59</T>
            <G>0.21</G>
        </DNA>

...
```

The extra blank lines come from the `FOREACH` and `END` directives. These do not produce any text by themselves, so they show up as blank lines in the output. To get rid of the unnecessary blank lines and other spaces, just add dashes at the end of each directive, as shown in template 10.4.

Summary

- To convert non-XML data to the XML format, one can use the same techniques that apply to any kind of processing of text data. The XML document is just another kind of output format.

```
use Template;

while (<>) {
  chomp;
  if (/Motif #([0-9]+):/) {
    $label = $1;
    $i = 0;
  } elsif ($label && /^[0-9]+/) {
    split;
    $motifs{$label}[$i]{A} = $_[1];
    $motifs{$label}[$i]{G} = $_[2];
    $motifs{$label}[$i]{C} = $_[3];
    $motifs{$label}[$i]{T} = $_[4];
    $i++;
  }
}
$tt = Template->new();
$vars = {
  MotifData => {%motifs},
};
$tt->process("motif.tt", $vars);
```

Program 10.23 Using pattern matching to extract data and then formatting it with Perl templates

- The Perl Template Toolkit simplifies the production of XML documents by using a WYSIWIG style.
- The Perl Template Toolkit has its own language for iteration and selecting an item of a hash or array. The Template Toolkit language is much simpler than Perl because it has fewer features.

10.2.5 Transforming XML to XML

The last type of transformation task is transforming from XML to XML. In theory, this kind of transformation is just a special case of transforming from XML to text. However, there are new issues that arise in this case, and there are some new tools that are designed for this case.

```
<MotifData>
[% FOREACH label IN MotifData.keys.sort %]
    <Motif id='[% label %]'>
    [% FOREACH position IN MotifData.$label %]
        <DNA>
            <A>[% position.A %]</A>
            <C>[% position.C %]</C>
            <T>[% position.T %]</T>
            <G>[% position.G %]</G>
        </DNA>
    [% END %]
    </Motif>
[% END %]
</MotifData>
```

Template 10.3 Perl template for formatting Perl hashes and arrays

```
<MotifData>
[% FOREACH label IN MotifData.keys.sort -%]
    <Motif id='[% label %]'>
[% FOREACH position IN MotifData.$label -%]
        <DNA>
            <A>[% position.A %]</A>
            <C>[% position.C %]</C>
            <T>[% position.T %]</T>
            <G>[% position.G %]</G>
        </DNA>
[% END -%]
    </Motif>
[% END -%]
</MotifData>
```

Template 10.4 Perl template that avoids unnecessary blank lines

```
use XML::Parser;

$p = new XML::Parser(Handlers => { Start => \&start });

print("<HealthStudyUS>\n");
$p->parsefile($ARGV[0]);
print("</HealthStudyUS>\n");

sub start {
  $tag = $_[1];
  %attributes = @_;
  if ($tag eq "Interview") {
    print("  <Interview");
    print(" Date='$attributes{Date}'");
    $WeightUS = $attributes{Weight} * 2.2;
    print(" Weight='$WeightUS'");
    $HeightUS = $attributes{Height} * 0.39;
    print(" Height='$HeightUS'");
    print("/>\n");
  }
}
```

Program 10.24 Transforming XML attributes

The most efficient technique for transforming from XML to XML is to use the handlers style. Consider the task of reading the health study database and changing the height and weight from centimeters and kilograms to inches and pounds. The input document looks like this:

```
<HealthStudy>
<Interview Date='2000-1-15' BMI='18.66' .../>
<Interview Date='2000-1-15' BMI='26.93' .../>
...
```

The output document should contain only the date, height, and weight attributes of each interview. The result of running program 10.24 using the example database is

```
<HealthStudyUS>
  <Interview Date='2000-1-15' Weight='101.794' Height='24.18'/>
  <Interview Date='2000-1-15' Weight='151.69' Height='24.57'/>
  <Interview Date='2000-2-1' Weight='203.566' Height='25.35'/>
  <Interview Date='2000-2-1' Weight='110.77' Height='26.13'/>
</HealthStudyUS>
```

Program 10.24 deals only with information in attributes. When information is in XML content, one must use additional handlers. Suppose that one has the same task as in program 10.15, but the output must be in XML. The output should look like this:

```
<WeightList>
  <Weight>46.27</Weight>
  <Weight>68.95</Weight>
  <Weight>92.53</Weight>
  <Weight>50.35</Weight>
</WeightList>
```

The solution is shown in program 10.25.

XML allows data to be in either attributes or content. Attributes are much simpler to process, but they are more limited than content. Content can have markup while attributes cannot. Generally speaking, one should use attributes for simple data values and one should use content for more complex data values.

One common transformation task is to convert from one of these two formats to the other. Consider the task of converting the health study from content attributes to ordinary attributes. In program 10.26 the `$printContent` variable is used by the `start` handler to inform the `char` and `end` handlers that the content information is to be printed. The `end` handler turns this variable "off."

While the handlers style of parsing and processing XML documents is efficient, programs can get very complicated as the transformation task involves data and attributes on more than one level. As an exercise, try to modify the program above so that it converts the weight and height from kilograms and centimeters to pounds and inches. To do this exercise, one must introduce one or more variables that allow the `start` handler to inform the `char` handler about which attribute is being printed so that the appropriate conversion can be performed. The problem with this style is that the handling of each element is spread over the three handlers. It would be better if all the processing for each type of element were handled in one place. Other

```
use XML::Parser;

$p = new XML::Parser(Handlers => { Start => \&start,
                                   End => \&end,
                                   Char => \&char });

print("<WeightList>\n");
$p->parsefile($ARGV[0]);
print("</WeightList>\n");

sub start {
  $tag = $_[1];
  if ($tag eq "Weight") {
    print("  <Weight>");
    $weightElement = 1;
  }
}
sub char {
  if ($weightElement) {
    print($_[1]);
  }
}
sub end {
  if ($weightElement) {
    print("</Weight>\n");
    $weightElement = 0;
  }
}
```

Program 10.25 Transforming XML content

```
use XML::Parser;
$p = new XML::Parser(Handlers =>
 { Start => \&start, End => \&end, Char => \&char });
$p->parsefile($ARGV[0]);
sub start {
  $tag = $_[1];
  if ($tag eq "HealthStudy") {
    print("<HealthStudy>\n");
  }
  elsif ($tag eq "Interview") {
    print("<Interview");
  }
  elsif ($tag eq "Date") {
    print(" Date='");
    $printContent = 1;
  }
  ...
}
sub char {
  if ($printContent) {
    print($_[1]);
  }
}
sub end {
  $tag = $_[1];
  if ($tag eq "HealthStudy") {
    print("</HealthStudy>\n");
  }
  elsif ($tag eq "Interview") {
    print("/>\n");
  }
  elsif ($printContent) {
    print("'");
    $printContent = 0;
  }
}
```

Program 10.26 Transforming XML content to XML attributes

parsing styles can help simplify the program, but all of them have disadvantages. A better approach that has become very popular is to use the XML Transformation Language that is introduced in the next chapter.

Summary

- Transformation from XML to XML using Perl can be done using any of the parsing styles.
- None of the styles are completely satisfactory when the transformation task is complicated.

10.3 Exercises

In the following exercises, write a Perl program that determines the specified information. The solutions to these exercises are available online at the book website `ontobio.org`. Additional exercises are also available at this site.

1. Using the health study database in section 1.1, find all interviews in the year 2000 for which the study subject had a BMI greater than 30. Print the information for each such interview using tab-delimited fields. Compare your answer with your solution to exercise 10.1.

2. Perform the same task as in exercise 10.1, but using a database in XML format as in section 1.2. Write your program first by using patterns to extract the information, and then by using the XML::Parser module.

3. Generalize exercise 10.2 to extract interviews for any year and any minimum BMI value. Write your program as a Perl procedure which has two parameters.

4. Given a BioML document as in figure 1.3, find all literature references for the insulin gene. Compare your answer with your solution to exercise 10.2.

5. As in exercise 10.3, find all PubMed citations dealing with the therapeutic use of glutethimide. For each citation print one line containing the MedlineID, the title, and the date of publication in tab-delimited format.

6. For the health study database in section 1.1, the subject identifier is a field named `SID`. Find all subjects in the database for which the BMI of the subject increased by more than 4.5 during any period of time. For each subject, print the subject identifier, the amount that the BMI increased, and the period of time. Print the results in XML format. If this condition is satisfied more than once by a subject, then print the maximum increase in the BMI for this subject. Hint: Collect information about each subject in a hash or array.

7. Read the GO association database, and compute the number of associations of the term GO:0003673. Unlike exercise 10.6, determine the actual number of `go:association` elements for the specified GO term. Do not use the `n_associations` attribute.

8. A file contains BioML data as in figure 1.3. For each gene in this file, compute the total length of all exons that it contains.

11 *The XML Transformation Language*

The XML Transformation Language (XSLT) (W3C 2001d) is one of the most popular, as well as the most commonly available, transformation languages for XML documents. Although this language was originally intended for use by the XML Stylesheet Language (XSL), one can use XSLT for many other useful transformations, including data transformations for bioinformatics. In fact, XSLT is used mostly for transformation today. While there are many XML transformation languages, XSLT has the advantage of being rule-based and being itself written in XML. This chapter introduces this style of programming.

11.1 Transformation as Digestion

XSLT is very different from the procedural style of programming that dominates mainstream programming languages. XSLT is rule-based. An XSLT rule is called a *template*, and an XSLT program is just a set of templates. The templates are separate from one another (i.e., one template can never contain another), and the order in which they appear in the program does not matter. The whole XSLT program is called a *transformation program* or a *transform*.

Consider the document in figure 11.1 that shows some protein interaction data from a microarray experiment. Suppose that one would like to change the names (tags) of some of the elements. Specifically, suppose that instead of `Protein` we want to use `P`, and instead of `Substrate`, use `S`. Transform 11.1 shows the XSLT program for doing this task. To understand how this program functions, consider how enzymes digest molecules such as proteins. Proteins are long chains of amino acids, and each enzyme is capable of splitting the chain at one or more specific points in the chain, which match the active site of the enzyme. This process is shown symbolically in figure 11.2.

```
<Array>
 <Protein id="Mas375">
  <interaction substrate="Sub89032">
   <BindingStrength>5.67</BindingStrength>
   <Concentration unit="nm">43</Concentration>
  </interaction>
  <interaction substrate="Sub89033">
   <BindingStrength>4.37</BindingStrength>
   <Concentration unit="nm">75</Concentration>
  </interaction>
 </Protein>
 <Protein id="Mtr245">
  <interaction substrate="Sub89032">
   <BindingStrength>0.65</BindingStrength>
   <Concentration unit="um">0.53</Concentration>
  </interaction>
  <interaction substrate="Sub80933">
   <BindingStrength>8.87</BindingStrength>
   <Concentration unit="nm">8.4</Concentration>
  </interaction>
 </Protein>
 <Substrate id="Sub89032"/>
 <Substrate id="Sub89033"/>
</Array>
```

Figure 11.1 Example of a document specifying some values obtained by a microarray experiment.

Each template acts like an enzyme that acts upon one or more kinds of elements in the XML document. The kinds of elements that the template can "attack" is specified by the `match` attribute. Most commonly, the `match` condition is either the tag of the elements that the template can attack or a "wild card" that allows the template to attack any element. If there are both specific and generic templates, then the specific ones take precedence.

Since elements and attributes can have the same names, XSLT distinguishes them by prefixing attribute names with an @ sign. Thus `chromosome` is the name of an element, but `@start` is the name of an attribute. The wild card notation for elements is `node()`, and the wild card notation for attributes is `@*`. The templates in transform 11.1 use both of the wild card notations.

Enzymes can only attack locations on a protein chain that are "exposed." In the same way, templates only attack the highest-level elements that can be matched. Lower-level elements become exposed only when the contain-

```
<?xml version="1.0"?>
<xsl:transform version="1.0"
 xmlns:xsl="http://www.w3.org/1999/XSL/Transform">

<!-- Change all occurrences of Protein to P -->
<xsl:template match="Protein">
  <P>
    <xsl:apply-templates select="@*|node()"/>
  </P>
</xsl:template>

<!-- Change all occurrences of Substrate to S -->
<xsl:template match="Substrate">
  <S>
    <xsl:apply-templates select="@*|node()"/>
  </S>
</xsl:template>

<!-- Don't change anything else -->
<xsl:template match="@*|node()">
  <xsl:copy>
    <xsl:apply-templates match="@*|node()"/>
  </xsl:copy>
</xsl:template>

</xsl:transform>
```

Transform 11.1 An XML transformation program that changes the name of protein elements from "Protein" to "P", and similarly changes "Substrate" to "S". All other elements are unchanged.

ing elements have been "digested." Digestion and the subsequent exposing of child elements to attack by other templates is accomplished by using the `xsl:apply-templates` command. One can be selective about exactly which of the child elements will be exposed by using a `select` criterion. Figure 11.3 illustrates how the hierarchical structure relates to the templates. Note that the context changes as a result of the "digestion" of an element.

The last template in transform 11.1 is saying: "by default, copy all elements and attributes, and then apply appropriate templates to the attributes and child elements that are in each element." This template is a handy one to include in any XSLT program that is modifying some of the features of an XML document, but which is leaving most of the features unchanged.

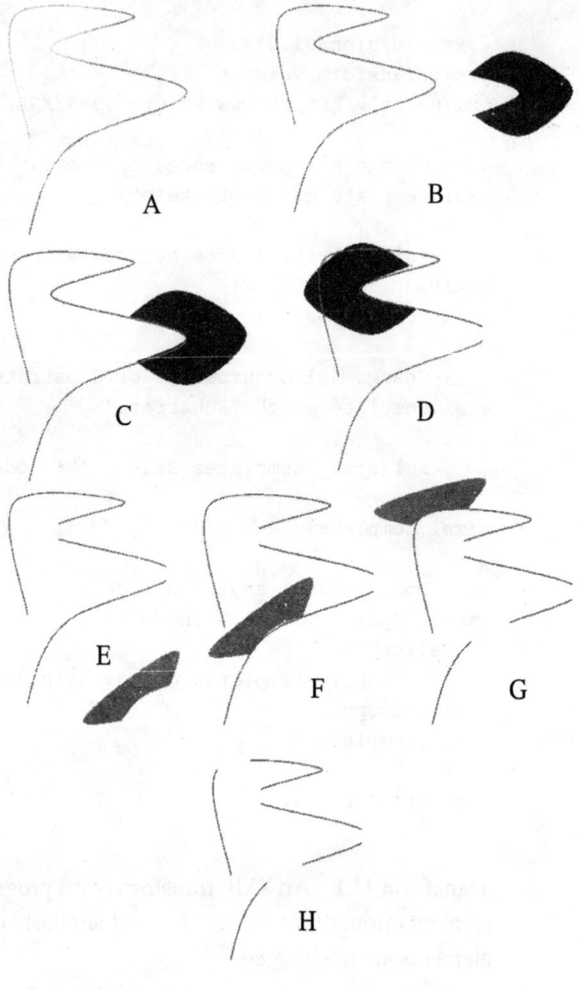

Figure 11.2 Abstract depiction of the process of digestion. The original chain is shown in A. An enzyme (dark gray region) attacks the chain (B, C, D) in two locations, splitting the chain each time. A second enzyme (light gray region) attacks two of the subchains (E, F, G). The end result is five subchains (H).

Summary

- An XSLT program consists of templates.
- A template either matches a specific kind of element or attribute or it uses a wild card to match many kinds of elements and attributes.
- A template performs an action on the matching elements and attributes.

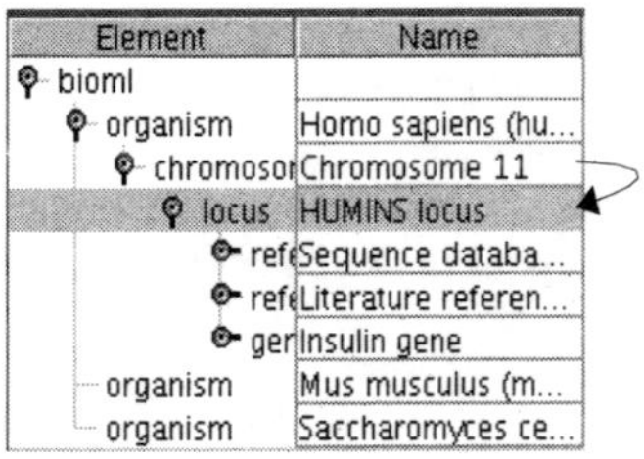


```
<xsl:template match="chromosome">
  ...
  <xsl:apply-templates select="locus"/>
</xsl:template>

<xsl:template match="locus">
  ...
</xsl:template>
```

Figure 11.3 The digestion process during XML transformation. The first template digests a `chromosome` element and then releases the `locus` child elements to the second template. The corresponding action on the hierarchy is to change the context from the `chromosome` element to the `locus` as shown in the screen image.

- After transforming the matching element or attribute, a template can apply other templates to continue the transformation.

11.2 Programming in XSLT

Every XSLT template acts within a *context*. When one is using an XML editor, the context is the element (or attribute) that is currently highlighted. One selects context in an XML editor by clicking the mouse on the desired element in the hierarchy or by clicking on a data entry box for an attribute. In the same way, the `xsl:apply-templates` changes the transformation context from one element or attribute to another.

The context for an XML editor is selected by a person using a mouse. The selection can occur wherever the person expresses an interest. The context for an XML transformation is more systematic. Elements are normally selected by reading the document from the beginning to the end, just as one would

```
...
<!-- Change all occurrences of Protein to P -->
<xsl:template match="Protein">
  <xsl:sort select="@id"/>
  <P>
    <xsl:apply-templates select="@*|node()"/>
  </P>
</xsl:template>

<!-- Change all occurrences of Substrate to S -->
<xsl:template match="Substrate">
  <xsl:sort select="@id"/>
  <S>
    <xsl:apply-templates select="@*|node()"/>
  </S>
</xsl:template>
...
```

Transform 11.2 A modification of the program in transform 11.1 in which the proteins and substrates have been sorted by their ids

read a novel. This order is called the *document order*. However, the order in which elements are selected during the transformation can be changed by using a `xsl:sort` element. In transform 11.2 a transformation is performed that not only changes some element names but also changes the order of those elements.

The `apply-templates` command serves to change the context of the transformation from one element or attribute to another one. The `for-each` is another command that accomplishes the same effect. The only difference between them is that `apply-templates` causes another template to become active in a new context while the `for-each` command stays inside the same template. This is illustrated in transform 11.3 which changes the tag of `interaction` elements within `Protein` elements to `I`.

While both `apply-templates` and `for-each` have the same effect, there are some differences. The `for-each` command is a traditional technique for controlling the actions performed by a computer program, and those who have programming experience will find it a familiar command. By contrast, `apply-templates` is a rule-based command that uses a matching or "lock-and-key" mechanism which is much more flexible and powerful.

The power of the `apply-templates` rule-based command is illustrated by transform 11.2. In this program, child elements of a `Protein` other than

```
...
<xsl:template match="Protein">
  <P>
    <xsl:apply-templates select="@*"/>
    <xsl:for-each select="interaction">
      <I>
        <xsl:apply-templates select="@*|node()"/>
      </I>
    <xsl:for-each/>
  </P>
</xsl:template>
...
</xsl:transform>
```

Transform 11.3 A modification of the program in transform 11.1 in which the `interaction` elements contained in `Protein` elements are changed to `I` elements. All other kinds of child element in a `Protein` element are lost.

`interaction` elements would be lost. This would not occur if the transformation of the `interaction` elements were done using another template. The only `interaction` elements that will be transformed by the `for-each` command are the ones that are child elements of a `Protein` element.

Nevertheless, the `for-each` command is useful, especially when one is performing numerical calculations. This is the topic of the next section.

Summary

- A transformation action occurs in a context: the element or attribute being transformed.
- The context is normally chosen in the same order in which the elements or attributes appear in the document, but which can be changed by using a `sort` command.
- The context is changed by using either an `apply-templates` (rule-based) command or a `for-each` (traditional iteration) command.

11.3 Navigation and Computation

Like any programming language, one can perform numerical computations using XSLT. This is specified using a notation similar to that in traditional

```
...
<xsl:template match="Protein">
  <P>
    <xsl:attribute name="averageBindingStrength">
      <xsl:value-of
      select="sum(interaction/BindingStrength) div
              count(interaction/BindingStrength)"/>
    </xsl:attribute>
    <xsl:apply-templates select="@*|node()"/>
  </P>
</xsl:template>
..
```

Transform 11.4 A modification of the program in transform 11.1 to compute the average binding strength of all interactions with a protein. The average binding strength is shown as another attribute in each `P` element.

programming languages such as Perl, but XSLT adds a new feature to computation: *navigation*.

Navigation is the process of conducting vehicles from one place to another. The original meaning was concerned with ships on the sea. Nowadays it is more commonly applied to the directions for driving a car from one place to another. In the case of XML documents, one navigates from one element to another. Instead of streets one navigates over elements, and instead of turning from one street to another, one traverses either "down" from an element to a child element, or "up" from an element to its parent element.

The template in transform 11.4 shows how to perform both navigation and computation. The objective is to compute the average binding strength of all interactions of a protein. The `value-of` command evaluates the expression in its `select` attribute. The `interaction/BindingStrength` part of this expression is the navigation using XPath as in section 8.1. It specifies that one should select all `interaction` elements in the context and then select all `BindingStrength` elements within the `interaction` elements. The slash means that one navigates from a parent element to a child element. This notation emulates the notation used for navigating among directories and files (except that in Windows, a backward slash is used instead of a forward slash).

An `attribute` command inserts an attribute into the current element (in this case a `P` element). The `sum` is the numerical sum of all matching elements, and the `count` is the number of all matching elements. The `div`

```
...
<xsl:template match="interaction">
  <interaction>
    <xsl:attribute name="protein">
      <xsl:value-of select="../@id"/>
    </xsl:attribute>
    <xsl:apply-templates select="@*|node()"/>
  </interaction>
</xsl:template>
..
```

Transform 11.5 A template that adds the `id` of the containing element as a new attribute

operator is short for "division." Programming languages often use the slash to denote division. Obviously one cannot use the same notation because that would conflict with the use of slash to denote navigation.

Navigating from a child to a parent uses the same notation as in directories. In transform 11.5, an attribute is added to the `interaction` element that has the identifier of the corresponding `Protein` element.

The XSLT language inherits all of the operators that are available in XPath, such as the ones in table 8.1. Two operators that seem to be missing are the maximum and minimum operators. In fact, both of these can be computed by using the `xsl:sort` command. This is explained in the next section.

Summary

- XSLT navigation is the process of traveling from one element or attribute to another one in the document.
- Navigation is specified using the same notation as in directory trees.
- Computations are specified using operators, such as the ones shown in table 8.1

11.4 Conditionals

Conditionals are used to define special cases. For example, in section 1.1 the health study record defined normal weight, overweight, and obesity in terms of ranges for the body mass index (BMI). In XSLT these ranges would be written like this:

```
<xsl:choose>
  <xsl:when test="@bmi&lt;25">
    Normal
  </xsl:when>
  <xsl:when test="@bmi&lt;30">
    Overweight
  </xsl:when>
  <xsl:otherwise>
    Obese
  </xsl:otherwise>
</xsl:choose>
```

By using sorting and conditionals one can compute the maximum and minimum. Here is the computation of a maximum:

```
<xsl:for-each select="interaction/BindingStrength">
  <xsl:sort data-type="number" select="."/>
  <xsl:if test="position()=last()">
    <xsl:value-of select="."/>
  </xsl:if>
</xsl:for-each>
```

This computation sorts all the binding strengths in increasing numerical order. It then selects just the last (largest) one. Note the use of the "." to denote the current element. Alternatively, one could have sorted in descending order and selected the first one as follows:

```
<xsl:for-each select="interaction/BindingStrength">
  <xsl:sort data-type="number"
            order="descending" select="."/>
  <xsl:if test="position()=1">
    <xsl:value-of select="."/>
  </xsl:if>
</xsl:for-each>
```

Conditionals can appear either as elements using `xsl:choose` or `xsl:if` as above or within match and select attributes. For example,

```
<xsl:value-of select="BindingStrength[position()=1]"/>
```

will select just the first `BindingStrength` element. One can abbreviate the test above as

```
<xsl:value-of select="BindingStrength[1]"/>
```

but this should only be used in simple cases like this one. Do not expect such abbreviations to work for more complicated expressions.

Summary

- Conditionals are used for special cases.
- The `xsl:if` conditional element is used to restrict to a single special case.
- The `xsl:choose` conditional element is used for handling several special cases.
- Conditional can be specified in `match` and `select` attributes by writing the condition in brackets after the match criterion.

11.5 Precise Formatting

XSLT is not limited to producing only XML files. The output file can have any format, although XSLT is primarily intended for XML. Of course, the input document must necessarily be an XML document. This book, for example, was written in XML and then translated to the LaTeX typesetting language using XSLT.

The most common formatting issue is the formatting of numbers. The `format-number` function is used for formatting numbers. For example,

```
<xsl:value-of
  select="format-number(3674.9806, '#,##0.0##')"/>
```

will print `3,674.981`. The # symbol represents a digit that will be omitted if it is insignificant. Zero represents a digit (not just `0`) that will always be printed even if it is insignificant. As another example,

```
<xsl:value-of
  select="format-number(3674.9805, '#,##0.000')"/>
```

will print `3,674.980`.

When XSLT is producing an XML file, it attempts to place the elements so that one can read the output document without using any special tools. In particular, the elements are successively indented to show the boundaries of the elements. Of course, it is better to view XML documents using an XML editor.

Although XSLT is usually pretty good about guessing your intentions, it is not always obvious whether your output file is supposed to be an XML file or it just has some of the XML features. To tell XSLT exactly what you intend, include one of the following at the beginning of your XSLT program:

- For XML output files:

```
<xsl:output method="xml" media-type="text/xml"/>
```

- For HTML output files:

```
<xsl:output method="html" media-type="text/html"/>
```

- For output files that are neither XML nor HTML:

```
<xsl:output method="text" omit-xml-declaration="yes"/>
```

If your output file is not an XML document, then you may want to exercise more precise control over the output formatting by using the `xsl:text` element. Consider these two templates:

```
<xsl:template match="Protein">
  Protein information:
  <xsl:apply-templates select="@*|node()"/>
</xsl:template>

<xsl:template match="Protein">
  <xsl:text>Protein information:</xsl:text>
  <xsl:apply-templates select="@*|node()"/>
</xsl:template>
```

The first template would produce generous amounts of space before and after the `Protein information:` text in the output file, while the second would write nothing more than just the `Protein information:` text.

Since XSLT is designed to produce XML documents, it automatically changes the left angle bracket from < to `<`. XSLT also automatically changes the ampersand character from `&` to `&` These two characters have a special meaning in XML documents. If XSLT is being used to produce a non-XML document, then one may want these two characters to be left alone. To force XSLT to write left angle brackets and ampersands verbatim, use `disable-output-escaping` attribute in each element where this behavior is desired.

Summary

- The `format-number` function allows one to specify the format of a number.
- The `xsl:output` element tells XSLT the kind of document that is being produced so it can format the output document appropriately.
- The `xsl:text` element is used for controlling the amount of space in the output document and also for informing XSLT whether or not to escape the XML special characters.

11.6 Multiple Source Documents

When the amount of source information is large, it is convenient to break up a large file into several smaller files. There are two strategies for dealing with such collections of files:

1. The collection of files is a single XML document that was split into pieces for convenience. In this strategy, all of the pieces must form a document that conforms to a single DTD.
2. The collection of files is a collection of different XML documents that are used for a single purpose. This strategy allows the individual files in the collection to use different DTDs.

Suppose that one has performed five experiments and that the data are stored in five separate files, called `experiment1.xml` through `experiment5.xml`. The `experiment1.xml` file might look like this:

```
<Experiment date="2003-09-01">
  <Observation id="A23">
  ...
</Experiment>
```

The first strategy can be accomplished by using the notion of an XML entity as discussed in section 1.4. A separate "main" file is created that looks like this:

```
<?xml version="1.0"?>
<!DOCTYPE ExperimentSet SYSTEM "experiment.dtd"
[
  <!ENTITY experiment1 SYSTEM "experiment1.xml">
  <!ENTITY experiment2 SYSTEM "experiment2.xml">
  <!ENTITY experiment3 SYSTEM "experiment3.xml">
  <!ENTITY experiment4 SYSTEM "experiment4.xml">
  <!ENTITY experiment5 SYSTEM "experiment5.xml">
]>
<ExperimentSet>
  &experiment1;
  &experiment2;
  &experiment3;
  &experiment4;
  &experiment5;
</ExperimentSet>
```

The five files will automatically be incorporated into the main file. This is done by the XML processor, not by XSLT, and there is nothing in the XSLT transformation program that mentions anything about these files. Note that only the main file mentions the `DOCTYPE`. This strategy requires that the files being combined form an XML document that conforms to the overall DTD.

To accomplish the second strategy use the `document` function. For example,

```
  <xsl:for-each select="document('experiment1.xml')">
    <xsl:apply-templates/>
  <xsl:for-each>
  <xsl:for-each select="document('experiment2.xml')">
    <xsl:apply-templates/>
  <xsl:for-each>
```

will process the `experiment1.xml` and `experiment2.xml` documents. Unlike the first strategy, the second strategy processes each document independently. So they could have different DTDs.

Summary

- XSLT can process multiple input source files by using XML entities to include one file in another.
- Alternatively, XSLT can process multiple files by using the `document` function.

11.7 Procedural Programming

Although XSLT is a rule-based language, one can also program in XSLT using the traditional procedural style. In particular, this means that one can declare and use variables and procedures, and one can pass parameters to procedures.

A variable is declared using the `xsl:variable` command. For example,

```
<xsl:variable name="x" select="BindingStrength[1]"/>
```

will set the variable x to the first `binding_strength` element in the current context. This command has approximately the same meaning as

```
$x = $BindingStrength[0];
```

in Perl. Note that XSLT starts counting at 1 while Perl normally starts counting at 0.

An XSLT variable is used (evaluated) by writing the `$` character before the variable name. This convention is almost the same as in Perl, except that Perl variables are not declared so they always appear with a preceding character such as `$`. Another difference is that Perl distinguishes between variables that represent collections of values from variables that represent single values (called "scalars" in Perl). XSLT makes no such distinction.

Procedures in XSLT are just templates that have a name. They are called by using the `xsl:call-template` command. The following template computes the average of all `binding_strength` elements in the current context:

```
<xsl:template name="BindingStrengthAverage">
  <xsl:value-of select="sum(BindingStrength) div
                         count(BindingStrength)"/>
</xsl:template>
```

The procedure is called as follows:

```
<xsl:call-template name="BindingStrengthAverage"/>
```

Procedures often have parameters, and these are specified in XSLT by using `xsl:param` in the procedure. For example, the following will compute the average of any set of elements:

```
<xsl:template name="average">
  <xsl:param name="elements"/>
  <xsl:value-of
    select="sum($elements) div count($elements)"/>
</xsl:template>
```

When a procedure is called, the parameters are specified using the `xsl:with-param` command as follows:

```
<xsl:call-template name="average">
  <xsl:with-param name="elements"
    select="BindingStrength"/>
</xsl:call-template>
```

In general, a computation procedure consists of the following parts:

1. The procedure declaration. This consists of the name of the procedure and the names of the parameters.
2. The procedure body. This is the part that performs the actual computation. It usually consists of a conditional element having two parts:
   (a) The computation performed on each subelement
   (b) The computation performed after all subelements have been processed

The previous example shows the computation of the average, so it is natural to consider how one might compute the variance using XSLT. The first step is writing the procedure declaration. In this case there are three relevant parameters. The first is the set of elements whose variance is to be computed. The second is the *accumulator*. It is the variable that is used for computing the sum of squares of the elements. It is called the accumulator because it accumulates the sum by successively adding terms until the entire sum has been computed. The last parameter is the *iterator*. Its purpose is to indicate which term is to be added to the accumulator. Here is the declaration for a procedure to compute the variance:

```
<xsl:template name="variance">
  <xsl:param name="elements"/>
  <xsl:param name="ssq"/>
  <xsl:param name="i"/>
```

The name of the procedure is `variance`. The set of elements to be used for the computation is called `elements`. The accumulator is `ssq` and the iterator is `i`. The second step is to write the procedure body. This consists of a conditional element for the two cases. It looks like this:

```
  <xsl:choose>
    <xsl:when test="$i > count($elements)">
  <!-- The final computation goes here. -->
    </xsl:when>
    <xsl:otherwise>
  <!-- The computation on each subelement goes here. -->
    </xsl:otherwise>
  </xsl:choose>
```

Since the iterator starts at 1, the computation is complete when the iterator exceeds the total number of elements to be processed. It does not matter whether the final computation is written first or second. So it could also be written this way:

```
  <xsl:choose>
    <xsl:when test="$i &lt;= count($elements)">
  <!-- The computation on each subelement goes here. -->
    </xsl:when>
    <xsl:otherwise>
  <!-- The final computation goes here. -->
    </xsl:otherwise>
  </xsl:choose>
```

The computation on each subelement consists of three steps:

1. Add the next square to the accumulator.

2. Increase the iterator by 1.

3. Continue the computation.

Here is the program:

```
<xsl:call-template name="variance">
  <xsl:with-param name="elements" select="$elements"/>
  <xsl:with-param name="ssq"
    select="$ssq + $elements[position()=$i] *
                   $elements[position()=$i]"/>
  <xsl:with-param name="i" select="$i + 1"/>
</xsl:call-template>
```

The first `xsl:with-param` command adds the square of the next element to the accumulator. The second command increases the iterator by 1. The call to the procedure continues the computation. The two commands can be written in either order as they take effect only after the computation is continued. So the following program does the same computation:

```
<xsl:call-template name="variance">
  <xsl:with-param name="i" select="$i + 1"/>
  <xsl:with-param name="ssq"
    select="$ssq + $elements[position()=$i] *
                   $elements[position()=$i]"/>
  <xsl:with-param name="elements" select="$elements"/>
</xsl:call-template>
```

The final computation divides the sum of squares by the number of elements and subtracts the square of the average:

```
<xsl:variable name="avg"
  select="sum($elements) div count($elements)"/>
<xsl:value-of
  select="$ssq div count($elements) - $avg * $avg"/>
```

Putting these together gives the following procedure for computing the variance:

```
<xsl:template name="variance">
  <xsl:param name="elements"/>
  <xsl:param name="ssq"/>
  <xsl:param name="i"/>
  <xsl:choose>
    <xsl:when test="$i > count($elements)">
      <xsl:variable name="avg"
        select="sum($elements) div count($elements)"/>
      <xsl:value-of
        select="$ssq div count($elements) - $avg * $avg"/>
```

```
      </xsl:when>
      <xsl:otherwise>
        <xsl:call-template name="variance">
          <xsl:with-param name="ssq"
            select="$ssq + $elements[position()=$i] *
                           $elements[position()=$i]"/>
          <xsl:with-param name="i" select="$i + 1"/>
        </xsl:call-template>
      </xsl:otherwise>
    </xsl:choose>
</xsl:template>
```

Here is an example of how this procedure would be called:

```
<xsl:variable name="trialvariance">
  <xsl:call-template name="variance">
    <xsl:with-param name="elements" select="trial"/>
    <xsl:with-param name="ssq" select="0"/>
    <xsl:with-param name="i" select="1"/>
  </xsl:call-template>
</xsl:variable>
<xsl:value-of
  select="format-number($trialvariance, '###0.##')"/>
```

As this example suggests, XSLT can be used for numerical computations provided the computations are not too complicated. When the computations get complex, it would be better to use software tools and languages that are designed for such computations (such as Perl).

Summary

- XSLT can be used for traditional procedural programming.
- Variables are declared by using an `xsl:variable` element.
- Procedures are templates that have a name. The parameters of a procedure are declared by using `xsl:param` elements.
- Procedures are called by using an `xsl:call-template` element. Parameters are passed to the procedure by using `xsl:with-param` elements.
- Although one could implement complex numerical algorithms in XSLT, it is probably easier to use programming languages and tools that are designed for such algorithms.

11.8 Exercises

The following exercises use the BioML example in figure 1.3. Each exercise is solved with one or two templates that transform the kinds of elements mentioned in the exercise. Each of the solutions is an XSLT program having the following form:

```
<xsl:transform version='1.0'
 xmlns:xsl='http://www.w3.org/1999/XSL/Transform'>

<!-- The answer to the exercise goes here. -->

<!--
  This template copies all elements and attributes
  that do not appear in the template(s) above.
-->
<xsl:template match="@*|node()">
  <xsl:copy>
    <xsl:apply-templates select="@*|node()"/>
  </xsl:copy>
</xsl:template>

</xsl:transform>
```

1. Copy the `locus` name attribute so that it is an attribute of `gene`.
2. Remove all `locus` elements, and move any `reference` elements from being child elements of the `locus` element to being child elements of the `gene` element.
3. Change the BioML example to have a striping layer between `organism` and `chromosome` as shown in figure 1.14.
4. Change the `reference` elements to be either `isStoredIn` or `isCited By` depending on whether the reference is to a sequence database or to a literature reference.
5. Infer the EMBL sequence number and the organism of each gene, adding them to the `gene` element as attributes.
6. For each gene, compute the total length of all exons that it contains. Compare your solution with your solution to exercise 11.8.

12 *Building Bioinformatics Ontologies*

Unstructured data, such as natural language text, and semistructured data, such as tables and graphs, are adequate mechanisms for individuals to communicate with one another using traditional print media and when the amount of published material is relatively small. However, the amount of biomedical knowledge is becoming much too large for traditional approaches. While printed research publications are still very important, other forms of biomedical information are now being published electronically. Formal ontologies can increase the likelihood that such published information will be found and used, by making the data easier to query and transform. Given this situation, it is not surprising to learn that ontologies for biology and medicine are proliferating. Unfortunately, as we have seen in chapters 2 and 4, there are a many web-based ontology languages. Furthermore, even if one has selected an ontology language, there are many ways to build an ontology. This chapter discusses how to deal with the diversity of ontology languages and how to build high-quality ontologies.

However, before beginning to develop an ontology, one should examine the purpose and motivation for embarking on this activity. The first section is concerned with the questions that should be answered in this regard. Once one has a clear understanding of the purpose of the ontology, there are four major activities that must be undertaken: choosing an ontology language, obtaining a development tool, acquiring domain knowledge, and reusing existing ontologies. These activities are explained in a series of sections devoted to each of the topics. Although the topics are presented in a particular order, they do not have to be undertaken in that order, and may even be performed in parallel.

Having explained the major activities required for ontology development, the chapter turns to the issue of how to ensure that the ontology being developed is a high quality one. An ontology is a precise formulation of the concepts that form the basis for communication. Accordingly, the main parts of an ontology are the concepts and the relationships between them. The next two sections consider each of these in turn. Section 12.6 discusses the concept hierarchy and explains how to ensure that it is properly designed, and the design of the relationships is discussed in section 12.7. Once an ontology has been designed, it should then be validated to ensure that it satisfies its purpose. If flaws are discovered during validation, then the ontology must be modified. Ontologies are also modified after they have been published, as the field evolves. Techniques for ontology validation and modification are presented in section 12.8. Finally, the chapter ends with some exercises.

12.1 Purpose of Ontology Development

Before embarking on a project to develop an ontology it is important to have a firm understanding of the purpose of your ontology and the community that it is intended to serve. It is commonplace for ontology development projects to have no explicitly stated purpose beyond the acquisition of the domain knowledge. The assumption seems to be that the ontology should not be dependent on any particular purpose. This situation is unfortunate because it has been known at least since the middle of the nineteenth century that the design of an ontology depends on its purpose and viewpoint (Whewell 1847). This important fact has been forgotten and painfully rediscovered frequently since then. The purpose of the ontology should include the following:

1. *Why* the ontology is being developed. One of the most common reasons for building a formal ontology is to make shared information more usable. However, there are other reasons why one would build a formal ontology. It can be very useful for managing information used by small groups of people or even by a single individual. This book, for example, was written in XML, using an ontology that was built specifically for the needs of this project. Yet another reason why one might build a formal ontology is to analyze a domain, making explicit the assumptions being made by the community. In this case, the very act of formalizing the domain can be valuable irrespective of any other uses of the ontology. Finally, ontologies are often needed as part of a larger project, as in the example at the beginning of the chapter.

2. *What* will be covered by the ontology. This is also called its *scope*. A clear definition of the scope will prevent the ontology development effort from expanding unnecessarily. Many ontologies have already been developed. If an existing ontology has overlapping scope, then one should consider reusing it. When an ontology is being developed as part of a larger project, the scope will be dependent on the project scope.

3. *Who* will be using the ontology. If an ontology will only be used by a few persons, possibly only one person, then its design will be very different from an ontology that will be used by a much larger community. Indeed, if an ontology will only be used by one person for a short time, then it is possible to avoid writing it down explicitly. The authors of this book built a formal ontology to help with the writing of the book. This ontology was very useful even though it was used by only two persons.

4. *When* and for how long the ontology will be used. An ontology that will be used for a few weeks will generally have a much different design than one that is intended to be used for decades. Generally speaking, the longer an ontology will be used, the more effort one should invest in its design.

5. *How* the ontology is intended to be used. An ontology intended for information retrieval may be different from one intended to be used for scientific experimentation.

When a design choice is made, it is helpful to document the rationale for the choice and to refer back to the original purpose of the ontology. A design rationale should include the alternatives that were considered as well as the reason for the choice that was made. When an ontology development project involves a substantial amount of effort, then the statement of purpose will take the form of a statement of *project requirements*. Such a statement can be regarded as the contract which the developers have agreed to fulfill.

In this chapter we will use a medical chart ontology as an example of ontology development. Another example is developed in the exercises at the end of the chapter. The purpose of an ontology has a significant influence on how it should be developed. We begin by giving an informal description of the purpose of this ontology: *A hospital would like to make its medical chart information more easily available in its medical information system. The plan is to develop an ontology that will be useful for the next decade. The medical chart information will be used only by medical personnel who have permission to access the information. The information will be used both for immediate diagnostic decisions and for statistical data mining to detect long-term trends. The ontology must cover medically relevant events for patients and must also allow for personnel to make notes about patients and events. Events include tests, drug prescriptions, and operations performed. All events must be categorized using standard categories.*

The purpose of this ontology is summarized in the following table:

Why	Assist medical practice
What	Relevant medical events
Who	Medical personnel
When	Ten years
How	Diagnosis and trend analysis

Requirements for software development are often expressed using *use case diagrams*. Use case diagrams are part of the Unified Modeling Language (UML) (UML 2004). Although use case diagrams are intended for developing software, the technique can also be used for ontology development. Use case diagrams are primarily useful for specifying who will be using the ontology and how it will be used. They are not an effective way to specify why the ontology is being developed, what will be covered, and how long the ontology will be used. A use case diagram shows the relationships among *actors* and *use cases*. Actors represent anything that interacts with the system. An actor is usually a role played by a person, but it can also represent an organization or another computer system. A use case represents an interaction with the system. An ontology is not a computer system, but one can identify the actors that interact with it, as well as the components with which the actors interact. The requirements for the medical chart ontology could be represented diagrammatically as in figure 12.1. This diagram was created using the ArgoUML tool (Tigris 2004).

Summary

- Before developing an ontology, one should understand its purpose.
- The purpose of the ontology should answer the following questions:

  1. Why is it being developed?
  2. What will be covered?
  3. Who will use it?
  4. How long will it be used?
  5. How will it be used?

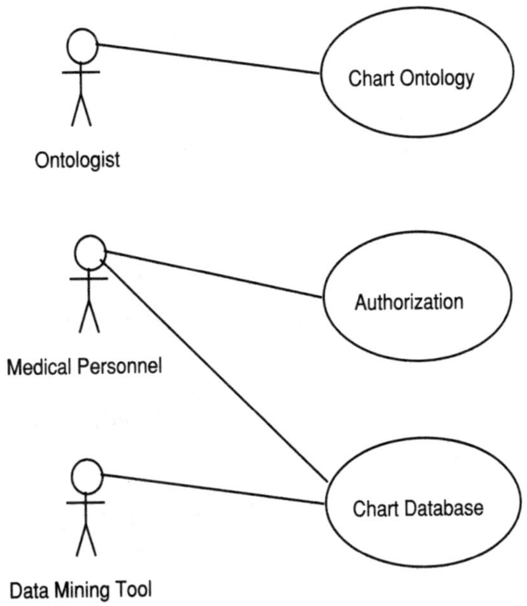

Figure 12.1 Use case diagram for the medical chart ontology.

- Use case diagrams can specify who will use the ontology and how it will be used.

12.2 Selecting an Ontology Language

Ontology languages can be given a rough classification into three categories:

- *Logical languages.* These languages express knowledge as logical statements. One of the best-known examples of such an ontology language is the Knowledge Interchange Format (KIF) (Genesereth 1998).

- *Frame-based languages.* These languages are similar to database languages. Frame-based languages organize data using record-like structures called *frames*. As we saw in section 1.1, a database record consists of a series of data items called "fields." In the same way, a frame consists of a series of data items called *slots*. One of the best-known frame-based languages is KL-ONE (Brachman and Schmolze 1985).

- *Graph-based languages.* These include semantic networks and conceptual graphs. Knowledge is represented using nodes and links between the nodes. XML and the Semantic Web languages are the best-known examples of graph-based languages.

Perhaps because of the strong analogy between hypertext and semantic networks, most recent ontology languages have been graph-based.

Deciding what approach to use for building an ontology is not an easy one. In this book, the emphasis is on the major web-based approaches as follows:

- **XML DTD** is the most basic as well as the most widely supported. However, it has serious limitations as an ontology language.

- **XSD** is quickly gaining acceptance, and conversion from XML DTD to XSD has been automated. However, it shares most of the limitations of XML DTDs.

- **XML Topic Maps** is a language for defining abstract subjects, called *topics*, and the relationships between them. Topic maps directly support higher-order relationships, which is not the case for the other languages in this list. On the other hand, topic maps do not have the complex data structures of XSD or the sophisticated semantics of RDF and OWL. Unfortunately, there are very few tools available for XML Topic Maps, so the development of ontologies using this language will not be discussed in this chapter.

- **RDF** has been gaining in popularity. There are fewer tools available for RDF than there are for XML, but new tools are continually becoming available. Unfortunately, there is no easy path for converting from XML DTDs or schemas to RDF (or vice versa). RDF has some inference built in, and RDF semantics is compatible with modern rule engines (either forward- or backward-chaining). It is also well suited to high-performance graph matching systems.

- **RDF specified with an XML DTD** is an approach that is compatible with XML DTD, XSD, RDF and the OWL languages. The Gene Ontology (GO) has used this technique. In this approach the DTD is designed so that it complies with RDF as well as with the OWL languages. Since an XML DTD can easily be converted to XSD, this makes the documents compatible with all major Web based ontology languages except for Topic Maps. However, only the most rudimentary features of RDF and OWL can be used by this approach.

- **OWL Lite.** This limited form of OWL was intended more for developers than for serious use. It allows a developer a first step on the way to supporting the more substantial OWL-DL and OWL Full languages.

- **OWL-DL.** For applications that fit the description logic approach, this is a very effective ontology language. Unfortunately, many domains do not fit.

- **OWL Full.** This is the richest and most flexible of the web-based ontology languages. It is also the least supported. Inference using OWL Full can be slow or can fail entirely. This is not a flaw of existing tools, but rather is a fundamental aspect of this language.

The major ontology languages can be divided into these main groups:

- XML DTD and XSD.

- XML Topic Maps.

- RDF and the three OWL languages.

Ontologies within a single group are mostly compatible with one another. XSD has more features than XML DTD, and it is easy to convert from a DTD to a schema. Similarly RDF and the OWL languages differ from one another mainly in what features are supported. Converting ontologies from one of these language groups to another can be difficult. Converting from the first group to one of the other two is especially problematic. Topic Maps, RDF, and OWL require that all relationships be explicit, while XML relationships are mostly implicit. As noted in the list above, there is an approach that combines the first and third groups. Developing an ontology using this technique is relatively easy, but it has the disadvantage that one is making no use of the expressiveness of RDF and OWL.

Note that in the discussion of ontology languages above, the concern was with conversion of ontologies from one ontology language to another, not transformation of data from one ontology to another. Data transformation, which we discussed at length in chapters 9 through 11, can involve transforming data within the same ontology language group as well as between language groups. Transformation can also involve data that are not web-based or data that are not based on any formal ontology. While making a good choice of an ontology language can make the transformation task easier, developing correct transformation programs can still be difficult.

The choice of ontology language will be highly dependent on the purpose of the ontology. As a first step one should choose among the four major approaches: one of the three main groups or the combination of the first and third groups. For the medical chart example, the intention is for the ontology to be used for 10 years. This is a good argument for either RDF or OWL because it is difficult to convert from an XML DTD or schema to RDF or OWL. The fact that notes can be added to other entries is another argument in favor of RDF or OWL, both of which are designed for annotating resources that are stored elsewhere. So there are very good reasons why one would choose RDF or OWL for the medical chart ontology. Since the purpose of the development project focuses on data representation and not on logic and reasoning, none of the more sophisticated features of OWL are needed. Therefore this ontology should use the combined approach. This will make the medical charts accessible to tools that are based on XML alone. This is an advantage in the shortterm. In the longterm, the compatibility with RDF allows one to take advantage of Semantic Web tools as they become available.

Summary

- The major ontology languages used today can be classified as follows:
  - Basic XML
    - XML DTD
    - XSD
  - XML Topic Maps
  - Semantic Web
    - RDF
    - OWL
      1. OWL Lite
      2. OWL-DL
      3. OWL Full
- It is possible to use an approach that is compatible with XML DTD, XSD, RDF, and the OWL languages.

12.3 Ontology Development Tools

Having chosen an ontology language or approach, the next step is to choose a suitable development tool. Unfortunately, such tools seldom refer to themselves as "ontology development tools." They are more commonly called "ontology editors." This is unfortunate because it fails to distinguish between the development of the ontology itself and the creation of the knowledge base. In bioinformatics, the data being stored in the knowledge base are often obtained by automated means, either directly or by means of a transformation (as shown in chapters 10 and 11). Semantic Web ontologies, however, often presume that the knowledge base will be manually constructed. Indeed, SW data are often referred to as "annotations." As a result, Semantic Web ontology editors have focused on the annotation task rather than the ontology development task. Nevertheless, many editing tools (especially XML editors) do include the ability to develop the ontology as well as the documents.

The following is a rough classification of the approaches and tools that can be used for ontology development:

1. **No explicit ontology.** One way to develop an ontology is to write the data in XML using a text editor or a specialized XML editor, without any DTD. XML can be parsed and processed without a DTD. This technique works only when the data have a relatively simple structure and the data will be used by a small number of people and processes.

2. **DTD generator.** A DTD can be automatically constructed from XML files using a DTD generator program. Some XML editors (such as XML Spy) include DTD generation as one of their features. There are also standalone DTD generators. This kind of tool allows one to transition from using no DTD to using a generated DTD. It is not necessary to obtain and install a DTD generator as there are online services that will perform the generation for you, such as the Hit Software website (Software 2004). This site also has services for generating XML schemas and converting DTDs to XML schemas. If the BioML document in figure 1.3 is processed by this website, one obtains a DTD that looks like this:

```
<!ELEMENT bioml ( organism ) >

<!ELEMENT chromosome ( locus ) >
<!ATTLIST chromosome name CDATA #REQUIRED >
<!ATTLIST chromosome number NMTOKEN #REQUIRED >

<!ELEMENT db_entry EMPTY >
<!ATTLIST db_entry entry NMTOKEN #REQUIRED >
```

```
<!ATTLIST db_entry format NMTOKEN #REQUIRED >
<!ATTLIST db_entry name CDATA #REQUIRED >

<!ELEMENT ddomain EMPTY >
<!ATTLIST ddomain end NMTOKEN #REQUIRED >
<!ATTLIST ddomain name CDATA #REQUIRED >
<!ATTLIST ddomain start NMTOKEN #REQUIRED >

<!ELEMENT dna ( #PCDATA ) >
<!ATTLIST dna end NMTOKEN #REQUIRED >
<!ATTLIST dna name CDATA #REQUIRED >
<!ATTLIST dna start NMTOKEN #REQUIRED >

<!ELEMENT exon EMPTY >
<!ATTLIST exon end NMTOKEN #REQUIRED >
<!ATTLIST exon name CDATA #REQUIRED >
<!ATTLIST exon start NMTOKEN #REQUIRED >

<!ELEMENT gene ( dna, ddomain+, exon, intron ) >
<!ATTLIST gene name CDATA #REQUIRED >

...
```

If the DTD generated by this tool is not exactly what one had in mind, then it is easy to modify it. The most common modification is to relax some of the constraints. For example, one might change some of the mandatory (`#REQUIRED`) attributes to optional (`#IMPLIED`) attributes.

3. **XML editor.** There are many XML editors, and some of them allow one to create DTDs and XML schemas. For a survey of these tools, see (XML 2004).

4. **RDF editor.** Many RDF editors are now available. For a survey of the RDF editors that were available as of 2002, see (Denny 2002a,b).

5. **OWL editor.** There are very few of these. The few that do exist were originally developed for another ontology language and were adapted for OWL. The best known OWL editor is Protégé-2000 from Stanford Medical Informatics (Noy et al. 2003). Protégé is an open source ontology and

knowledge-base editor available at `protege.stanford.edu`. It is written in Java, and can be used to create customized knowledge-based applications. It was not originally developed for OWL, but it has an OWL "plug-in" called ezOWL (ezOWL 2004)that allows one to edit OWL ontologies. Two of the authors of Protégé have written a nice article on ontology development using Protégé, called "Ontology 101" (Noy and McGuinness 2001).

6. **Computer-aided software engineering (CASE) tool.** Ontologies are a form of software and it is possible to use software engineering tools and techniques to build ontologies (Baclawski 2003). There are differences between ordinary software development and ontology development, but it is possible to reconcile these differences (Baclawski et al. 2001; Kogut et al. 2002). The DAML UML Enhanced Tool (DUET) translates from the Unified Modeling Language (UML), the standard modeling language for software, to DAML, the predecessor of OWL (DUET 2002). If one is accustomed to using CASE tools, then it is convenient to use a CASE tool. However, if one has no experience with CASE tools, it is better to use a tool specifically designed for ontology development.

Summary

The following are the main groups of approaches and tools for ontology development:

- No explicit ontology or tool
- Automatic generation of the ontology from examples
- XML editor
- RDF editor
- OWL editor
- CASE tool adapted for ontology development

12.4 Acquiring Domain Knowledge

All ontologies are ultimately based on "domain knowledge." Acquiring and formalizing knowledge about the domain is necessary for ontologies that are intended to be used by a larger community than just a local group of people. If information is going to be posted on the web, then it is especially important to have a formal ontology of some kind.

Unfortunately, even if one is an expert in a field it can still be an effort to formalize one's knowledge about the subject. The two main ways to acquire domain knowledge are glossaries and examples of use. For ontology development projects, the statement of purpose or project requirements can also be a source of relevant domain knowledge. A *glossary* or *dictionary* is a listing of the terminology in the domain, together with definitions. A *usage example* is a test case for an ontology. It is a statement using terminology from the field. It is best to express usage examples using simple statements in the form of subject-verb-object. More complex statements that have subordinate clauses are also useful, provided that the subordinate clauses are in the form of subject-verb-object. Glossaries and usage examples are often combined. A definition is an example of a use of the term, and it is common for a glossary to include examples of how each term is used.

To understand the structure of the definition of a word, recall that in biology the genus is the next most general concept above the species level. One "defines" a species by giving its genus and then specifying the features of the species that distinguish it within its genus. Words are defined the same way. One first gives the next most general term, which is called the "genus proximus," and then one specifies the features of the word that distinguish it within the genus proximus. Consider this definition of "lumbar puncture:"

```
A lumbar puncture is a procedure whereby spinal fluid
is removed from the spinal canal for the purpose of
diagnostic testing.
```

The genus proximus in this case is "procedure." The subordinate clauses "spinal fluid is removed from the spinal column" and "for the purpose of diagnostic testing" distinguish "lumbar puncture" from other "procedures."

While the ideal form for a statement is subject-verb-object, it is not always easy or natural to express domain knowledge this way. One sometimes uses verbs as concepts, and nouns or noun phrases are sometimes necessary for expressing a relationship. In the lumbar puncture definition one finds the phrase "for the purpose of diagnostic testing." To express this using the subject-verb-object format, one must use a statement such as "Lumbar puncture has the purpose of diagnostic testing." Although this statement is awkward, it is much better for ontology development.

Definitions of nouns are usually expressed by comparing them with more general nouns. While verbs can also be defined this way, it is not as useful as it is for nouns. In practice, most verbs are defined in terms of the corresponding noun, which is defined in the usual way. This fails to address the role that a verb plays, which is to express relationships between concepts. Consider the following definition of "inject:"

```
give an injection to; "We injected
the glucose into the patient's vein"
```

The definition begins by defining the verb in terms of the corresponding noun. The usage example that follows is the more interesting part of the definition. It suggests that "inject" relates some agent (e.g., some medical practitioner) with a substance (in this case, glucose). One can also specify the location where the injection occurs (e.g., the patient's vein).

Here are some usage examples for the medical chart ontology:

```
George is a patient.
George is in the infectious disease ward.
George was admitted on 2 September 2004.
Dr. Lenz noted that George was experiencing nausea.
George's temperature 38.9 degrees C.
Nausea is classified using code S00034.
```

Summary

- Ontologies are based on domain knowledge.
- The following are the main sources of domain knowledge for ontology development:
  1. Statement of purpose of the ontology
  2. Glossaries and dictionaries
  3. Usage examples

12.5 Reusing Existing Ontologies

If an ontology already exists for some of the terminology in your domain, then it is sometimes better to use the existing ontology than to construct it anew. However, there are risks involved that must be balanced against the advantages. Here are some of the reasons why existing ontologies might not be appropriate:

1. The ontology may have inappropriate features or the wrong level of detail. This could happen because the ontology was constructed for a different purpose or in a different context.

2. The ontology may be in an incompatible language. For example, it is difficult, in general, to convert a database schema to one of the XML ontology languages. Similarly, it is difficult to convert most XML DTDs and schemas to RDF or OWL because DTDs and schemas may not explicitly specify relationships. For example, the hospital that is developing the medical chart ontology may already have a relational database that includes medical chart information. However, relational database schemas are difficult to convert to any of the XML ontology languages.

3. Existing ontologies, especially database schemas, can have artifacts that were introduced to improve performance rather than being fundamental to the domain. Aside from the fact that such artifacts are conceptually irrelevant, they might actually result in worse performance because performance is highly dependent on the environment in which the database is being used.

Having determined that an existing ontology is at least partially suitable for reuse, there are a number of ways to incorporate it in another ontology. The simplest technique is to download the ontology and copy (i.e., "cut and paste") all or part of it.[1] One can then modify it if necessary. However, this technique can result in very large and unwieldy ontologies. Maintaining an ontology developed in this way can become very time-consuming if several ontologies have been reused.

There are two alternatives to copying:

1. **Include.** This is nearly the same as cutting and pasting except that it occurs every time that the document is processed. The inclusion is specified by giving the URL of the ontology to be included. The ontology is downloaded and substituted like any other included document into the place where the inclusion was requested. An example of this is shown in section 1.4 where five XML documents containing experimental data are

1. Of course, one should be careful to ensure that doing this does not violate the copyright.

merged to form a single XML document. The merger occurs each time the main XML document is processed. In particular, this means that changing one of the five XML documents would also change the merged document. By contrast, if one constructed a new document by cutting and pasting, then changes to any of the five XML documents would not be reflected in the new document.

2. **Import.** This is similar to inclusion except that the information in an imported ontology remains in its original namespace. To refer to a resource in the imported document, it is necessary to qualify it with the namespace of the imported ontology. Distinguishing imported resources with namespaces helps prevent ambiguity when the same concept name is used in more than one imported ontology.

Each ontology language has its own special way to include and to import:

1. **XML.** One can only include into an XML document. There is no XML import mechanism. The mechanism for inclusion is called an *entity*, and there are two main kinds of entity, distinguished by the character that signals the inclusion. One can use either an ampersand or a percent sign. The percent sign is used for including material into a DTD. The ampersand is used for including material into the XML document. This was discussed in section 1.4.

2. **XSD.** Both include and import are supported by XSD. The include mechanism uses the `include` element. Because included elements are in the same namespace, there is the possibility of the same name being used for two different purposes. The `redefine` element can be used to eliminate such an ambiguity. Alternatively, one can choose to use the `import` element which prevents ambiguities by keeping resources in different namespaces.

3. **RDF.** There is no special mechanism for importing or including RDF ontologies beyond what is already available for XML documents.

4. **OWL.** One can import an OWL ontology into another one by using a property named `owl:imports`. In addition, it is possible to declare that a resource in one namespace is the same as a resource in another. This allows one to introduce concepts from one namespace to another one namespace. This is similar to the `redefine` element of XSD.

We have already seen examples of including one XML file in another in section 1.4. SBML is an example of an XSD schema that imports another. The following shows how SBML imports the MathML schema:

```
<xsd:import
 namespace=
  "http://www.w3.org/1998/Math/MathML"
 schemaLocation=
  "http://www.w3.org/Math/XMLSchema/mathml2/mathml2.xsd"/>
```

Note that the URI of the MathML namespace is not the same as its URL.

Summary

- Reusing existing ontologies can save time and improve quality.
- However, reusing an existing ontology is not always appropriate. One must balance the risks against the advantages.
- There are three techniques for reusing an ontology:
  1. Copy the ontology
  2. Include the ontology
  3. Import the ontology

12.6 Designing the Concept Hierarchy

The most important part of any ontology is its concept hierarchy. The concepts are the subjects and objects that appear in whatever statement of purpose, glossaries, and usage examples one has available. Consider the medical chart ontology. From the purpose and usage examples, one finds the following terms: medical personnel, chart, event, patient, note, test, prescription, operation, category, and admission. The following are examples of how these concepts could be organized in a hierarchy for two ontology languages:

1. *XML DTD.* The concepts are organized according to how they will be arranged in the document. The hierarchy is specified by giving the content model for each concept in terms of other concepts. The top-level concept of the medical chart ontology is the chart concept, which consists of a sequence of events for a patient, which can be admission, tests, prescriptions, or operations. Events are categorized using standard categories,

and one can attach notes to events. Notes are made by medical personnel. The following are the content models:

```
<!--
  A chart consists of one patient
  and a sequence of events.
-->
<!ELEMENT Chart (Patient,Event*)>

<!--
  An event is an admission, test, prescription
  or operation.  It also has at least one category
  and may have any number of notes.
-->
<!ELEMENT Event
   ((Admission|Test|Prescription|Operation),
    Category+, Note*)
>
```

An XML DTD is developed by carefully restating requirements and definitions using simple sentences to relate concepts to other concepts. These restatements appear in the comments of the DTD above. Use only verbs and verb phrases such as *consists of, contains, has, is, may, must, is a sequence of, has exactly one, has at least one* and *has at most one*. These expressions are easily translated into content models. The content models can be converted back into sentences which can be verified against the requirements and definitions.

2. *RDF.* The concepts are organized according to subclass relationships. This is very different from the hierarchical organization of an XML document. In the medical chart ontology, we know that "a chart consists of one patient and a sequence of events," so that there are relationships between the chart concept and the patient and event concepts. However, neither of these relationships is a subclass relationship. We also know that "an event is an admission, test, prescription, or operation," so there are relationships between these concepts. Unlike the relationships with chart, these four relationships do represent subclass relationships. How can one determine that a relationship is a subclass relationship?

There are a number of ways to help distinguish subclass relationships from other hierarchical relationships. When a concept A is a subclass of concept B, then the statement "Every instance of A is also an instance of B" should be valid. To test whether two concepts are related by subclass, try this statement in each direction to see if one of them is valid. Applying this criterion to the chart and event concepts gives the statements "Every chart is also an event" and "Every event is also a chart," neither of which is valid. So neither the chart concept nor the event concept is a subclass of the other. Similarly, applying this criterion to events and tests gives the statements "Every event is also a test" and "Every test is also an event." The first of these statements is not valid, as there are other kinds of event than just tests, but the second statement is valid, so `Test` is a subclass of `Event`.

Subclasses have an important feature called *inheritance.* An instance of a subclass automatically inherits every characteristic of the superclass. Exceptions are never allowed. For example, one definition of a eukaryotic cell is a cell with at least one nucleus. However, if one specifies that erythrocytes (red blood cells) are a subclass of eukaryotic cells, then one has a problem because normal mature erythrocytes do not have nuclei. Generally speaking, if one wishes to allow a particular subclass to have some exception, then it is best not to use a subclass relationship in this case. Alternatives to the subclass relationship are discussed in subsection 12.7.1.

Other sentences that help distinguish subclass relationships are "A is a kind of B," "A is a subset of B," or "A is a B." One must be careful with the last of these, as the same statement is often used to specify that a particular instance belongs to a class.

Here is how the concept hierarchy for the medical chart ontology would be defined using RDF:

```
<rdfs:Class rdf:ID="Chart"/>
<rdfs:Class rdf:ID="Patient"/>
<rdfs:Class rdf:ID="Event"/>
<rdfs:Class rdf:ID="Admission">
  <rdfs:subClassOf rdf:resource="#Event"/>
</rdfs:Class>
<rdfs:Class rdf:ID="Test">
  <rdfs:subClassOf rdf:resource="#Event"/>
</rdfs:Class>
```

```
<rdfs:Class rdf:ID="Prescription">
  <rdfs:subClassOf rdf:resource="#Event"/>
</rdfs:Class>
<rdfs:Class rdf:ID="Operation">
  <rdfs:subClassOf rdf:resource="#Event"/>
</rdfs:Class>
<rdfs:Class rdf:ID="Category"/>
<rdfs:Class rdf:ID="Note"/>
```

When discussing concepts it is important to distinguish the concept from the instances belonging to the concept. In the case of charts and events, the same words are used for the concept and for an instance of the concept. To make the distinction clear one can use the word "concept" or "class" when one is referring to the concept. Thus "the chart class" refers to the concept while "a chart" or "George's chart" refer to instances. Capitalization is another common technique for distinguishing concepts from instances. Thus "Chart" refers to the chart concept, while "chart" refers to an instance chart. However, capitalization is used for other purposes in many languages, so it is not very reliable.

As we discussed in section 1.5, there are several ways to develop a concept hierarchy. One can begin with the most general concepts and then successively specialize them. This is called *top-down* development. The XML DTD for the medical chart ontology was developed by starting with the Chart concept and then specializing it. Conversely, one can start with the most specific concepts and successively group them in progressively larger classes. This is called *bottom-up* development. Neither development technique is intrinsically better than the other. In practice, one uses a combination of top-down and bottom-up techniques. In fact, there is evidence that human beings tend to start in the middle of the hierarchy, generalizing and specializing from there (Rosch and Lloyd 1978). This middle level is called the *basic* level.

If a concept hierarchy is large or is going to be in use for a relatively long time, then one should make an effort to have a design of as high a quality as possible. The hierarchy should be as uniform as possible, classes must be distinguished from instances, concepts should be elaborated to the appropriate level of detail, and one should specify whether classes can overlap one another. For the rest of this section, we discuss these criteria in more detail.

12.6.1 Uniform Hierarchy

To help understand large hierarchies one should try to make them as uniform as possible. While uniformity is a subjective notion, there are some objective criteria that one can use to help make a taxonomy more uniform:

1. Every level of a hierarchy should represent the same level of generality. In other words, all subclasses of one class should be at the same conceptual level. Of course, this depends on the purpose of the ontology. One way to classify animals would be to divide them into three subclasses: humans, domesticated animals, and wild animals. For people in everyday life, these three subclasses are on the same conceptual level. However, this is not the case when animals are classified genetically.

2. Every class that has subclasses should be subdivided into at least two and no more than a dozen subclasses. Subdividing into a single class suggests that the ontology is either incomplete or that the subclass is superfluous. Subdividing into a large number of subclasses makes it difficult for a person to understand or to navigate the taxonomy.

Unfortunately, these two criteria can conflict with each other. The taxonomy of living beings is a good example of this. The most general concept is subdivided into domains, which are subdivided into kingdoms, which are subdivided into phyla, continuing until one reaches individual species. The notion of a phylum, for example, serves to identify a level in the taxonomy, and every phylum represents the same level of generality throughout the hierarchy. However, the price that one pays for this uniformity is that some subclassifications consist of a single subclass while others consist of a large number of subclasses.

When the number of subclasses is large, one can introduce new levels into the hierarchy. In the taxonomy of living beings, additional levels are sometimes used to reduce the number of classes in a subclassification, such as "subphyla," "superphyla," "suborders," "superfamilies," and so on. Unfortunately, there is no easy way to deal with classes that have only one subclass. In the case of the taxonomy of living beings, one can argue that the single subclass is the only one that is currently known, leaving open the possibility that others may have existed in the past or may be discovered in the future. The species *H. sapiens* is the only species in the genus *Homo*. However, there were other species in this genus in the past.

12.6.2 Classes vs. Instances

One important design issue is whether a concept should be represented as a class or an instance. Up to now we have been tacitly assuming that concepts are always classes and that it is obvious what it means to be an instance of the class. Unfortunately, in general, there is no clear division between classes and instances, and a concept can be either one. Choosing between them represents a design choice that is dependent on the purpose of the ontology.

Consider, for example, the concept of a disease. Acute promyleocytic leukemia is a particular form of leukemia. Should it be regarded as a subclass of leukemia or an instance of leukemia? The answer depends on the purpose of the ontology. If there are usage examples in which there are instances of acute promyleocytic leukemia (such as particular cases of the disease), then acute promyleocytic leukemia should be a class. It should also be a class if it has subclasses. However, if there are no subclasses and no instances, then acute promyleocytic leukemia should be an instance. This would be appropriate for ontologies that are concerned with understanding the causes, symptoms, and progression of diseases rather than with observations of occurrences of diseases.

In general, instances represent the most specific concepts that are being represented. If a concept has a subclassification, then the concept must necessarily be a class. However, the absence of a subclassification does not mean that a concept is an instance. The determination in this case relies heavily on the usage examples.

12.6.3 Ontological Commitment

Ontology development efforts can sometimes be afflicted with an ailment that frequently occurs in software development efforts. This is the tendency to expand the scope of the development beyond the original purpose and intent. This problem is known by many names, such as "featuritis" and "scope creep." The accumulation of new features is usually gradual, with each addition being so small that it generally gets overlooked in terms of its impact on the project. However, the cumulative effect can be considerable. In ontology development, the term *ontological commitment* refers to the level of elaboration and detail that has been chosen for each concept.

Consider, for example, the notion of the temperature of a human being. This is commonly used for diagnostic purposes. There are, however, several ways that the temperature can be measured, so one should also specify how the measurement was performed (e.g., orally, rectally, etc.). But one might not stop there. Body temperature normally fluctuates with a circadian rhythm, so the time of day should also be considered. One could continue this elaboration forever.

As the example suggests, there is no limit to the degree of detail for any concept. Aside from the additional development cost and effort that results from scope creep, larger ontologies are harder to understand and to use. In addition, overelaboration can result in overlapping scope with other ontologies. This is not a problem in itself, but it can become a problem when the designs of the overlapping concepts are significantly different and it is necessary to make use of both ontologies.

All ontological commitments should be documented with a rationale for why the commitment was made. Documenting such commitments is much harder than it seems. The problem is that one may not be aware of the assumptions that are being made. Realizing that one is making implicit assumptions can be a difficult exercise. The best way to discover such assumptions is to have a well-stated purpose and scope for the ontology. Ontological commitments most commonly occur at the "boundaries" of the project scope. It is best to keep the ontology as simple as possible and to elaborate all concepts only as required. Staying within the scope not only limits the amount of work required, it also furnishes a good rationale for ontological commitments.

12.6.4 Strict Taxonomies

In many taxonomies, such as the taxonomy of living beings, there is an implicit assumption that subclassifications are nonoverlapping. The mathematical term for this situation is that the subclasses are *disjoint*. In the medical chart ontology, the four subclasses of the Event class are disjoint. For example, a test cannot also be an admission event. Specifying that these two classes are disjoint will help detect errors that would otherwise be missed.

Another way to look at this distinction is whether an instance can belong to two different subclasses of another class. When this happens it may be an indication that the taxonomy is inaccurate. For example, in the taxonomy of living beings originated by Linnaeus there were just two kingdoms: Plant and Animal. When microscopic living beings were discovered, it soon became apparent that there were living beings that could be regarded as being both plants and animals. Rather than allow the kingdoms to overlap, a new kingdom Protista, was added to deal with these new kinds of living being.

As more was learned about them, the taxonomy continued to change, but the disjointness condition was maintained.

Ontology languages differ from one another with respect to how disjointness is specified and whether it is implicitly assumed. In some ontology languages, subclasses are necessarily disjoint, unless one specifies otherwise. Other ontology languages presume that subclasses may overlap unless one specifies that they are disjoint. XML DTDs do not have a mechanism for allowing a particular element to belong to more than one type of element. Each particular element has exactly one tag. Thus XML DTDs do not allow any overlap among element types. By contrast, RDF and OWL allow instances to belong to more than one class, as long as the classes have not been explicitly specified to be disjoint (which can be specified in OWL, but not in RDF).

Summary

- XML hierarchies are concerned with the structure of the document.
- RDF and OWL hierarchies are concerned with the subclass relationships.
- Concept hierarchies can be developed in several ways:
  1. From the most general to the most specific (top-down)
  2. From the most specific to the most general (bottom-up)
  3. Starting at an intermediate, basic level (middle-out)
- Developing high quality concept hierarchies is difficult. The following techniques have been helpful:
  1. Maintain a uniform structure throughout the hierarchy.
  2. Carefully distinguish instances from classes.
  3. Keep the hierarchy as simple as possible, elaborating concepts only when necessary.
  4. Specify whether or not the hierarchy is strict (nonoverlapping).

12.7 Designing the Properties

One can think of the class hierarchy as being the skeleton of the ontology. It forms the structure on which the rest of the ontology is based. The properties form the rest of the ontology, and are analogous to the rest of the organism:

muscles, organs, skin, and so on. While one can have an ontology consisting of just a class hierarchy, such as the many classic taxonomies, they are just as lifeless as a skeleton by itself. Properties are essential for giving an ontology real meaning.

Properties can be classified in several ways:

1. **Attribute vs. relationship.** In XML, attributes are very different from child elements, both syntactically and semantically, as discussed in section 2.2. RDF and OWL eliminate this distinction, as explained in section 4.2.

2. **Data vs. resource.** XSD is in two parts. The first part deals with data structures (made up of XML elements) and the second deals with datatypes (such as numbers and dates, which do not involve XML elements). In XML, data structures are built using child elements. For example, a Medline citation such as figure 2.1 is an elaborate data structure using many elements. A simple datatype value, on the other hand, can be represented in XML using either XML attributes or XML elements. For example, the fact that George's height is 185 can be expressed either as an attribute:

```
<Person name="George" height="185"/>
```

   or as a child element:

```
<Person name="George">
  <height>185</height>
</Person>
```

3. **Intrinsic or extrinsic.** A property is *intrinsic* if it is a fundamental feature of the entity. For example, the chemical formula and structure of an enzyme is intrinsic. Properties that are not intrinsic are said to be *extrinsic*. For example, the name of an enzyme is extrinsic. To tell whether a property is intrinsic or extrinsic ask whether changing the property would change the entity. Changing the name of an enzyme would not normally be regarded as changing it in a fundamental way. Changing its chemical formula, however, would normally be seen as changing the enzyme.

   Although being intrinsic or extrinsic is an important feature of a property, none of the major ontology languages have such a notion. At the moment, one can only state this in the informal description of the property. However, it is useful to classify properties this way since it affects the design of the ontology. We will see examples in the rest of this section.

There is a notion of *subproperty* that is similar to the notion of *subclass*. Mathematically, both are subset relationships. As a result, properties can be organized as a hierarchy, just as classes can be so organized. However, the subproperty relationship is not an important one in RDF and OWL ontologies, and it is not used very often. Unlike classes, which are often defined by their position in the class hierarchy, properties are most often defined by their relationships with classes. Usage examples are especially important for determining which properties will be needed in an ontology. The goal is to be able to express every usage example using a property. In the rest of this section we discuss the main issues involved in the design of properties: domain, range, and cardinality constraints. However, before discussing the features of properties, we first address whether one should introduce a property at all.

12.7.1 Classes vs. Property Values

Although properties and classes are very different notions, there are cases in which one must decide whether to use one or the other. In the medical chart ontology, the Event concept was subclassified into four subclasses: Admission, Test, Prescription, and Operation. In the XML DTD these were represented as child elements of the Event element. In RDF or OWL, these were represented as subclasses of the Event class. An alternative to subclassification is to use an XML attribute or an RDF property which can take four possible values. For example, if Test is a class, then one would define an instance of a test using a `Test` element, whereas if Test is a property value, then an instance of a test would look like this:

```
<Event eventType="Test">
  ...
</Event>
```

The `eventType` attribute is defined in the XML DTD like this:

```
<!ATTLIST Event
  eventType
    (Admission|Test|Prescription|Operation) #REQUIRED>
```

Alternatively, if one is using OWL, then the eventType property is defined in the ontology as follows:

```
<owl:ObjectProperty rdf:ID="eventType">
  <rdfs:range>
    <owl:Class rdf:ID="EventType">
      <owl:oneOf parseType="Collection">
        <EventType rdf:ID="Admission"/>
        <EventType rdf:ID="Test"/>
        <EventType rdf:ID="Prescription"/>
        <EventType rdf:ID="Operation"/>
      </owl:oneOf>
    </owl:Class>
  </rdfs:range>
</owl:ObjectProperty>
```

Choosing between subclassing and property values can be difficult, and it can be dependent on the purpose of the ontology. As a result, different ontologies, ostensibly about the same domain, may use different designs. There are a number of criteria that can help one make this design decision:

1. *Intrinsic vs. extrinsic properties.* Generally speaking, extrinsic property values should be designed as property values rather than subclasses. For example, people are often classified into groups according to their ages, such as Infant, Toddler, Adolescent, and Adult. However, the age of a person is constantly changing so it is normally considered to be extrinsic. Thus the age group is better handled using a property value rather than as a subclass.

2. *Exceptions.* Subclasses have the important characteristic of *inheritance.* This was already explained in the introduction to section 12.6. If one wishes to allow exceptions, then it is better to use property values to distinguish cases.

3. *Subclasses have new properties.* When a subclass will have additional features, such as an additional property, then it is better to use subclassing rather than property values. For example, in the Event classification of the medical chart ontology, one would expect that each of the subclasses will have properties unique to the subclass. For example, a prescription instance will have a drug and administration schedule, which other events, such as an admission, would not have. However, this criterion is not as

compelling as the others mentioned above. If there is some good reason for not using subclasses (such as exceptions or extrinsic properties), then the introduction of new properties is not a sufficient reason for using subclasses. On the other hand, having additional kinds of properties is not a requirement for using subclasses. There are many taxonomies, including biological taxonomies, where subclasses do not introduce any new properties.

Possibly the most subtle issue concerning subclasses is the issue of how the instances act or are acted upon. The classic example of the resulting confusion is the question of whether square is a subclass of rectangle. From a logical point of view, it seems obvious that squares are a proper subset (and therefore subclass) of rectangles. However, according to one view of cognition and concepts, objects can only be defined by specifying the possible ways of acting on them (Indurkhya 1992). For instance, Piaget showed that the child *constructs* the notion of object permanence in terms of his or her own actions (Piaget 1971).

Accordingly, the concept square, when defined to be the set of all squares without any actions on them, is not the same as the concept square in which the objects are allowed to be shrunk or stretched. In fact, it has been found that children's concepts of square and rectangle undergo several transformations as the child's repertoire of operations increases (Piaget and Inhelder 1967; Piaget et al. 1981). This suggests that one should model "squareness" as a property value of the `Rectangle` class, called something like `isSquare`, which can be either true or false.

In general, concepts in the real world, which ontologies attempt to model, do not come in neatly packaged, mind-independent hierarchies. There are many actions that can potentially be performed on or by objects. The ones that are relevant to the purpose of the ontology can have a strong affect on how the ontology should be designed (Baclawski 1997b). For still more examples of how complex our everyday concept hierarchies can be, see (Indurkhya 2002; Lakoff 1987; Rosch and Lloyd 1978).

12.7.2 Domain and Range Constraints

It is sometimes convenient to think of a property as being analogous to a mathematical function. A mathematical function maps each element of a *domain* to an element of a *range*. For example, the square-root function has the set of nonnegative numbers as its domain and range. However, ontological properties differ from functions in two important ways:

1. A mathematical function is allowed to have only one value on each domain element. A property may, in general, have many values on each domain element. One sometimes says that properties are "multivalued." For example, every positive number actually has two squareroots. The property that maps every nonnegative number to all of its square roots has the nonnegative numbers as its domain and all real numbers as its range. This property is not a mathematical function.

2. A mathematical function must take a value on every element of its domain. A property need not have any values for some domain elements. Mathematically, properties are only *partial functions* in general.

The *domain* of a property is the set of entities that are allowed to have that property. For example, the `supervisor` property applies only to people. The *range* of a property is the set of entities that may be values of the property. For example, a `height` is a nonnegative number. When designing an ontology it is useful to choose appropriate domains and ranges for properties. They should be neither too specific nor too general. If a domain or range is too limiting, then acceptable statements may be disallowed. If a domain or range is too general, then meaningless statements will be allowed.

A more subtle ontology design issue is to ensure that the property is attached to the right set in the first place. For example, it may seem obvious that the body temperature is a property of a person. However, this fails to consider the fact that a person's body temperature varies with time. This may be important when one is recording more than one temperature measurement as in the medical chart ontology. As a result, it would be more appropriate for the domain of the body temperature to be an event rather than a person.

In XML and XSD, the domain of an attribute is the set of elements that use the attribute. In the BioML DTD, for example, virtually every element can have a `name` attribute, but not every element can have a `start` attribute. XML DTDs have only a limited capability for specifying ranges of attributes. The most commonly used ranges are `CDATA` (arbitrary text) and `NMTOKEN` (which limits the attribute to names using only letters, digits, and a few other characters such as underscores). XSD has a much more elaborate capability for specifying attribute ranges, as discussed in section 2.4.

In RDF, the domain and range of a property are specified using domain and range statements. For example, the height of a person would be declared as follows:

```
<rdf:Property rdf:ID="personHeight">
  <rdfs:domain rdf:resource="#Person"/>
  <rdfs:range rdf:resource="xsd:decimal"/>
</rdf:Property>
```

Similarly, only persons can have or be supervisors, so the supervisor property would be declared as follows:

```
<rdf:Property rdf:ID="supervisor">
  <rdfs:domain rdf:resource="#Person"/>
  <rdfs:range rdf:resource="#Person"/>
</rdf:Property>
```

In OWL, one can define domains and ranges in the same way as in RDF. For example, in the medical chart ontology, each event may be authorized by a member of the staff. This is specified in OWL as follows:

```
<owl:ObjectProperty rdf:ID="authorizedBy">
  <rdfs:domain rdf:resource="#Event"/>
  <rdfs:range rdf:resource="#Staff"/>
</owl:ObjectProperty>
```

In addition, OWL has the ability to specify *local ranges* relative to a domain by means of *owl:someValuesFrom*. For example, suppose that admissions may only be authorized by a doctor. In other words, when an event is in the `Admission` subclass of `Event`, then the range of `authorizedBy` is the subclass `Doctor` of `Staff`. This is specified in OWL as follows:

```
<owl:Class rdf:about="#Admission">
  <rdfs:subClassOf>
    <owl:Restriction>
      <owl:onProperty rdf:resource="#authorizedBy"/>
      <owl:allValuesFrom rdf:resource="#Doctor"/>
    </owl:Restriction>
  <rdfs:subClassOf>
</owl:Class>
```

Note that this does not require that admissions be authorized. It only states that when an admission has been authorized, then it must be authorized by a doctor. To require authorization, one must impose a cardinality constraint. This is covered in the next subsection.

Many design methodologies treat classes as the most important design notion, and relegate properties to a subsidiary role in which properties belong to their domain classes. For example, the methodology in (Noy and McGuinness 2001) regards properties as "slots" belonging to a class. XML and XSD, as well as most software engineering methodologies, take this point of view. OWL and RDF use an alternative point of view in which classes and properties have the same status (Baclawski et al. 2001). This design methodology is called *aspect-oriented modeling*, and it is supported by the most recent version of UML (UML 2004).

12.7.3 Cardinality Constraints

Cardinality is another word for the number of elements of a set. When used in the context of ontologies, the word refers to the number of values that can be taken by a property or attribute. For example, in BioML every chromosome can have at most one name. In terms of cardinalities, one would say that the name property has *maximum cardinality* equal to 1. As another example, every genus must have at least one species; in other words, the *minimum cardinality* is equal to 1. One can impose both a minimum and a maximum cardinality, and when these two are the same number, then it is called an exact *cardinality* constraint. In such a case, the number of values will always be the same. For example, if one requires that every patient have exactly one name, then one says that the name property has *cardinality* equal to 1. The most commonly imposed cardinality constraints are shown in the following table:

Constraint	Description	XML DTD
No cardinality constraint	*Any number* of values	*
Minimum cardinality is 1	*At least* one value	+
Maximum cardinality is 1	*At most* one value	?
Cardinality is 1	*Exactly* one value	
Maximum cardinality is 0	*Not applicable*	N/A

The last column shows the symbol used in XML DTDs, and Perl patterns. One can impose cardinality constraints with numbers other than 0 or 1, but this is rarely done. The last row in the table is used by properties that are defined on a general domain, but which do not apply for some subsets. One cannot specify this in an XML DTD or Perl pattern. The various ontology languages differ a great deal with respect to their ability to specify cardinality constraints.

XML DTD. An attribute can only appear once in an element. In other words, every attribute has maximum cardinality equal to 1. If an attribute is `#IMPLIED`, then the attribute is optional. One can specify that an attribute is `#REQUIRED`. This is the same as requiring that the cardinality be equal to 1. The number of occurrences of a child element is specified using the character in the last column of the table above.

XSD. As in XML DTDs, an attribute can only appear once in an element. For child elements, the number of occurrences is specified using `minOccurs` and `maxOccurs`.

RDF. In this case, properties can have any number of values, and one cannot impose any cardinality constraints.

OWL. This language has the most elaborate cardinality constraints, and they can be either global (i.e., applying to the property no matter how it is used) or local (i.e., applying only to specific uses of the property). The global cardinality constraints are `owl:FunctionalProperty` and `owl:InverseFunctionalProperty`. If a property is declared to be an `owl:FunctionalProperty`, then it is mathematically a partial function, that is, it can take at most one value on each domain element. This is the same as stating that this property has a maximum cardinality equal to 1. If a property is declared to be an `owl:InverseFunctionalProperty`, then its inverse property is a partial function.

The local cardinality constraints are:

1. **owl:someValuesFrom.** A constraint of this kind specifies that the property has a minimum cardinality equal to 1 for a specified domain. In addition, the minimum cardinality must be fulfilled by using values from a particular class. For example, in the medical chart ontology, suppose that admission events must be authorized by a doctor. This is specified as follows:

```
<owl:Class rdf:about="#Admission">
  <rdfs:subClassOf>
    <owl:Restriction>
      <owl:onProperty rdf:resource="#authorizedBy"/>
      <owl:someValuesFrom rdf:resource="#Doctor"/>
```

```
    </owl:Restriction>
  <rdfs:subClassOf>
</owl:Class>
```

This constraint requires that at least one authorization of the admission must be by a doctor. However, there could be multiple authorizations of the same admission, and some of them could be by staff members who are not doctors. To constrain all of the authorizations for a particular admission to be by doctors, one should use `owl:someValuesFrom` as in the previous section.

2. **owl:minCardinality.** This restricts the number of values that a property can have. For example, if every admission must be authorized, then specify the following:

```
<owl:Class rdf:about="#Admission">
  <rdfs:subClassOf>
    <owl:Restriction>
      <owl:onProperty rdf:resource="#authorizedBy"/>
      <owl:minCardinality
        rdf:datatype="&xsd;nonNegativeInteger"
        >1</owl:minCardinality>
    </owl:Restriction>
  <rdfs:subClassOf>
</owl:Class>
```

3. **owl:maxCardinality.** This is the reverse of a minimum cardinality restriction. It specifies the maximum number of values.

4. **owl:cardinality.** This is equivalent to two restrictions: a minimum and a maximum, both specifying the same number. For example, if every event should be authorized by exactly one staff member, then use the following:

```
<owl:Class rdf:about="#Admission">
  <rdfs:subClassOf>
    <owl:Restriction>
      <owl:onProperty rdf:resource="#authorizedBy"/>
      <owl:cardinality
        rdf:datatype="&xsd;nonNegativeInteger"
```

```
          >1</owl:minCardinality>
      </owl:Restriction>
    <rdfs:subClassOf>
  </owl:Class>
```

Summary

- Properties can be classified in several ways:
  - Attribute vs. relationship
  - Property values are data or resources
  - Intrinsic vs. extrinsic
- Subclassification and property values can sometimes be used interchangeably. Choosing between the two design possibilities can be difficult.
- One should specify the domain and range of every property. They should be neither too general nor too specific.
- Cardinality constraints are important for ensuring the integrity of the knowledge base.
- Depending on the ontology language, one can specify other constraints, but these are less important.

12.8 Validating and Modifying the Ontology

Validation is the process to determine whether a work product satisfies its requirements. One should always validate an ontology, but the amount of effort one should devote to validation depends on the size of the community being served by the ontology. Validation can be performed after the ontology has been developed, but it is usually better to validate while the ontology is being built. There are several techniques that can be used to validate an ontology:

1. Verify that the purpose has been fulfilled.
2. Check that all usage examples can be expressed in the ontology.
3. Create examples from the ontology and check that they are meaningful.

4. Check that the formal ontology is consistent.

The fulfillment of the purpose is verified by examining the design rationales. Every design decision should be supported by a design rationale, and each rationale should be complete and convincing. A rationale is complete if all design alternatives have been considered.

Checking that the usage examples are expressible is done by expressing them in the ontology language that has been chosen. For example, in the medical chart ontology, the usage examples would look like the following:

```
<Patient rdf:ID="p08639" name="George"/>
<Admission patient="#p08639"
    ward="#InfectiousDiseaseWard">
  <date rdf:datatype="xsd:date">2004-09-02</date>
</Admission>
<Note patient="#p08639" author="#Lenz"
      classification="#S00034">
  Patient is experiencing nausea
</Note>
<Test patient="#p08639">
  <temperature
    rdf:datatype="xsd:decimal">38.9</temperature>
</Test>
```

Creating examples from the ontology is the reverse of the process above. Instead of starting with meaningful usage examples and expressing them, one expresses examples and checks that they are meaningful. The examples can be either specific or generic. Some of the issues one should consider are the following:

1. **Disjointness.** Can an instance belong to two different classes? This is testing the disjointness property discussed in subsection 12.6.4. For example, can an event be both a prescription and a test?

2. **Cardinality.** Can a property be left unspecified? Can a property have more than one value? This kind of example tests whether the cardinality constraints are valid and whether cardinality constraints should be added (see subsection 12.7.3). For example, could a test not include a measurement of body temperature? Could a test have two measurements of body temperature?

3. **Dynamics.** Can a property value change? This tests whether the property is intrinsic or extrinsic and can affect the class hierarchy as discussed in subsection 12.7.1. For example, can a patient change his or her name?

Consistency checking is the only validation activity that can currently be done using automated tools. Obviously, consistency is fundamental to any formal system that supports logical inference. If a formal system is inconsistent, then every statement can be proven true (and also proven false, since true = false in an inconsistent formal system). Nearly all XML parsers have the ability to check consistency, but one must usually request that this be done. XML parsers refer to consistency checking as *validation*, even though this term more properly refers to much more than just consistency checking.

There are many RDF and OWL tools that can be used for checking consistency:

- **ConsVISor** (Kokar et al. 2001) web service
- **Euler** (Euler 2003) downloadable software
- **F-OWL** (FOWL 2003) downloadable software
- **Pellet** (Pellet 2003) web service
- **vOWLidator** (vOWLidator 2003) web service
- **WonderWeb** (WonderWeb 2004) web service

Consistency checkers vary with respect to how they explain the problems that are found. In addition to finding inconsistencies, most of the tools also give advice about situations that are not inconsistencies but which could be indicative of an error. Such a situation is called a *symptom* by analogy with the medical notion of a symptom of a disease (Baclawski et al. 2004). The ConsVISor consistency checker is unique in having the ability to produce output that itself conforms to an ontology.

When flaws in the ontology design are revealed during validation, the ontology must be modified. Ontologies are also modified after they are published. This can happen because new concepts have been introduced, existing concepts change their meaning, or concepts can be related in new ways. When concepts and relationships change, it is tempting to modify the ontology to reflect those changes. However, the danger is that programs and data that depend on the ontology will no longer be compatible. Ontology modification is also called ontology evolution. Certain modifications are relatively benign and are unlikely to have much effect on programs. Adding new attributes and relaxing constraints are usually innocuous. Some of the more substantial modifications include:

- Modifying property domains and ranges. A set of properties on subclasses can be changed to a single property on a superclass, or a property on one class can be shifted to become a property on a related class.
- Reification. This is the process whereby concepts that are not classes are given class status. For example, a relationship can be reified to become a class. Reifying a relationship will replace the relationship with a class and two relationships.
- Unreification. This is the reverse of reification.

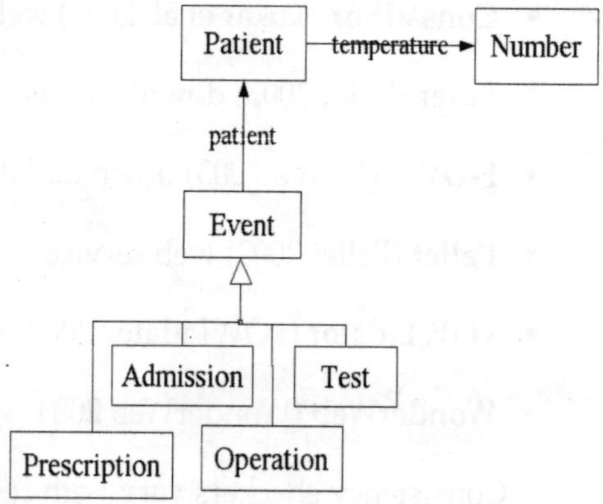

Figure 12.2 Part of the medical chart ontology.

As an example of ontology evolution, consider the medical chart ontology. In the part of the design shown in figure 12.2, the `temperature` of `Patient` is unrelated to any event. The problem is that the body temperature of a patient could be important information for events. A prescription could be based, in part, on a temperature measurement, but the current body temperature could differ from the temperature when the test was performed. To deal with these possibilities, the design should be modified as shown in figure 12.3.

As another example of ontology evolution, it is a common practice to measure the body temperature of a patient at many significant events, not just during tests. To allow for such a measurement for other events, one could

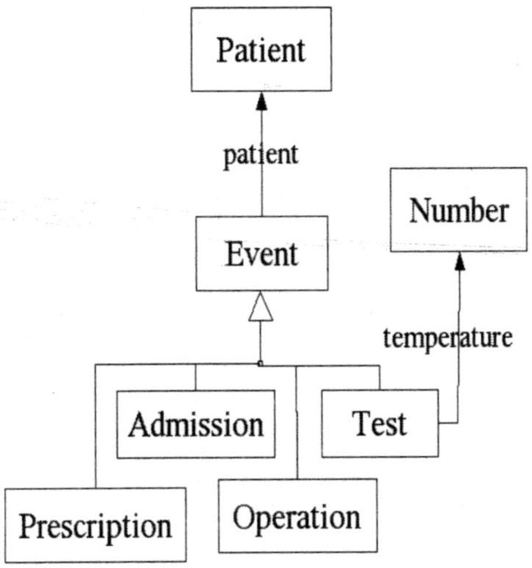

Figure 12.3 Medical chart ontology evolution. The temperature property is now connected to the `Test` class.

either connect it to all of the relevant subclasses of event. Alternatively, one could simply connect it with the `Event` class as in figure 12.4 so that every event can potentially have such a measurement.

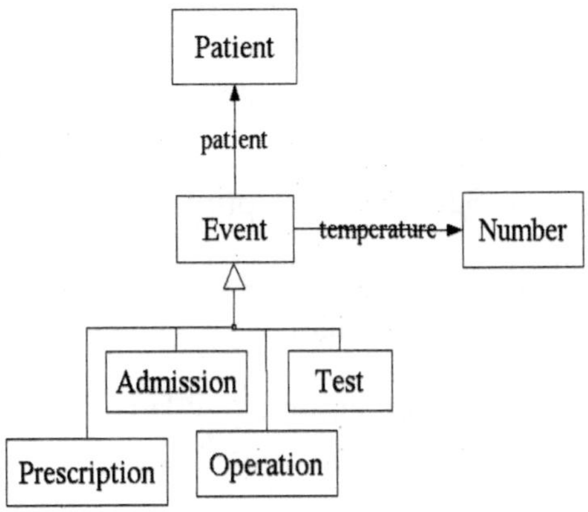

Figure 12.4 Another medical chart ontology modification. The temperature property is now connected to the `Event` class.

If an ontology is intended to be in use for a relatively long period of time, one must expect that the ontology will evolve. When the ontology is being used by a large community, one should develop standardized procedures for any additions or modifications to the ontology.

Summary

- Ontology validation consists of the following activities:
  1. Verify the fulfillment of the purpose.
  2. Check that all usage examples are expressible.
  3. Create examples that are consistent with the ontology, and determine whether they are meaningful.
  4. Check that the ontology is formally consistent.
- Ontologies evolve over time due to changing requirements and circumstances.

12.9 Exercises

Most of the exercises are based on the development of an ontology for single nucleotide polymorphism (SNP).

1. The informal description of the purpose of the SNP ontology is the following: *A small group of researchers would like to formalize their understanding of single nucleotide polymorphisms. The ontology will only be used for a few weeks. The ontology is only concerned with giving a high-level view of SNPs, which does not deal with the details.*

   Summarize the purpose succinctly in a table as in section 12.1.

2. Add consistency checking to the use case diagram in figure 12.1 for the medical chart ontology.

3. Choose an ontology language for the SNP ontology, and give a design rationale for your choice.

4. There already exist ontologies that deal with SNPs. For example, the SNP database (SNPdb) ontology in (Niu et al. 2003) is written in OWL and gives detailed information about the methods for finding SNPs. Is it appropriate to reuse SNPdb by importing it?

5. Build a concept hierarchy for the SNP ontology.

Part III

Reasoning with Uncertainty

The formal ontologies and languages developed in the first two parts of the book are based on deductive reasoning and deterministic algorithms. There is no room for uncertainty. Reality, unfortunately, is quite different, and any endeavor that attempts to model reality must deal with this fact. This part of the book compares and contrasts deductive and inductive reasoning, and then proposes how they can be reconciled.

The first chapter compares deductive reasoning with inductive reasoning, taking a high level point of view. There are several approaches to reasoning about uncertainty, and these are surveyed. The most successful approach to uncertainty is known as Bayesian analysis, and the rest of the book takes this point of view. Bayesian networks are a popular and effective mechanism for expressing complex joint probability distributions and for performing probabilistic inference. The second chapter covers Bayesian networks and stochastic inference.

Combining information from different sources is an important activity in many professions, and it is especially important in the life sciences. One can give a precise mathematical formulation of the process whereby probabilistic information is combined. This process is known as "meta-analysis," and it is a large subject in its own right. The third chapter gives a brief introduction to this subject.

The book ends by proposing a means by which inductive reasoning can be supported by the World Wide Web. Because Bayesian networks express reasoning with uncertainty, we refer to the inductive layer of the web as the *Bayesian Web.* Although this proposal is speculative, it is realistic. It has the advantage of allowing uncertainty to be formally represented in a web-based form. It offers the prospect of assisting scientists in some important tasks such as propagating uncertainty through a chain of reasoning, performing stochastic inference based on observations, and combining information from different sources.

13 *Inductive vs. Deductive Reasoning*

Deductive reasoning, also known more briefly as "logic," is the process for which facts can be deduced from other facts in a completely unambiguous manner using axioms and rules. Modern digital computer programs are fundamentally logical. They function in a manner that is unambiguously deterministic.

Reality is unlike a computer in many respects. It is much larger and far more complex than any computer program could ever be. Furthermore, most of what takes place is governed by rules that are either unknown or only imperfectly known. The lack of full knowledge about reality manifests itself as ambiguity and nondeterminism. There is no reason to suppose that reality is actually ambiguous or nondeterministic. Despite this situation, people manage to function effectively in the world.

There are two important mechanisms that people use to function in the world. The first is the ability to restrict attention to a small part of all of reality. The second is to accept that information is uncertain. These two mechanisms are related to one another. In theory, if one were capable of omniscience, then reality would be as unambiguous and deterministic as a computer program. However, since people are not capable of such a capacity, we are forced to suppress nearly all of what occurs in the world. The suppressed details manifest themselves in the form of uncertainties, ambiguities, and nondeterminism in the details that we do choose to observe. The former mechanism is called by various names such as "abstraction" and "relevance." Ontologies are fundamental to specifying what is relevant. The latter mechanism is fundamental to scientific reasoning. The combination of these two mechanisms is the subject of this part of the book.

When scientific reasoning is relatively simple, it is easy to ignore the role of the ontology, leaving it informal or implicit. However, medicine and biology are becoming too complex for informal descriptions of the context in which reasoning is taking place. This is especially important for combining inferences made in different contexts.

13.1 Sources and Semantics of Uncertainty

There are many sources of uncertainty. Many measurements, for example, are intrinsically uncertain. This is apparent with sensors for which accuracy of the measurement seems to be mainly a question of how much effort one wishes to expend on the measurement. However, even at the level of subatomic particles, there are limits to what one can measure, as stated by the Heisenberg uncertainty principle.

A more common source of uncertainty is the fact that a stochastic model will not include all possible variables that can affect system behavior. There is effectively no limit on how elaborate a model can be. However, models that have too many variables become unwieldy and computationally intractable. The model designer must make choices concerning which variables will be the most relevant and useful. The remaining variables are then ignored. The cost of ignorance is nondeterminism and what appear to be measurement errors, but what are in fact the result of unmodeled variables.

Yet another source of uncertainty is subjectivity. Probabilities are sometimes used as a means of expressing subjective assessments such as judgment, belief, trust, and so on. Some philosophers take the position that probability can only be applied to events according to their relative frequencies of occurrence, and they reject the interpretation of probability as a degree of belief. This point of view is called *frequentism*. Most researchers and philosophers today accept that probabilities can be applied both to relative frequencies of occurrence as well as to other degrees of belief. This interpretation of probability is called *Bayesian analysis*.

Some of the major philosophical works (specifically, Spinoza's *Ethics* (Spinoza 1998), Leibniz's *Monadology* (Leibniz 1998), and Wittgenstein's *Tractatus* (Wittgenstein 1922)) propound some version of logical atomism. In other words, they conceptualize the world using objects and their attributes, and they propose relationships identified by words that link the the mental concepts of objects and attributes to the corresponding physical objects. They also specify how more complex objects can be constructed from more elementary objects. Unfortunately, this point of view ignores issues such as observational uncertainty, belief, judgment, and trust, all of which affect our perception of the world. To address these issues, one must not only accept that our knowledge about the world is uncertain, it is also necessary to quantify and formalize the notion of uncertainty so that statements about the world will have more than just a vague, informal meaning.

Many mathematical theories have been introduced to give a formal semantics to uncertainty. On can classify these approaches into two main classes (Perez and Jirousek 1985):

1. *Extensional.* An approach of this kind treats uncertainty as a generalized truth-value attached to formulas. Uncertainty is computed by applying rules or procedures. When rules are used, the computation of uncertainty is analogous to the determination of facts by a rule-based system, as in chapter 3. When procedures are used, uncertainty is computed in a manner that is more closely related to programming with traditional programming languages such as Perl.

2. *Intensional.* This approach attaches uncertainty to "states" or "possible worlds." These approaches are also described as being *declarative* or *model-based.* Intensional approaches have been implemented using computer systems, as, for example, data mining and stochastic control systems.

Note that uncertainty is represented using probabilities regardless of whether a system is extensional or intensional. In this chapter we discuss some of the major theories of uncertainty, classifying them according to the extensional/intensional dichotomy. For an excellent discussion and comparison of these two approaches to uncertainty, see (Pearl 1988). Much of the material in this chapter is based on the treatment in Pearl's book.

Summary

- There are many sources of uncertainty, such as measurements, unmodeled variables, and subjectivity.

- Formalizations of uncertainty can be classified as either extensional or intensional.

- Extensional uncertainty is a generalized truth-value.

- Intensional uncertainty assigns probabilities to possible worlds.

13.2 Extensional Approaches to Uncertainty

One theory of uncertainty that has achieved some degree of popularity is the theory of *fuzzy logic*Zadeh (Zadeh 1965, 1981). Fuzzy logic is an extensional approach to uncertainty. In fuzzy logic one associates a number between 0 and 1 with each statement. This number is called a *truth-value* or *possibility* to distinguish it from probabilities used in probability theory. Truth-values are either given a priori as ground facts, or they are computed. Statements may be combined with other statements using the operations AND, OR, and NOT, just as in classic Boolean logic. Fuzzy logic is a generalization of Boolean logic in that if all statements are either fully true or fully false (i.e., their truth-values are either 1 or 0), then combining the statements using Boolean operations will always produce the same result as in Boolean logic. Statements that are entirely true or false are called *crisp statements*, and Boolean logic is called *crisp logic*. The truth-value of general statements combined using the Boolean operations is determined by a function called the *t-norm*. The t-norm is a function from a pair of truth-values to a single truth-value. It is the function that computes the truth-value of the AND of two fuzzy statements. The most commonly used t-norm is the minimum, also called the Gödel t-norm.

Because fuzzy logic depends on the choice of a t-norm, there are many different kinds of fuzzy logic. Truth-values computed using one t-norm are not compatible with truth-values computed using a different t-norm. One can define rules for fuzzy logic, and these rules can be fuzzy in the sense that each rule is assigned a strength between 0 and 1. The strength specifies the degree of confidence in the rule.

In rule-based systems, one begins with a collection of known facts and rules. The rule engine then infers new facts using the rules. It can do this either in a forward-chaining manner where all facts are inferred or a backward-chaining manner in which one infers only the facts needed to answer a particular query (see chapter 3 for how this works). Fuzzy logic, as in other extensional systems, is similar except that it is the truth-values that propagate, not the facts. Like rule-based systems, one can use either forward-chaining or backward-chaining.

Note that the term "fuzzy" is often used for any notion of uncertainty, not just for the specific class of theories due to Zadeh. For example, there is a notion of "fuzzy Bayesian network," which is unrelated to fuzzy logic.

There are many other extensional approaches to uncertainty. MYCIN (Shortliffe 1976) is an expert system that was developed for the purpose of medical diagnosis. Like fuzzy logic, MYCIN propagates uncertainty using rules, each of which has a strength (called the "credibility" of the rule). It differs from fuzzy logic primarily in the formulas used for propagating certainty levels.

Summary

- Fuzzy logic associates a generalized truth-value between 0 and 1 to each statement.
- A statement is crisp if it is fully true or fully false.
- There are many fuzzy logics, one for each choice of a t-norm.
- The generalized truth-value of a statement is computed by propagating uncertainty when the statement is inferred.
- Extensional logics differ in the formulas used for propagating uncertainty.

13.3 Intensional Approaches to Uncertainty

The dominant intensional approach to uncertainty is probability theory. Probability theory assigns a number between 0 and 1 (inclusive) to statements. Probabilistic statements are called *events*. Events can be combined to form new events using Boolean operations, and the probability assigned to events must satisfy the axioms of probability theory. In particular, there is a universal event that contains all others and that has probability 1. This universal event has various names, such as the probability space or sample space.

A *discrete random variable* is a set of disjoint events such that each event is assigned a value of the domain of the random variable, and such that the union of all these events is the universal event. For example, the states of a traffic light L are {green, yellow, red, failed}. The events are $(L = green)$, $(L = yellow)$, $(L = red)$, $(L = failed)$. The probabilities are $Pr(L = green)$, and so on. These probabilities define the *probability distribution* of the random variable. The sum of the probabilities over all possible values of the random variable is equal to 1. This is a consequence of the fact that the universal event has probability 1.

A *continuous random variable* is somewhat more difficult to define because the probability of such a variable taking any particular value is 0. There are two ways to define the probability distribution of such a variable:

1. The probability on intervals or regions. For example, a temperature T defines events such as $(t_1 \leq T \leq t_2)$, the set of temperatures between t_1 and t_2.

2. The ratio of the probability on a small interval or region divided by the size of the interval or region. This is called the *probability density*.

As with discrete random variables, the total probability must be 1. In terms of the probability density this means that the integral over all possible values is equal to 1.

Both discrete and continuous random variables have total probability 1. However, it is sometimes convenient to relax this requirement. An assignment of nonnegative weights to the possible values of a variable is called a *distribution*. In mathematics, a distribution is called a *measure*. If a distribution has a total weight that is some positive number other than 1, then one can convert the distribution to a probability distribution by dividing all weights by the total weight. This is called *normalization*. However, some distributions cannot be normalized because their total weight is either 0 or infinite.

The simplest example of a distribution is the *uniform distribution*. A random variable has the uniform distribution when every value has the same weight as every other value. A continuous random variable is uniform when its density is a positive constant. Uniform distributions on infinite sets or regions cannot be normalized.

When there are several random variables, their probabilistic structure is completely defined by the intersections of their events. Thus the events defined by the random variables L and T include such events as

$$(L = green \text{ and } T \leq 5).$$

The probabilities of these events define the *joint probability distribution* (JPD) of the random variables L and T. A *stochastic model* is another name for a collection of random variables. The probabilistic structure of the stochastic model is the JPD of the collection of random variables. One could give a strong argument that the stochastic model is the fundamental construct, and that the probability space is secondary. However, it is convenient to treat the probability space as fundamental and the random variables as derived from it (as measurable functions on the probability space).

Given two events A and B, a *conditional probability* of A given B is any number c between 0 and 1 (inclusive) such that $Pr(A \text{ and } B) = cPr(B)$.

Note that when $Pr(B) = 0$, every number between 0 and 1 is a conditional probability given B. The notation for a conditional probability is $Pr(A \mid B)$. The event B is the "condition" or "input," while the event A is the "consequence" or "output." However, this terminology does not mean that B and A have a cause-and-effect relationship.

Conditional probability is the most basic form of inference for probability theory. If one knows that the event B has occurred, then the probability of A changes from $Pr(A)$ to $Pr(A \mid B)$. If the probability of A does not change, then one says that A and B are *independent* or *statistically independent*. More generally, as one finds out more about what has happened, then the probability continually changes. Much more elaborate forms of stochastic inference are developed in chapter 14.

The defining formula for the conditional probability of A given B is

$$Pr(A \text{ and } B) = Pr(A \mid B)Pr(B).$$

If $Pr(B)$ is nonzero, one can solve for the conditional probability:

$$Pr(A \mid B) = \frac{Pr(A \text{ and } B)}{Pr(B)}.$$

This is sometimes used as the definition of the conditional probability. By reversing the roles of A and B, the defining formula for the conditional probability of B given A is

$$Pr(A \text{ and } B) = Pr(B \mid A)Pr(A).$$

The left-hand side of this equation is the same as the left-hand side of the defining formula for the conditional probability of A given B. Therefore

$$Pr(A \mid B)Pr(B) = Pr(B \mid A)Pr(A).$$

If $Pr(B)$ is nonzero, then one can solve for $Pr(A \mid B)$ to obtain

$$Pr(A \mid B) = \frac{Pr(B \mid A)Pr(A)}{Pr(B)}.$$

This is known as *Bayes' law*. It is named after the English mathematician Thomas Bayes who proved a special case of it.

In spite of its simplicity, Bayes' law is powerful. For example, suppose that A is a disease and B is a symptom of the disease. $Pr(A)$ is probability of the disease in the population and $Pr(B)$ is the probability of the symptom. If we know how often the symptom occurs when a person has the disease, then one knows $Pr(B \mid A)$. Bayes' law then gives the probability that a person has the disease when the symptom is observed. In other words, Bayes' law gives important information which can be used for the diagnosis of diseases based on symptoms. Specific examples of the use of Bayes' law for diagnosis are given in section 14.2.

As we discussed in section 13.1, The applicability and interpretation of probability theory has been the subject of both philosophical and mathematical analysis since the time it was originally founded. More recently, other ways of expressing uncertainty have emerged, as we discussed in section 13.2. The question is which of these approaches to uncertainty is the best.

The current methodology for such comparing approaches to uncertainty was introduced by De Finetti (De Finetti 1937) who formulated subjective probability in terms of betting against an adversary. This formulation is called the *Dutch book* argument. This made it possible to extend the applicability of probability to questions such as "Will the sun rise tomorrow?" or "Was there life on Mars?" It also can be used to prove the power of probability theory in general, and Bayesian analysis in particular. The argument is that if one knows that an agent consistently follows a non-Bayesian belief system in a known way, then one can arrange the bets so that the Bayesian *always* wins (not just on average). If the departure from Bayesian analysis is inconsistent, then the Bayesian can only win on average.

Although stated in financial terms, the Dutch book argument applies equally well to any activity which involves some form of utility, whether it is financial or not, and the associated risk in trying to increase this utility. It follows that Bayesian analysis is a minimal requirement for rational inference in experimental science.

There are significant advantages to probability theory as a mechanism for expressing uncertainty. It is the only approach that is empirically grounded, and it can be used either empirically or subjectively. Furthermore, Bayesian analysis will always win over a non-Bayesian analysis whenever one quantifies the risks associated with decisions based on the events in question.

However, probability theory has significant disadvantages. It is much more computationally complex than the extensional approaches. Specifying a general JPD is a formidable task as the number of random variables increases. Even for random variables that can take only two values, if there are 20 random variables, then a joint probability distribution has over 10^6 probabilities. Accordingly, it is very common to assume that the random variables satisfy various simplifying properties. The most common simplification is to assume that the random variables are statistically independent. This assumption works well for many random processes, such as card games and queues, but it does not capture the more complex processes that occur in biological systems, where interdependence is very common.

Another disadvantage of probability theory is that probabilistic inference is not the result of a propagation process as it is for the extensional approaches. This makes the intensional approach incompatible with rule-based reasoning systems.

Thus in spite of the advantages of probability theory pragmatically and theoretically, other approaches to uncertainty have been introduced that are more computationally tractable and more compatible with logical reasoning systems. However, new techniques and algorithms for probabilistic reasoning have now been introduced that have made it much more tractable as well as more compatible with rule-based systems. These techniques are discussed in chapter 14.

Summary

- Probability theory is the dominant intensional approach to uncertainty.
- Probabilistic statements are called events.
- A random variable is a collection of events distinguished from one another by a value of the variable.
- A stochastic model is a set of random variables together with their joint probability distribution.
- Conditional probability is the most basic from of inference in probability theory.
- Bayes' Law is the basis for diagnostic inference and subjective probabilities.
- The Dutch book argument shows that Bayesian analysis is always better than non-Bayesian analysis.
- Probability theory has long been regarded as being too computationally complex to be the basis for modeling the uncertainty of large systems, but new techniques have been introduced that are changing this.

14 *Bayesian Networks*

Stochastic modeling has a long history, and it is the basis for the empirical methodology that has been used with great success by modern scientific disciplines. Stochastic models have traditionally been expressed using mathematical notation that was developed long before computers and GUIs became commonly available. A Bayesian network (BN) is a graphical mechanism for specifying the joint probability distribution (JPD) of a set of random variables (Pearl 1998). As such, BNs are a fundamental probabilistic representation mechanism for stochastic models. The use of graphs provides an intuitive and visually appealing interface whereby humans can express complex stochastic models. This graphical structure has other consequences. It is the basis for an interchange format for stochastic models, and it can be used in the design of efficient algorithms for data mining, learning, and inference.

The range of potential applicability of BNs is large, and their popularity has been growing rapidly. BNs have been especially popular in biomedical applications where they have been used for diagnosing diseases (Jaakkola and Jordan 1999) and studying complex cellular networks (Friedman 2004), among many other applications.

This chapter divides the subject of BNs into three sections. The sections answer three questions: What BNs are, How BNs are used, and How BNs are constructed. The chapter begins with the definition of the notion of a BN (section 14.1). BNs are primarily used for stochastic inference, as discussed in section 14.2. BNs are named after Bayes because of the fundamental importance of Bayes' law for stochastic inference. Because BNs require one to specify probability distributions as part of the structure, statistical methods will be needed as part of the task of constructing a BN. Section 14.3 gives an overview of the statistical techniques needed for constructing and evaluating BNs.

14.1 The Bayesian Network Formalism

A *Bayesian network* is a graphical formalism for specifying a stochastic model. The random variables of the stochastic model are represented as nodes of a graph. We will use the terms "node" and "random variable" interchangeably. The edges denote dependencies between the random variables. This is done by specifying a *conditional probability distribution* (CPD) for each node as follows:

1. If the node has no incoming edges, then the CPD is just the probability distribution of the node.

2. If the node has incoming edges, then the CPD specifies a conditional probability of each value of the node given each combination of values of the nodes at the other ends of the incoming edges. The nodes at the other ends of the incoming edges are called the *parent* nodes. A CPD is a function from all the possible values of the parent nodes to probability distributions (PDs) on the node. Such a function has been called a *stochastic function* in (Koller and Pfeffer 1997).

It is also required that the edges of a BN never form a directed cycle: a BN is *acyclic*. If two nodes are not linked by an edge, then they are independent. One can view this independence property as defined by (or a consequence of) the following property of a BN: The JPD of the nodes of a BN is the product of the CPDs of the nodes of the BN. This property is also known as the chain rule of probability. This is the reason why the BN was assumed to be acyclic: the chain rule of probability cannot be applied when there is a cycle. When the BN is acyclic one can order the CPDs in such a way that the definitions of conditional probability and statistical independence can be applied to get a series of cancellations, such that only the JPD remains.

In section 13.3 we mentioned that it is sometimes convenient to use unnormalized distributions. The same is true for BNs. However, one must be careful when using unnormalized BNs because normalization need not produce a BN with the same graph. Furthermore, unnormalized BNs do not have the same independence properties that normalized BNs have.

Some of the earliest work on BNs, and one of the motivations for the notion was to add probabilities to expert systems used for medical diagnosis. The Quick Medical Reference Decision Theoretic (QMR-DT) project (Jaakkola and Jordan 1999) is building a very large (448 nodes and 908 edges) BN. A simple example of a medical diagnosis BN is shown in figure 14.1. This BN has four random variables:

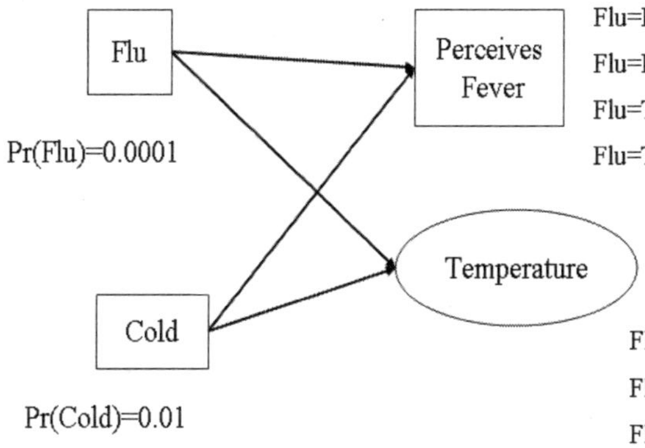

	Perceives Fever	
	TRUE	FALSE
Flu=F & Cold=F	0.01	0.99
Flu=F & Cold=T	0.10	0.90
Flu=T & Cold=F	0.90	0.10
Flu=T & Cold=T	0.95	0.05

	Temperature	
	Mean	Std Dev
Flu=F & Cold=F	37.0	0.5
Flu=F & Cold=T	37.5	1.0
Flu=T & Cold=F	39.0	2.0
Flu=T & Cold=T	39.5	2.5

Figure 14.1 Example of a BN for medical diagnosis. Rectangles represent discrete random variables and the oval represents a continuous random variable.

1. Flu, meaning that a patient has influenza.
2. Cold, meaning that a patient has one of a number of milder respiratory infections.
3. Perceives Fever (PF), meaning that the patient perceives that he or she has a fever.
4. Temperature, the continuous random variable representing a measurement of the patient's body temperature.

Note that three of the random variables are Boolean, the simplest kind of discrete random variable, and that the fourth random variable is continuous. Two of the nodes have no incoming edges, so their CPDs are just PDs, and because the nodes are Boolean, they can be specified with just one probability. We assume that $Pr(Flu) = 0.0001$, corresponding to the fact that the annual incidence rate for influenza (serious enough to require hospitalization) in the United States was 1.0 per 10,000 in 2002 (NCHS 2003). The common cold is much more common. We will use $Pr(Cold) = 0.01$, although the actual incidence rate is higher than this.

The CPD for the PF node has two incoming edges, so its CPD is a table that gives a conditional probability for every combination of inputs and outputs. For example, the CPD might be the following:

	not(PF)	PF
not(Flu) and not(Cold)	0.99	0.01
not(Flu) and (Cold)	0.90	0.10
(Flu) and not(Cold)	0.10	0.90
(Flu) and (Cold)	0.05	0.95

The CPD for the Temperature (T) node has two incoming edges, so its CPD will have four entries as in the case above, but because T is continuous, it must be specified using some technique other than a table. For example, one could model it as a normal distribution for each of the four cases as follows:

	Mean	Std Dev
not(Flu) and not(Cold)	37.0	0.5
not(Flu) and (Cold)	37.5	1.0
(Flu) and not(Cold)	39.0	1.5
(Flu) and (Cold)	39.2	1.6

As an example of one term of the JPD, consider the probability of the event (Flu) and not $(Cold)$ and (PF) and $(T \leq 39.0)$. This will be the product of the four probabilities: Pr(Flu), Pr(not(Cold)) = (1-Pr(Cold)), Pr(PF|Flu and not(Cold)), and Pr(T≤39.0|Flu and not(Cold)). Multiplying these gives

$$(0.0001)(.99)(.90)(0.5) = 0.004455.$$

Although the BN example above has no directed cycles, it does have *undirected* cycles. It is much harder to process BNs that have undirected cycles than those that do not. Some BN tools do not allow undirected cycles because of this.

Many of the classic stochastic models are special cases of this general graphical model formalism. Although this formalism goes by the name of *Bayesian* network, it is a general framework for specifying JPDs, and it need not involve any applications of Bayes' law. Bayes' law becomes important only when one performs inference in a BN, as discussed below. Examples of the classic models subsumed by BNs include mixture models, factor analysis, hidden Markov models (HMMs), Kalman filters, and Ising models, to name a few.

BNs have a number of other names. One of these, *belief networks*, happens to have the same initialism. BNs are also called probabilistic networks, directed graphical models, causal networks, and "generative" models. The last two of these names arise from the fact that the edges can be interpreted as specifying how causes generate effects. One of the motivations for introducing BNs was to give a solid mathematical foundation for the notion of causality. In particular, the concern was to distinguish causality from correlation. A number of books have appeared that deal with these issues such as one by Pearl (Pearl 2000) who originated the notion of BNs. For causation in biology, see (Shipley 2000). Other books that deal with this subject are (Glymour and Cooper 1999; Spirtes et al. 2001).

Summary

- A BN is a graphical mechanism for specifying JPDs.
- The nodes of a BN are random variables.
- The edges of a BN represent stochastic dependencies.
- The graph of a BN must not have any directed cycles.
- Each node of a BN has an associated CPD.
- The JPD is the product of the CPDs.

14.2 Stochastic Inference

The main use of a BN is to perform inference. This is done by observing some of the random variables. One can then query the BN to determine how the distributions of other random variables are affected. Specifying known facts is done by giving the values of some of the random variables. The values can be given as actual crisp values or as a PD on the values. The nodes that have been given values are termed the *evidence*. One can then choose one or more of the other nodes as the *query* nodes. The answer to the query is the JPD of the query nodes given the evidence.

Inference in a BN is analogous to inference performed by a rule engine. Recall from section 3.1, that in a rule engine one specifies a collection of if-then rules, called the rule base. One can then input a collection of known facts (typically obtained by some kind of measurement or observation). The rule engine then explicitly (as in a forward-chaining rule engine) or implicitly (as in a backward-chaining rule engine) infers other facts using the rules. The set of specified and inferred facts form the knowledge base. One can then query the knowledge base concerning whether a particular fact or set of facts has been inferred. A BN is analogous to a rule base. The evidence presented to a BN is analogous to the facts initially specified in a rule engine. Both BNs and knowledge bases can be queried.

Event A	Pr(A)
PF and not(Flu) and not(Cold)	(0.9999)(0.99)(0.01) = 0.0099
PF and not(Flu) and (Cold)	(0.9999)(0.01)(0.10) = 0.0010
(PF and Flu) and not(Cold)	(0.0001)(0.99)(0.90) = 0.0001
(PF and Flu) and (Cold)	(0.0001)(0.01)(0.95) = 0.0000
not(PF) and not(Flu) and not(Cold)	(0.9999)(0.99)(0.99) = 0.9800
not(PF) and not(Flu) and (Cold)	(0.9999)(0.01)(0.90) = 0.0090
(not(PF) and Flu) and not(Cold)	(0.0001)(0.99)(0.10) = 0.0000
(not(PF) and Flu) and (Cold)	(0.0001)(0.01)(0.05) = 0.0000

Table 14.1 A joint probability distribution as the result of a stochastic inference.

Consider first the case of BN inference starting with no evidence at all. In this case, the PS of the query nodes is computed by summing the terms of the JPD over all of the random variables that are not in the query set. For continuous random variables, one must integrate over the probability density. Consider the diagnostic BN in figure 14.1. Suppose one would like to know the probability that a patient reports a fever. Integrating over the temperature node produces this JPD (rounding all results to four decimal places):

Next, by summing the columns, one obtains the distribution of the PF node (except for some roundoff error): $Pr(PF) = 0.011$, $Pr(not(PF)) = 0.989$. The process of summing over random variables is *computing the marginal distribution*.

Now suppose that some evidence is available, such as that the patient is complaining of a fever, and that the practitioner would like to know whether the patient has influenza. This is shown in figure 14.2. The evidence presented to the BN is the fact that the random variable PF is true. The query is the value of the Flu random variable. The evidence is asserted by conditioning on the evidence. This is where Bayes' law finally appears and is the reason why BNs are named after Bayes. To compute this distribution, one first selects the terms of the JPD that satisfy the evidence, compute the marginal distribution, and finally normalize to get a PD. This last step is equivalent to dividing by the probability of the evidence.

To see why this is equivalent to Bayes' law, consider the case of two Boolean random variables A and B joined by an edge from A to B. The probability distribution of a Boolean random variable is determined by just one prob-

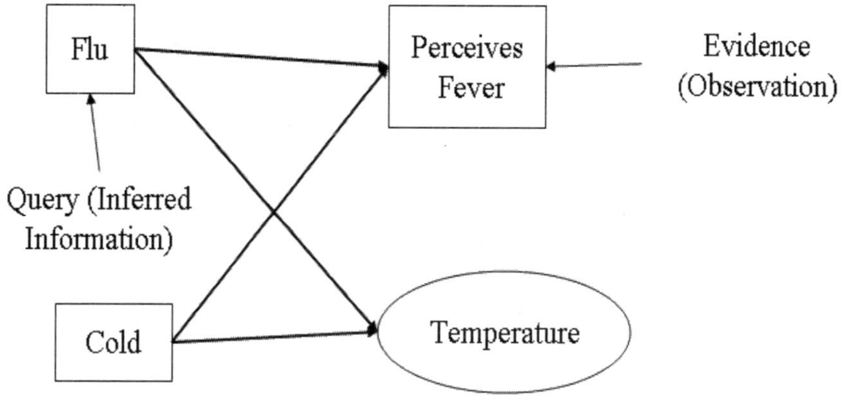

Figure 14.2 Example of diagnostic inference using a BN. The evidence for diagnosis is the perception of a fever by the patient. The question to be answered is whether the patient has influenza.

ability, so it is essentially the same as the probabilistic notion of an *event*. Let A and B be the two events in this case. The BN is specified by giving $Pr(A)$, $Pr(B \mid A)$ and $Pr(B \mid \text{not } A)$. Suppose that one is given the evidence that B is true. What is the probability that A is true? In other words, what is $Pr(A \mid B)$? The JPD of this BN is given by the four products $Pr(B \mid A)Pr(A)$, $Pr(\text{not } B \mid A)Pr(A)$, $Pr(B \mid \text{not } A)Pr(\text{not } A)$, and $Pr(\text{not } B \mid \text{not } A)Pr(\text{not } A)$. Selecting just the ones for which B is true, gives the two probabilities $Pr(B \mid A)Pr(A)$ and $Pr(B \mid \text{not } A)Pr(\text{not } A)$. The sum of these two probabilities is easily seen to be $Pr(B)$. Dividing by $Pr(B)$ normalizes the distribution. In particular, $Pr(A \mid B) = Pr(B \mid A)Pr(A)/Pr(B)$, which is exactly the classic Bayes' law.

Returning to the problem of determining the probability of influenza, the evidence requires that we select only the terms of the JPD for which PF is true, then compute the marginal distribution. Integrating over the temperatures in the first column of table 14.1 gives the following: We are only interested in the Flu node, so we sum the rows above in pairs to get: Normalizing gives $Pr(Flu) = 0.009$. Thus there is less than a 1% chance of having the flu even if one is complaining of a fever. Perceiving a fever has the effect of increasing the probability of having the flu substantially over the case of no evidence, but it is still relatively low.

The most general form of BN inference is to give evidence in the form of a PD on the evidence nodes. The only difference in the computation is

Event A	Pr(PF and A)
not(Flu) and not(Cold)	(0.9999)(0.99)(0.01) = 0.0099
not(Flu) and (Cold)	(0.9999)(0.01)(0.10) = 0.0010
(Flu) and not(Cold)	(0.0001)(0.99)(0.90) = 0.0001
(Flu) and (Cold)	(0.0001)(0.01)(0.95) = 0.0000

Table 14.2 Intermediate result during a stochastic inference

Event A	Pr(PF and A)
not(Flu)	0.0109
(Flu)	0.0001

Table 14.3 Final result of a stochastic inference

that instead of selecting the terms of the JPD that satisfy the evidence, one multiplies the terms by the probability that the evidential event has occurred. In effect, one is weighting the terms by the evidence. The probabilistic basis for this process is given in chapter 15. We leave it as an exercise to compute the probability of the flu as well as the probability of a cold given only that there is a 30% chance of the patient complaining of a fever.

BN inference is substantially more complex when the evidence involves a continuous random variable. We will consider this problem later. Not surprisingly, many BN tools are limited to discrete random variables because of this added complexity.

In principle, there is nothing special about any particular node in the process of BN inference. Once one has the JPD, one can assert evidence on any nodes and compute the marginal distribution of any other nodes. However, BN algorithms can take advantage of the structure of the BN to compute the answer more efficiently in many cases. As a result, the pattern of inference does affect performance. The various types of inference are shown in figure 14.3. Generally speaking, it is easier to infer in the direction of the edges of the BN than against them. Inferring in the direction of the edges is called *causal inference*. Inferring against the direction of the edges is called *diagnostic inference*. Other forms of inference are called *mixed inference*.

So far we have considered only discrete nodes. Continuous nodes add some additional complexity to the process. There are several ways to deal with such nodes:

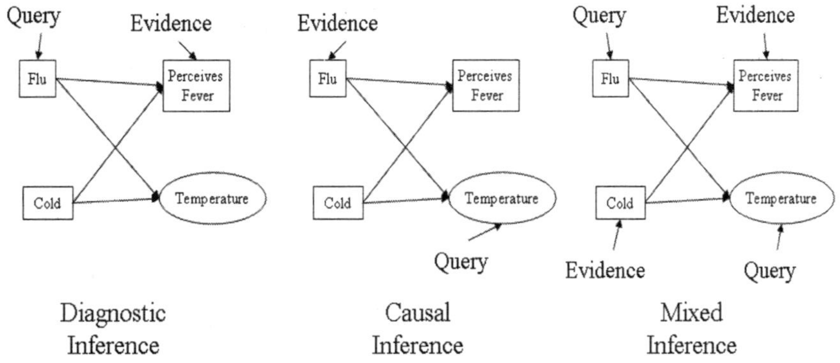

Figure 14.3 Various types of inference. Although information about any of the nodes (random variables) can be used as evidence, and any nodes can be queried, the pattern of inference determines how easy it is to compute the inferred probability distribution.

1. Partition. The possible values are partitioned into a series of intervals (also called bins). This has the disadvantage that it reduces the accuracy of the answer. However, it has the advantage that one only has to deal with discrete nodes. Many BN tools can only deal with discrete random variables.

2. Restrict to one class of distributions. A common restriction is to use only normal (Gaussian) distributions. This choice is supported by the central limit theorem. As in the case of partitioning, it reduces the accuracy of the answer. The advantage of this assumption is that the number of parameters needed to specify a distribution can be reduced dramatically. In the case of a normal distribution, one needs only two parameters. There are many other choices for a class of distributions that can be used. There will always be a tradeoff between improved accuracy vs. the increase in computational complexity. Since there will be many sources of error over which one has no control, the improvement in accuracy resulting from a more complex class of distribution may not actually improve the accuracy of the BN.

3. Use analytic techniques. This is more a theoretical than a practical approach. Only very small BNs or BNs of a specialized type (such as connectionist networks) can be processed in this way.

The techniques above are concerned with the specification of PDs. A CPD is a function from the possible values of the parent nodes to PDs on the node. If there are only a few possible values of the parent nodes (as in the diagnostic example in figure 14.1), then explicitly listing all of the PDs is feasible. Many BN tools have no other mechanism for specifying CPDs. When the number of possible values of the parent nodes is large or even infinite, then the CPD may be much better specified using a function. In the infinite case, one has no choice but to use this technique. Curve-fitting techniques such as least-squares analysis can be used to choose the function based on the available data.

A BN with both discrete and continuous nodes is called a *hybrid BN*. The diagnostic BN example above is a hybrid BN. When continuous nodes are dependent on discrete nodes, inference will produce a compound (mixed) Gaussian distribution. Such a distribution is the result of a compound process in which one of a finite set of Gaussians is selected according to a PD, and then a value is chosen based on the particular Gaussian that was selected.

If a discrete node is dependent on continuous nodes, then the discrete node can be regarded as defining a *classifier* since it takes continuous inputs and produces a discrete output which *classifies* the inputs. The CPDs for this situation are usually chosen to be logistic/softmax distributions. Connectionist networks (also called neural networks) are an example of this.

BNs are not the only graphical representation for stochastic models. Undirected graphical models, also called Markov random fields (MRFs) or Markov networks, are also used, especially in the physics and vision communities.

One application of BNs is to assist in decision making. To make a decision based on evidence one must quantify the risk associated with the various choices. This is done by using a utility function. It is possible to model some utility functions by adding *value nodes* (also called *utility nodes*) to a BN and linking them with dependency edges to ordinary BN nodes and to other utility nodes. The result of a decision is an action that is performed, and these can also be represented graphically by adding *decision nodes* and edges to a BN augmented with utility nodes. A BN augmented with utility and action nodes is called an *influence diagram* (also called a *relevance diagram*) (Howard and Matheson 1981). An influence diagram can, in principle, be used to determine the optimal actions to perform so as to maximize expected utility.

Summary

- The main use of BNs is for stochastic inference.
- BN inference is analogous to the process of logical inference and querying performed by rule engines.
- Bayes' law is the foundation for BN inference.
- Evidence can be either hard observations with no uncertainty or uncertain observations specified by a probability distribution.
- Evidence can be given for any nodes, and any nodes can be queried.
- The nodes of a BN can be continuous random variables, but inference in this case is more complicated.
- BNs can be augmented with other kinds of nodes, and used for making decisions based on stochastic inference.

14.3 Constructing Bayesian Networks

We now consider the important question of how to construct BNs. While there are many tools for performing inference in BNs, the methodology commonly employed for developing BNs is rudimentary. A typical methodology looks something like this:

1. Select the important variables.
2. Specify the dependencies.
3. Specify the CPDs.
4. Evaluate.
5. Iterate over the steps above.

This simple methodology will work for relatively small BNs, but it does not scale up to the larger BNs that are now being developed. The following are some of the development techniques that can be used as part of the process of constructing BNs, and each of them is discussed in more detail in its own section:

1. **Requirements**. Without clearly stated requirements, it is difficult to determine whether a BN has been successfully developed.

2. **Machine learning**. The PDs and CPDs are most commonly found by using statistical methods. There are a large number of such techniques.

3. **Component-based techniques**. A BN could be built from standard components or modules.

4. **Ontologies**. Ontologies can be used as the basis for the graph structure of the BN.

5. **Design patterns**. A BN development methodology has been introduced in (Neil et al. 2000) that is based on design patterns.

6. **Validating and revising**. As with any development activity, one must validate BNs. When testing uncovers a problem with a BN, it is necessary to adjust its CPDs or its structure. Revising the structure of a BN can also improve the design of a BN.

14.3.1 BN Requirements

Before embarking on any development project, it is important to have an understanding of its purpose. We saw this already in section 12.1 for the development of ontologies. The purpose of the BN should include the following:

1. *Why* the BN is being developed. One of the most common reasons for building a BN is to support diagnostic inference. However, BNs can also be used for combining information from different sources at different times. Yet another reason why one might build a BN is to analyze a domain, making independence assumptions more explicit. This allows these assumptions to be tested.

2. *What* will be covered by the BN. This is also called its *scope*. A clear definition of the scope will prevent the development effort from expanding unnecessarily.

3. *Who* will be using the BN. As with ontology development, this will affect the amount of effort that should be devoted to the design of the BN.

Analyzing the requirements of a BN not only involves acquiring an understanding of the domain, it should also determine the required accuracy, performance, and interfaces. BN development typically ignores these requirements. Indeed, the notion of a BN interface is only now beginning to be understood. BN interfaces are discussed in subsection 14.3.3. In most cases, BN development projects have no explicitly stated purpose. When there is a stated purpose, it is usually too generic to be useful in the development process. Having a detailed stated purpose would not entirely determine the required accuracy, performance, and interfaces, but it would certainly help.

Because the amount of knowledge to be acquired may be very large, it is important for the knowledge to be well organized. It is also important to track the source of knowledge so that one can determine its trustworthiness. These issues can be addressed to some degree by using ontologies. An example of this is discussed in subsection 14.3.6.

Summary

The purpose of a BN should address these issues:

- Why the BN is being developed and how it will be used
- What will be covered by the BN
- Who will be using the BN
- The required accuracy, performance, and interfaces

14.3.2 Machine Learning

This subsection gives some background on current statistical methods for constructing PDs. It begins with an overview of techniques for empirically determining PDs, CPDs, and BNs from data. Such data are regarded as being used to "train" the probabilistic model, so the techniques are known as *machine learning* methods.

Machine learning is a very large area that would be difficult to survey adequately, so we give only an overview. Since a BN is just a way of representing a JPD, virtually any data-mining technique qualifies as a mechanism for constructing a BN. It is just a matter of expressing the end result as a BN. For example, one might be interested in the body mass index (BMI) of individuals in a research study. The individuals have various characteristics, such as sex and age. Computing the average BMI of individuals with respect to these two characteristics gives rise to a three-node BN as in figure 14.4. The CPD for the BMI node gives the mean and standard deviation of the BMI for each possible combination of sex and age in the study.

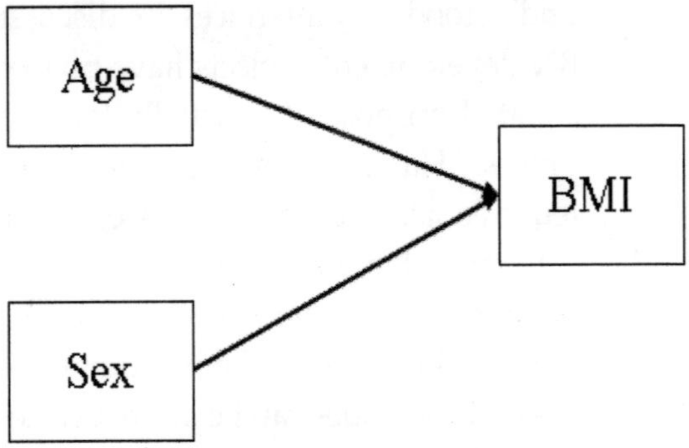

Figure 14.4 Bayesian network for the result of a research study of body mass index (BMI) as a function of age and sex.

It is common to assume that the CPDs are independent of each other, so they can be estimated individually. When data for both the parent nodes and the child node are available, estimating the CPD reduces to the problem of estimating a set of PDs, one for each of the possible values of the parent nodes. There are many techniques for estimating PDs in the literature. They can be classified into two categories:

1. **Frequentist methods**. These methods are associated with the statistician and geneticist Ronald Fisher, and so one sometimes sees at least some of these methods referred to as *Fisherian*. They also go by the name *maximum likelihood* (ML) estimation. The CPDs of discrete nodes that are dependent only on discrete nodes are obtained by simply counting the number of cases in each slot of the CPD table. This is why these techniques are called "frequentist." The CPDs for Gaussian nodes are computed by using means and variances. Other kinds of continuous node are computed using the ML estimators for their parameters.

2. **Bayesian methods**. These methods also go by the name *maximum a posteriori* (MAP) estimation. To perform such an estimation, one begins with a prior PD, and then modifies it using the data and Bayes' law. The use of an arbitrary prior PD makes these methods controversial. However, one can argue that the ML technique is just the special case of MAP for which the prior PD is the one which represents the maximum amount of ignorance possible. So one is making an arbitrary choice even when one is using frequentist methods. If one has some prior knowledge, even if it is sub-

jective, it is helpful to include it in the estimation. As the amount of data and learning increase, the effect of the prior PD gradually disappears.

The estimation techniques discussed above assume that data about all of the relevant nodes were available. This is not always the case. When one or more nodes are not directly measurable, one can either remove them from the BN or attempt to estimate them indirectly. The latter can be done by using BN inference iteratively. One treats the unobservable nodes as query nodes and the observable nodes as evidence nodes in a BN inference process. One then computes the expectations of the unobservable nodes and uses these values as if they were actually observed. One can then use ML or MAP as above. This whole process is then repeated until it converges. This technique is known as *expectation maximization* (EM).

It is possible to use machine learning techniques to learn the structure of the BN graph as well as to learn the CPDs. These tend to have very high computational complexity, so they can only be used for small BNs. In practice, it is much better to start with a carefully designed BN and then modify it in response to an evaluation of the quality of its results.

Connectionist networks are a class of BNs that are designed for efficient machine learning. Such BNs are most commonly known as "neural networks" because they have a superficial resemblance to the networks of neurons found in vertebrates, even though neurons have very different behavior than the nodes in connectionist networks. Many kinds of connectionist network support incremental machine learning. In other words, they continually learn as new training data are made available.

Connectionist networks constitute a large research area, and there are many software tools available that support them. There is an extensive frequently asked questions list (FAQ) for neural networks, including lists of both commercial and free software (Sarle 2002). Although connectionist networks are a special kind of BN, the specification of a connectionist network is very different from the specification of a BN. Consequently, techniques for machine learning of connectionist networks may not apply directly to BNs or vice versa. However, BNs are being used for connectionist networks (MacKay 2004) and some connectionist network structures are being incorporated into BNs, as in (Murphy 1998).

Summary

Probability distributions are computed by using statistical techniques.

- Frequentist (ML) techniques make no a priori assumptions.
- Bayesian (maximum a priori) techniques start with a prior distribution and gradually improve it as data become available.
- EM is used for determining the distribution of a random variable that is not directly observable.
- Connectionist networks are special kinds of statistical models for which there are efficient machine learning techniques.

14.3.3 Building BNs from Components

Modern software development methodologies make considerable use of components that have already been developed and tested. This greatly reduces the amount of effort required to develop large systems. An approach to BN development that makes use of previously developed BN components has been proposed in (Koller and Pfeffer 1997). In addition to reducing the BN development effort, this approach may also be able to improve performance during inference. This approach is known as *object-oriented Bayesian networks* (OOBNs). The basic OOBN concept is called an "object." An OOBN object can be just a random variable, but it can also have a more complex structure via attributes whose values are other objects. An OOBN object can correspond to an entity or it can be a relationship between entities.

A *simple* OOBN object corresponds to a BN node. It has a set of input attributes (i.e., the parent nodes in the BN) and an output attribute (i.e., its value). A *complex* OOBN object has input attributes just as in the case of a simple OOBN object, but a complex object can have more than one output attribute. The input and output attributes define the interface of the OOBN. The interface defines the formal relationship of the object to the rest of the BN. It can also have encapsulated attributes that are not visible outside the object. A complex object corresponds to several BN nodes, one for each of the outputs and encapsulated attributes. The notion of a complex object is a mechanism for grouping nodes in a BN. The JPD of an OOBN, as well as the process of inference, is exactly the same whether or not the grouping is used.

However, by grouping (encapsulating) nodes into objects, one gains a number of significant advantages:

1. Complex objects can be assigned to classes which can share CPDs. Reusing CPDs greatly simplifies the task of constructing a BN.

2. Classes can inherit from other classes which allows for still more possibilities for reuse.

3. Encapsulation can be used during inference to improve performance. This advantage is especially compelling. As shown in (Koller and Pfeffer 1997), if a BN has an OOBN structure, then the performance of inferencing can be improved by an order of magnitude or more compared with even a well-optimized BN inference algorithm.

Another feature of the OOBN methodology is the notion of an *object-oriented network fragment* (OONF). An OONF is a generalization of a BN which specifies the conditional distribution of a set of value attributes given some set of input attributes. If there are no input attributes, then an OONF is a BN. An OONF can be defined recursively in terms of other OONFs. An OONF can also be used as a *component* which can be "reused" multiple times in a single BN. Component-based methods are powerful development methodologies that allow one to build BNs from standard components that have been constructed independently.

Summary

The OOBN methodology introduces several notions to BN development:

- Components which can be used more than once
- Groupings of BN nodes with a formally defined interface
- Inference algorithms that take advantage of the OOBN structure to improve performance significantly

14.3.4 Ontologies as BNs

One of the earliest large BNs was the QMR-DT mentioned in section 14.1 which added probabilities to an expert system. The close connection between expert systems and ontologies would suggest that it ought to be possible to "add probabilities" to ontologies. Perhaps because of this analogy, an active research area has developed that is attempting to do this for ontologies, especially for ontologies based on description logic. See (Ding and Peng 2004; Koller et al. 1997).

Given an OWL-DL ontology, the corresponding BN has one node for each class. This node is a Boolean random variable which is true precisely when an entity belongs to the class. Edges are introduced when two classes are related. The most common relationship is the subclass relationship which means that one class is contained in another. Obviously this will result in a stochastic dependency. Other kinds of relationship can be expressed in terms of classes. For example, the age of a person (in years) gives rise to a collection of disjoint subclasses of the person class, one for each possible value of the age of a person.

Although this technique does seem to be a natural way to "add probabilities" to ontologies, it does not seem to produce BNs that are especially useful. The most peculiar feature of these BNs is that all of the classes are ultimately subclasses of a single universal class (called the Thing class), and the random variable for a class represents the probability that a randomly chosen thing is a member of the class. While this might make sense for some class hierarchies, the hierarchies of ontologies often contain a wide variety of types of entity. For example, a biomedical ontology would contain classes for research papers, journals, lists of authors, drugs, addresses of institutions, and so on. It is hard to see what kind of experiment would sometimes produce a drug, other times produce a list of authors, and still other times produce an address.

On the other hand, this technique can be the starting point for BN development, especially for diagnostic BNs. An example of this is discussed in subsection 14.3.6, where the ontology is used as the background for the development of a BN. The disadvantage of developing BNs by using ontologies in this way is that whatever formal connection exists between the ontology and the BN is quickly lost as the BN is modified. As a result, one cannot use any logical consequences entailed by the ontology during BN inference. Indeed, the ontology ultimately furnishes no more than informal documentation for the BN.

Summary

- It is possible to define a BN structure corresponding to an ontology.
- Such BNs are seldom useful in their original form, but can be used as the starting point for developing realistic BNs.

14.3.5 BN Design Patterns

The oldest and most commonly used design patterns for BNs are probabilistic analogs of the logical and arithmetic operations. For example, the so-called "noisy OR-gate" models the combining of evidence in favor of a single conclusion, as shown in figure 14.5 (Pearl 1988). The evidence nodes and conclusion are all modeled as random variables that are either true or false. Each evidence node that is true increases the likelihood that the conclusion is true, and vice versa. If all of the evidence nodes are false, then the conclusion is certain to be false. Each evidence node can contribute a different amount to the conclusion, and the evidence nodes contribute independently to the conclusion. figure 14.5 shows how the conditional probability table is defined for a noisy OR-gate. In this figure, the amount that an evidence node X contributes to the conclusion is $1 - q_X$. If this parameter is equal to 1, then the truth of the corresponding evidence node is conclusive: the conclusion is true with probability 1. In practice, evidence nodes will have much smaller contributions. Other "noisy" operations include noisy-AND, noisy-MAX and noisy-MIN operations (Pradhan et al. 1994).

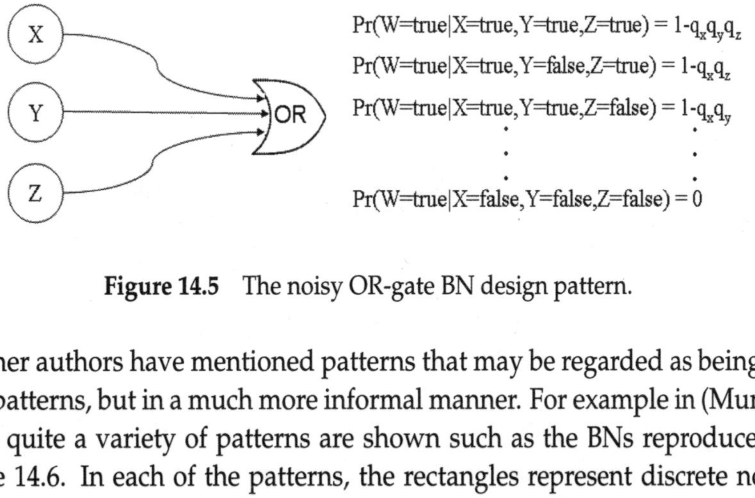

Figure 14.5 The noisy OR-gate BN design pattern.

Other authors have mentioned patterns that may be regarded as being design patterns, but in a much more informal manner. For example in (Murphy 1998) quite a variety of patterns are shown such as the BNs reproduced in figure 14.6. In each of the patterns, the rectangles represent discrete nodes and the ovals represent Gaussian nodes. The shaded nodes are visible (observable) while the unshaded nodes are hidden. Inference typically involves specifying some (or all) of the visible nodes and querying some (or all) of the hidden nodes.

A number of Design idioms for BNs were introduced by (Neil et al. 2000). The definitional/synthesis idiom models the synthesis or combination of many nodes into one node. It also models deterministic definitions. The cause-consequence idiom models an uncertain causal process whose consequences are observable. The measurement idiom models the uncertainty of a measuring instrument. The induction idiom models inductive reasoning based on populations of similar or exchangeable members. Lastly, the rec-

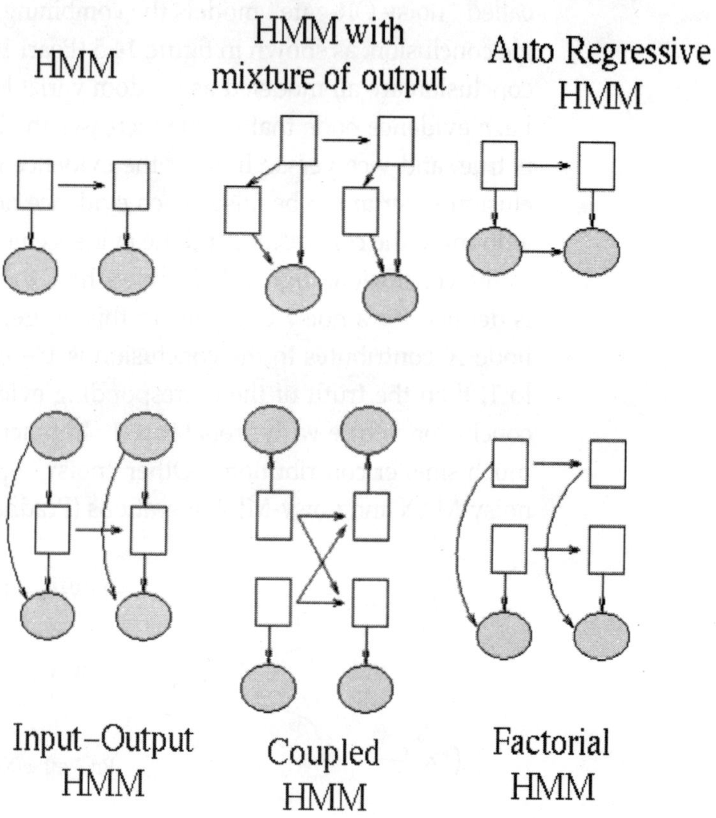

Figure 14.6 Various informal patterns for BNs. These examples are taken from (Murphy 1998).

onciliation idiom models the reconciliation of conflicting information. Note that reconciliation of conflicting information is not the same as combining compatible information, which is introduced in chapter 15.

Summary

- One methodology for designing BNs is to use design patterns or idioms.
- Many BN design patterns have been identified, but most are only informally specified.

14.3.6 Validating and Revising BNs

Testing and validation have always been accepted as an important part of any development activity. This is especially true for BNs. BNs have the advantage that there is a well-developed, sophisticated mechanism for testing hypotheses about PDs. However, this is also a disadvantage because statistical hypothesis testing generally requires more than one test case, and even with a large sample of test cases the result of a test can be equivocal. A BN test case is usually a specific example of a stochastic inference.

There are several techniques for validating BNs:

1. **Test cases.** When there are special cases whose answer is known, one can perform the inference and check that the result is close to the expected answer.

2. **Sensitivity analysis.** This technique determines the impact of inaccuracies of the CPD entries by systematically varying them and recording the effects on test cases. As one might expect, different CPD entries of BNs can have very different sensitivities (Henrion et al. 1996; Pradhan et al. 1996). Sensitivity analysis can also be used to focus attention on the probabilities that need to be determined more accurately.

3. **Uncertainty analysis.** In this technique, all of the probabilities are varied simultaneously by choosing each one from a prespecified distribution that reflects their uncertainties. One then records the effects on test cases. This technique can determine the overall reliability of a BN. However, it yields less insight into the effect of separate probabilities than is the case for sensitivity analysis.

4. **Consistency checking.** If a BN was developed using components as in the OOBN methodology, then one can check that the components have been used correctly. Software development tools make extensive use of this technique, which is called "type checking."

When an evaluation of a BN fails, the BN must be modified. Usually all that is necessary is to change one or more of the CPDs. However, sometimes testing uncovers previously unsuspected dependencies, and the BN structure must be changed. One can design a BN by starting with some simple structure and then revising the design to make it more accurate or simpler. This is the approach developed by Helsper and van der Gaag (2001, 2002). This methodology uses an ontology as the starting point, as in subsection 14.3.4. The authors have studied the use of their methodology within the domain of esophageal cancer.

The Helsper-van der Gaag methodology uses ontologies more as a background for the design process than as a formal specification for the BN structure. This is in contrast with the OOBN technique in subsection 14.3.3 in which the design not only specifies the BN completely but also affects the inference algorithm. In the Helsper methodology the ontology is used to provide an initial design for the BN in a manner similar to the way that this is done in subsection 14.3.4. This step in the methodology is called *translating*. However, this initial design is modified in a series of steps based on domain knowledge. Some of the modifications use the ontology, but most of them must be elicited from domain experts. The ontology "serves to document the elicited domain knowledge."

What makes the Helsper-van der Gaag methodology interesting are the systematic modification techniques that are employed. The methodology refers to this phase as *improving and optimizing*. The modifications must follow a set of *guidelines*, but these guidelines are only explained by examples in the articles.

One example of a modification operation used by Helsper and van der Gaag is shown in figure 14.7. In this operation, a node that depends on two (or more) other nodes is eliminated. This would be done if the node being eliminated is not observable or if it is difficult to observe the node. There are techniques for determining the CPDs for unobservable nodes such as the EM algorithm discussed in subsection 14.3.2. However, this algorithm is time-consuming. Furthermore, there is virtually no limit to what one could potentially model, as discussed in section 13.3. One must make choices about what variables are relevant, even when they could be observed, in order to make the model tractable.

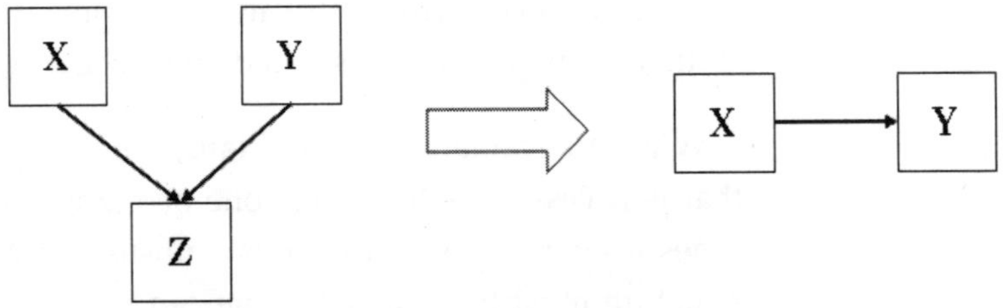

Figure 14.7 Modifying a BN by eliminating a node that other nodes depend on. The result is that the parent nodes become dependent on each other.

When a node is dependent on other nodes, the other nodes (which may otherwise be independent) become implicitly dependent on each other via the dependent node. In statistics this is known as Berkson's paradox, or "selection bias." The result of dropping a node is to make the parent nodes explicitly dependent on each other. This dependency can be specified in either direction, whichever is convenient and maintains the acyclicity of the BN.

The modification operation shown in figure 14.7 changes the JPD of the BN because one of the variables is being deleted. Furthermore, the new JPD need not be the same as the distribution obtained by computing the marginal distribution to remove the deleted variable, although it is approximately the same.

It is a general fact that the direction of a directed edge in a BN is probabilistically arbitrary. If one knows the JPD of two random variables, then one can choose either one as the parent node and then compute the CPD for the child node by conditioning. In practice, of course, the specification works the other way: the JPD is determined by specifying the CPD. For a particular modeling problem, the direction of the edge will usually be quite clear, especially when one is using a design pattern.

However, sometimes the direction of the dependency is ambiguous, and one of the modification operations is to reverse the direction. In this case the JPD is not changed by the operation. This situation occurs, for example, when two variables are Boolean, and one of them subsumes the other. In other words, if one of the variables is true, then the other one is necessarily true also (but not vice versa). Suppose that X and Y are two Boolean random variables such that X implies Y. Then we know that $Pr(Y = true \mid X = true) = 1$. This gives one half of the CPD of one of the variables with respect to the other, and the dependency can go either way. This is shown in figure 14.8

Summary

- It is important to test and validate BNs to ensure that they satisfy the requirements.

- The most commonly used techniques for validating BNs are

  1. specialized test cases,

  2. sensitivity analysis,

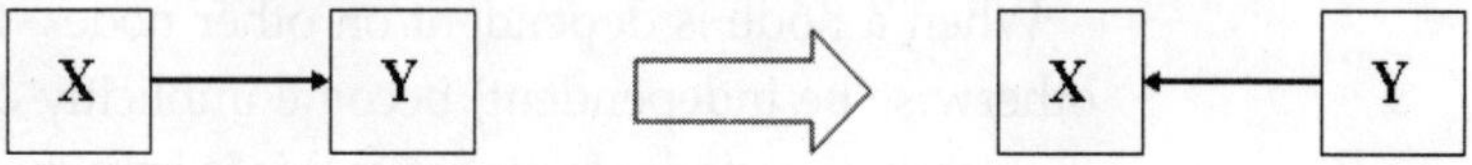

	Y	not Y
X	1	0
not X	p	1-p

	X	not X
Y	q	1-q
not Y	0	1

Figure 14.8 Modifying a BN by reversing the direction of a dependency when two Boolean nodes are related by subsumption.

3. uncertainty analysis,
4. consistency checking.

- A BN that fails a test must be modified.
- BN modification can be used as a normal part of BN development.
- BN modification operations have been identified and classified, and guidelines for when to apply them have been developed.

14.4 Exercises

1. In the diagnostic BN in figure 14.1, one can use either a temperature measurement or a patient's perception of a fever to diagnose influenza. Although these two measurements are a priori independent, they become dependent when one observes that the patient has the flu or a cold. In statistics this is known as Berkson's paradox, or "selection bias." It has the effect that a high temperature can reduce the likelihood that a patient reports being feverish and vice versa. Compute the JPD of the PF and T nodes in this BN given the observation that the patient has influenza.

2. Compute the probability that a patient has influenza using temperature measurements. For example, try 37°, 38°, 39°, and 40° C. These are all (in theory) exact measurements. In fact, a thermometer, like all sensors, can only give a measurement that is itself a random variable. Compute the probability of influenza given a temperature of 38.40° C, normally distributed with standard deviation 0.20° C.

15 *Combining Information*

Meta-analysis is the integration of data from disparate sources. While this can encompass a wide variety of phenomena, it is most commonly applied to data obtained from sensors that are observing the same (or at least overlapping) environments. The sensors can be different sensors or they can be the same sensor observing at different times. Scientific experimentation is also a form of sensing. In this case one is observing natural phenomena. Such observations are generally subject to uncertainty due to the lack of full knowledge about the phenomena being observed as well as the limitations of the measuring devices being used. One can reduce these uncertainties by making a series of independent observations. Meta-analysis is the process of combining the evidence afforded by the observations.

Meta-analysis goes by many names. The generic name is *combining information* (CI). In the social and behavioral sciences, it is also known as *quantitative research synthesis*. In medicine, it is often called *pooling of results* or *creating an overview*. Physicists refer to CI as *viewing the results of research*. For chemists, CI is used for determining physical constants based on empirical studies, and is called *critical evaluation* (NRC 1992). When applied to sensors, CI is most commonly called *data fusion*. Sensor data fusion has the most elaborate forms of CI, and the field of multi-sensor data fusion has a standard for data fusion, called the *JDL Model* that divides it into 4 levels (Steinberg et al. 1999).

Because of the uncertainty inherent in most forms of empirical knowledge, it is normally stated as a probability distribution (PD) on a set of possible states. For example, when one speaks of a temperature measurement of 30.5° ±0.4°C, one is asserting that the measurement has a normal distribution whose mean was 30.5°C and whose standard error was 0.4°C. Now suppose that one performs a second, independent, measurement of the same temperature, obtaining 30.2° ±0.3°C. One now has two independent normal distributions. Combining the two measurements is the same as combining the two distributions.

In this chapter the process of meta-analysis is formally defined and proven. Combining discrete distributions is covered first, followed by the continuous case. Stochastic inference is a special case of meta-analysis. More generally, one can combine two Bayesian networks (BNs). Conversely, the meta-analysis process can itself be expressed in terms of BNs. This is shown in section 15.3. The temperature measurement example above is an example of the combination of observations that are continuous distributions. PDs are not only a means of expressing the uncertainty of a observation, they can themselves be observations. In other words, PDs can have PDs. A large number of statistical tests are based on this idea, which is discussed in section 15.4. The last section introduces an interesting variation on information combination, called Dempster-Shafer theory.

15.1 Combining Discrete Information

We first consider the case of combining two discrete PDs. That means we have two independent random variables X and Y, whose values are discrete rather than continuous. For example, a patient might seek multiple independent opinions from practitioners, each of which gives the patient their estimates of the probabilities of the possible diagnoses. Combining these two discrete random variables into a single random variable is done as follows:

Discrete Information CombinationTheorem
Let X and Y be two discrete random variables that represent two independent observations of the same phenomenon. If there exists a value v such that both $Pr(X = v)$ and $Pr(Y = v)$ are positive, then there is a random variable Z that combines the information of these two observations, whose PD is

$$Pr(Z = v) = \frac{Pr(X = v)Pr(Y = v)}{\sum_w Pr(X = w)Pr(Y = w)}$$

Proof Since X and Y are independent, their JPD is given by $Pr(X = u, Y = v) = Pr(X = u)Pr(Y = v)$. The random variables X and Y are combined by conditioning on the event $(X = Y)$. This will be well defined if and only if $Pr(X = Y)$ is positive, which is the case when there is some value v such that $Pr(X = Y = v)$ is positive. When this is true, the distribution of the combined random variable Z is given by

$$Pr(Z = v) = Pr(X = v \mid X = Y) = \frac{Pr(X = v \text{ and } X = Y)}{Pr(X = Y)}.$$

Now $(X = v \text{ and } X = Y)$ is logically equivalent to $(X = v \text{ and } Y = v)$. Since X and Y are independent, $Pr(X = v \text{ and } Y = v) = Pr(X = v)Pr(Y = v)$. Therefore,

$$Pr(Z = v) = \frac{Pr(X = v)Pr(Y = v)}{Pr(X = Y)}.$$

The result then follows.

Information combination depends on two important criteria: the observations must be independent and they must be measuring the same phenomenon. As we saw in chapters 2 and 4, the essence of semantics is the determination of when two entities are the same. Information combination is also determined by sameness. Indeed, information combination can be regarded as the basis for the semantics of uncertainty.

Another important assumption of the information combination theorem above is that the event representing the sameness of two random variables is $(X = Y)$, i.e., where the two observations are exactly the same. However, it is conceivable that in some situations the sameness relationship could be more complicated. This is especially true when the values being observed are not sharply distinguishable. For example, not everyone uses the same criteria to characterize whether a person is obese or overweight. As a result, independent observations can be calibrated differently. Such observations should not be combined unless they can be recalibrated. In section 15.5, we consider a more general event for representing the sameness of two discrete random variables that addresses these concerns to some degree.

To illustrate how the information combination theorem can be applied, suppose that a patient is complaining of a severe headache. For simplicity, assume that the only possible diagnoses are concussion, meningitis, and tumor. One doctor concludes that the probabilities of the diagnoses are 0.7, 0.2, and 0.1, respectively. Another doctor concludes that the probabilities are 0.5, 0.3, and 0.2, respectively. Combining these two yields the probabilities 0.81, 0.14, and 0.05, respectively. Note that the most likely diagnosis becomes more likely, while the least likely one becomes less likely. The reason for this is that the diagnoses have been assumed to be *independent*. In practice, diagnoses will be based on tests and symptoms that are observed by both doctors. In addition, doctors have similar training and use the same criteria for making diagnoses. As a result, the diagnoses would not usually be independent.

For a more extreme example, suppose that the first doctor concludes that the probabilities are 0.9, 0.0, and 0.1; and the second doctor gives the probabilities as 0.0. 0.9, and 0.1. The combined distribution will conclude that the tumor has probability 1.0, while the other two diagnoses are impossible (Zadeh 1984). This seems to be wrong. However, it makes perfectly good sense. In the words of Sherlock Holmes in "The Blanched Soldier", "When you have eliminated all which is impossible, then whatever remains, however improbable, must be the truth." Each of the doctors has concluded that one of the diagnoses is impossible, so the third possibility must be the truth. On the other hand, one can question whether such totally different diagnoses would happen independently. In other words, it is unlikely that the doctors are independently observing the same phenomenon. Such observations are said to be *incompatible.*

However, there are other circumstances for which observations that are incompatible are actually independent and therefore fusable. For example, if one distribution represents the probability of occurrence of a rare disease, and another distribution represents the observation that a particular patient definitely has the disease, then the combination of the two distributions is simple: the patient has the disease.

An even more extreme example would be two observations in which all of the possibilities have been declared to be impossible in one or the other observation. The discrete information combination theorem gives no combined distribution in this case because the hypotheses are not satisfied. One says that such observations are *inconsistent*.

Summary

- The discrete information combination theorem gives the formula for fusing independent discrete random variables that measure the same phenomenon.

- Incompatible PDs can be combined but care must be taken to interpret the combined distribution properly.

- Inconsistent PDs cannot be combined at all.

15.2 Combining Continuous Information

One can also combine continuous random variables. The only difference is that one must be careful to ensure that the combined distribution can be rescaled to be a PD.

Continuous Information CombinationTheorem
Let X and Y be two continuous random variables that represent two independent observations of the same phenomenon. Let $f(x)$ and $g(x)$ be the probability density functions of X and Y, respectively. If f and g are bounded functions, and if $\int f(y)g(y)dy$ is positive, then there is a random variable Z that combines the information of these two observations, whose probability density function is

$$h(x) = \frac{f(x)g(x)}{\int f(y)g(y)dy}.$$

Proof The proof proceeds as in the discrete case except that one must check that $\int f(x)g(x)dx$ converges. Now $f(x)$ was assumed to be bounded. Let B be an upper bound of this function. Then $f(x)g(x) \leq Bg(x)$ for every x. Since $\int g(x)dx$ converges, it follows that $\int f(y)g(y)dy$ also converges. The result then follows as in the discrete case.

As with discrete random variables, information combination requires that the observations be independent and they measure the same phenomenon. Ensuring that the observations measure the same phenomenon can be difficult, as the observations can use different calibrations. Uncoordinated or miscalibrated observations should not be combined unless they can be recalibrated.

In both of the information combination theorems, the last step is to normalize the distribution. Consequently, one can combine *unnormalized distributions* as long as the combined distribution is normalizable. In particular, it makes sense to combine a uniformly distributed random variable U with a random variable X, even when the uniform distribution cannot be normalized. It is easy to see that the combination of U with X is the same as X. In other words, a uniform distribution adds no new information to any distribution.

Normal distributions are an especially important special case which follows easily from the general case:

Combining Normal Distributions
If X and Y are independent normally distributed random variables with means m,

n and variances v, w, respectively, then the combined random variable has mean

$$\frac{wm + vn}{v + w} = \frac{\frac{m}{v} + \frac{n}{w}}{\frac{1}{v} + \frac{1}{w}}$$

and variance

$$\frac{vw}{v + w} = \frac{1}{v} + \frac{1}{w}.$$

This result is easily extended to the combination of any number of independent normal distributions. The means are combined by means of a weighted average, using weights that are proportional to the inverse variances.

We can now combine the two temperature measurements 30.5° ±0.4°C and 30.2° ±0.3°C mentioned earlier. The variances are 0.16 and 0.09, so the combined mean is 30.3° ±0.24°C. The combined mean is closer to 30.2°C than to 30.5°C because the former measurement is more accurate.

The formula for combining normal distributions applies equally well to multivariate normal distributions. The only differences are that the mean is a vector and the variance is a symmetric matrix (often called the covariance). This formula is the basis for the Kalman filter (Maybeck 1979) in which a sequence of estimates is successively updated by independent observations. The Kalman filter update formula is usually derived by using an optimization criterion such as least squares. However, nothing more than elementary probability theory is necessary.

Information combination is commonly formulated in terms of a priori and a posteriori distributions. The a priori or prior distribution is one of the two distributions being combined, while the experiment or observation is the other one. The a posteriori distribution is the combined distribution. Although the formulation in terms of a priori and a posteriori distributions is equivalent to information combination, it can be somewhat misleading, as it suggests that the two distributions play different roles in the process. In fact, information combination is symmetric: the two distributions being combined play exactly the same role. One of the two distributions will generally have more effect on the result, but this is due to it having more accuracy, not because it is the prior distribution or the observation.

Another example of information combination is stochastic inference in a BN, as presented in section 14.2. The evidence is combined with the BN, and the distributions of the query nodes are obtained by computing the marginal distributions of the combined JPD. Since the evidence usually specifies information about only some of the nodes, a full JPD is constructed by using independent uniform distributions for all of the other nodes. As we noted earlier, a uniform distribution adds no information to an information combination process. In general, the evidence can be a BN. In other words, one can combine two independent BNs. However, the combined JPD need not have the same graph as the BNs that were combined, even when the original two BNs have the same graph. This is a consequence of Berkson's paradox. However, if the BNs are measurements from the same population and have the same graph, then the combined BN should also have the same graph. Indeed, if it does not, then this is evidence that the original BNs were not from the same population.

Summary

- The continuous information combination theorem gives the formula for fusing independent continuous random variables that measure the same phenomenon.
- The derivation of an a posteriori distribution from an a priori distribution and an observation is a special case of the information combination theorems.
- Stochastic inference in a BN is another special case of the information combination theorems.

15.3 Information Combination as a BN Design Pattern

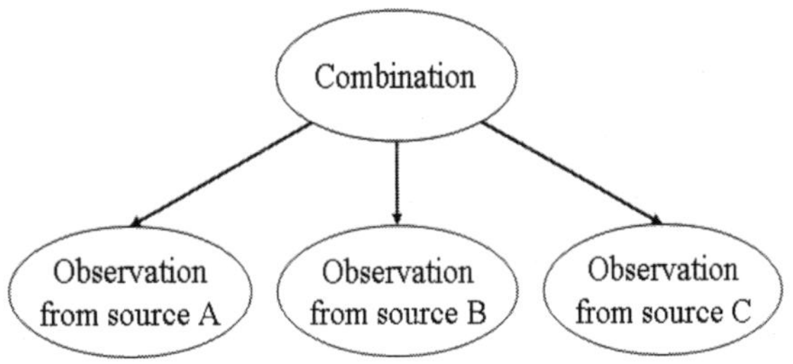

Figure 15.1 Information combination as a BN pattern. Two or more independent observations are combined to produce a single probability distribution.

The combination of independent sources of evidence can be expressed as the BN pattern shown in figure 15.1. This pattern differs from the reconciliation pattern discussed in subsection 14.3.5. In the reconciliation pattern, the dependency arrows are in the opposite direction. The conditional probability distributions that define the BN for information combination are shown in figure 15.2. The process of combining information from multiple sources is a special case of stochastic inference. The random variables to be combined (called X and Y in the figure) are given as evidence to the BN. Then query the BN to obtain the combined random variable Z.

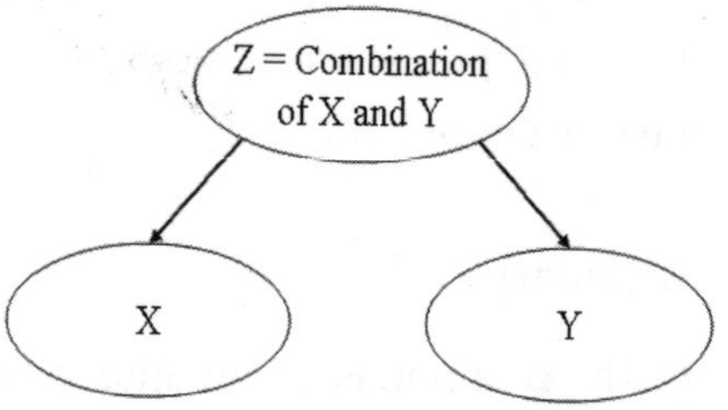

$$\Pr(X{=}x|Z{=}z) = \begin{cases} 1 \text{ if } x = z \\ 0 \text{ if } x \neq z \end{cases} \qquad \Pr(Y{=}y|Z{=}z) = \begin{cases} 1 \text{ if } y = z \\ 0 \text{ if } y \neq z \end{cases}$$

Figure 15.2 The conditional probability distributions that define the BN for combining two independent observations of the same phenomenon. The prior probability distribution on the Z is the uniform distribution.

Expressing information combination as a BN allows one to formulate more general information combination processes. For example, one can combine random variables that are dependent on each other or on common information, or one can combine random variables that are not directly observable, as shown in figure 15.3.

Summary

- The information combination process can be expressed as a BN.

- When expressed as a BN, information combination is a form of stochastic inference.

- The BN formulation of information combination allows one to formulate many information combination processes as well as other ways to combine information.

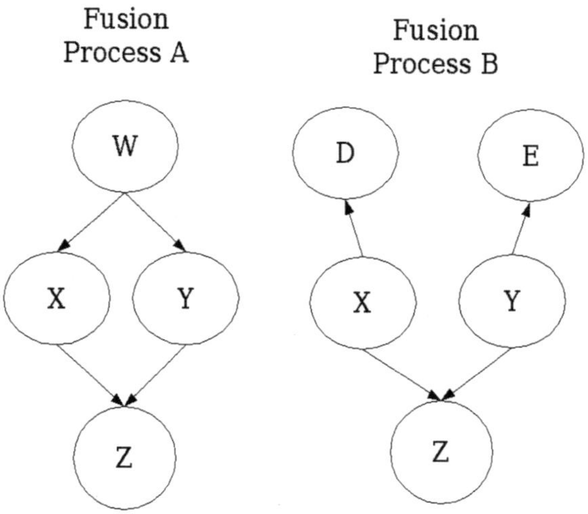

Figure 15.3 Examples of information combination processes. The process on the left side combines two random variables that have a mutual dependency. The process on the right side combines random variables that are not directly observable.

15.4 Measuring Probability

So far we have focused on PDs as a means of expressing an observation. The range of possibilities for what one can observe is very large, including concentrations, temperatures, pressures, and simple Boolean observations. However, some of the most important observations are measurements of PDs, and a large array of statistical tests (such as *t*-, chi-square, and *F*-tests) are concerned with such measurements. When observing a PD, one must be careful to distinguish the PD that is being measured from the one that is used for expressing the observation of the PD. It can get confusing because the observation is the PD of a PD.

To understand what this means, consider the problem of determining the body mass index (BMI) of individuals in a population. If one just focuses on the BMI measurement, then one will not capture the variation of BMI in the population. As data are accumulated, one will get more accurate measurements of the average BMI, and that is all. In practice, one is interested in the distribution of values in a population. In other words, one is measuring a PD. Although population distributions cannot be exactly normally distributed, since they are finite distributions, they can usually be approximated by a normal distribution.

A normal distribution is characterized by two parameters: the mean and the variance. Consequently, the measurement of a normal distribution is a measurement of these two numbers. When these are measured using a random sample, the mean has the t distribution, and the variance has the chi-square distribution. Both of these are approximately normally distributed for large samples. The mean and variance of the sample mean and sample variance are well-known:

Sample StatisticsTheorem

Let X be a random variable from a normally distributed population with mean μ and variance σ^2, For a random sample of N independent measurements of X:

1. *The sample mean $\overline{X}$ has mean μ and variance σ^2/N.*

2. *The sample variance s^2 has mean σ^2 and variance $\sigma^4/(N-1)$.*

When two populations are compared, one can compare them in a large variety of ways. Their means can be compared with a t-test, and their variances can be compared with either a chi-square test (to determine whether the difference of the variances is small) or an F-test (to determine whether the ratio of the variances is close to 1).

It is easy to experiment with these concepts either by using real data or by generating the data using a random number generator. In the following, a random number generator was used to generate two independent random samples of size 100 from a normal population with mean 10 and variance 16. For such a large sample size, the t and chi-square distributions are very close to being normal distributions. The estimates for the distributions (mean, variance) were (9.31, 13.81) and (10.55, 16.63). Now forget what these are measuring, and just think of them as two independent measurements. The uncertainty of each measurement is approximately normally distributed. The mean of the first measurement is the measurement itself, and the variance matrix is $\begin{pmatrix} 0.138 & 0 \\ 0 & 3.86 \end{pmatrix}$. The variance matrix of the second measurement is $\begin{pmatrix} 0.166 & 0 \\ 0 & 5.59 \end{pmatrix}$. The off-diagonal terms are zero because the two measurements are independent, and hence uncorrelated. Combining these two measurements can be done in two ways. The first is to apply the continuous information combination theorem. The combined distribution has mean (9.87, 14.97) and variance matrix $\begin{pmatrix} 0.075 & 0 \\ 0 & 2.28 \end{pmatrix}$. The second way to combine the two measurements is to treat them as a single combined random sample of size 200. In this case the distribution of the measurement has mean (9.93, 15.53) and variance matrix $\begin{pmatrix} 0.078 & 0 \\ 0 & 2.42 \end{pmatrix}$. The second way of combining the information is closer to the true value (10, 16), but both measurements are within one standard deviation of the true value, so they are both as accurate as one would expect. Fusing the samples will usually be more accurate because it makes use of more information about the measurements. The information combination technique uses only the distributions, not how the distributions were obtained. However, the advantage is small compared with the estimation error of either technique.

Summary

- PDs can be measured.
- Many standard statistical tests are based on measurements of PDs.
- Independent measurements of PDs can be combined, just like any other kinds of measurements.

15.5 Dempster-Shafer Theory

The analogy between the cosine similarity function for the vector space model in section 6.2 and the information combination theorem is intriguing. In both cases, one computes the product of weights that occur in the same position in a vector. In the vector space model the positions in the vector correspond to terms that can occur in a document. In the information combination theorem the positions are the possible values of the two random variables. In section 6.5 we discussed how the vector space model could be extended to deal with concept combinations. Surprisingly, a form of concept combination has been considered for information combination. This interesting development is called the Dempster-Shafer (D-S) theory of evidence (Shafer 1976).

D-S theory assumes that one has a set of entities which can be assigned probabilities, and the sum of the probabilities for all entities adds up to 1, exactly as in ordinary probability theory. The difference is that D-S theory allows entities to be combinations of other entities. Compared with the concept combinations considered in section 6.5, D-S combinations are very simple. A D-S combination is the set-theoretic union of its constituents. Thus, while there are many ways to combine the two terms "test" and "drug," D-S theory has only one way to combine them.

Consider the diagnosis problem of the patient suffering from a headache introduced at the beginning of this chapter. D-S theory, not only includes the three basic diagnoses, concussion, meningitis, and tumor, it also allows combined diagnoses such as concussion-meningitis. Note that such a combination is a separate point in the probability space, distinct from either concussion or meningitis: it does not represent an event such as (concussion OR meningitis). Try to imagine a new kind of trauma called a "concussion-meningitis" or "meningitis-concussion." Remember that there is only one way to combine entities in D-S theory, so these two must be regarded as being the same even though most interpretations would regard them as being different. The following are two examples of D-S distributions:

Diagnosis	Distribution P	Distribution Q
concussion	0.5	0.6
meningitis	0	0.2
tumor	0.3	0.2
concussion-meningitis	0.2	0

All entities in D-S theory are of three kinds: *elementary*, such as concussion and meningitis; a *combination*, such as concussion-meningitis; or the "empty entity," corresponding to the set-theoretic notion of an empty set. The empty entity plays a special role in D-S theory. We will say that a D-S distribution is *elementary* if all of its probability is on elementary entities. The distribution Q above is elementary, while distribution P is not.

The most important contribution of D-S theory is Dempster's rule of combination which specifies how to combine independent evidence. Unlike the discrete information combination theorem, Dempster's rule is postulated as an axiom; it is not proven based on some underlying theory.

Dempster's Rule of Combination

Let X and Y be two D-S distributions representing independent evidence for the same phenomenon. Define a combination distribution M by the formula $M(C) = \sum_{A \cap B = C} X(A)Y(B)$, for all nonempty C. If there exists some nonempty C such that $M(C) \neq 0$, then there exists a D-S distribution Z which combines X and Y and is defined by the formula

$$Z(C) = \frac{M(C)}{\sum_{nonempty D} M(D)},$$

for every nonempty C. In other words, Z is obtained from M by rescaling so that the probabilities add to 1.

If X and Y are elementary D-S distributions which can be combined, then the combination Z is also elementary and coincides with the distribution given by the information combination theorem. In fact, if either X or Y is elementary, then Z will also be elementary. Dempster's rule is therefore an extension of the information combination theorem. The combination of the D-S distributions P and Q defined above is given as follows:

Diagnosis	P	Q	Distribution M	Combination Z
concussion	0.5	0.6	0.42	0.81
meningitis	0	0.2	0.04	0.08
tumor	0.3	0.2	0.06	0.11
concussion-meningitis	0.2	0	0	0

The only complicated entry in the computation above is the value of M(concussion). This probability is the sum of two products: P(concussion)Q(concussion) and P(concussion-meningitis)Q(concussion). The rationale for including both of these in the combined probability for concussion is that both concussion and concussion-meningitis contribute to the evidence (or belief) in concussion because they both contain concussion.

There is some question about the role played by the empty entity. It is sometimes interpreted as representing the degree to which one is unsure about the overall observation. However, Dempster's rule of combination explicitly excludes the empty entity from any combined distribution. As a result, the only effect in D-S theory of a nonzero probability for the empty entity is to allow distributions to be unnormalized. The information combination theorems also apply to unnormalized distributions, as we noted in the discussion after the information combination theorems.

Summary

- D-S theory introduces a probabilistic form of concept combination.
- D-S distributions are combined by using Dempster's rule of combination.
- Dempster's rule of combination coincides with the discrete information combination theorem when the distributions are elementary.

16 *The Bayesian Web*

16.1 Introduction

The Semantic Web is an extension of the World Wide Web in which information is given a well-defined meaning, so that computers and people may more easily work in cooperation. This is done by introducing a formal logical layer to the web in which one can perform rigorous logical inference. However, the Semantic Web does not include a mechanism for empirical, scientific reasoning which is based on stochastic inference. Bayesian networks (BNs) are a popular mechanism for modeling uncertainty and performing stochastic inference in biomedical situations. They are a fundamental probabilistic representation mechanism that subsumes a great variety of other probabilistic modeling methods, such as hidden Markov models and stochastic dynamic systems. In this chapter we propose an extension to the Semantic Web which we call the Bayesian Web (BW) that supports BNs and that integrates stochastic inference with logical inference. Within the BW, one can perform both logical inference and stochastic inference, as well as make statistical decisions.

Although very large BNs are now being developed, each BN is constructed in isolation. Interoperability of BNs is possible only if there is a framework for one to identify common variables. The BW would make it possible to perform operations such as:

- Use a BN developed by some other group almost as easily as one now navigates from one webpage to another
- Make stochastic inference and statistical decisions using information from one source and a BN from another source

- Fuse BNs obtained from disparate sources by identifying variables that measure the same phenomenon
- Reconcile and validate BNs by checking mutual consistency

16.2 Requirements for Bayesian Network Interoperability

The most fundamental requirement of BN interoperability is to have a common interchange format. However, this alone would not be enough for one to automatically combine data and BNs from different sources. In this section we discuss the requirements for BNs to be fully interoperable in the sense discussed in the introduction.

The following are the requirements for BN interoperability and the proposed BW:

1. Interchange format. There already exists a format for representing BNs, called the XML Belief Network format (XBN) (XBN 1999). This XML file format was developed by Microsoft's Decision Theory and Adaptive Systems Group. An example is shown in section 16.4 below.
2. Common variables. It should be possible for the same variable to appear in different BNs. For example, whether a person has the flu should be the same variable no matter which BN it appears in. Being able to specify or to deduce that two entities are the same is a fundamental feature of the Semantic Web. Of course the context within which a BN is valid affects the meaning of the variable. For example, one might be interested only in the occurrence of the flu in Spain in 1918. This would be very different from the flu in Australia in 2004.
3. Annotation and reference makes it possible to specify the context of a BN. In so doing one also specifies the meaning of the variables. One should be able to refer to a BN and for a BN to refer to other information. In other words, the BN should itself be an entity about which one can make statements. Annotations are also important for authentication and trust.
4. Open hierarchy of distribution types. New probability distributions (PDs) and conditional probability distributions (CPDs) can be introduced by subclassing other distributions.
5. BN components. A BN can be constructed from known pieces. It can also be constructed by instantiating templates A BN component is a partially specified BN.

6. Meta-Analysis. Multiple BNs can be combined to form new BNs. This is a very different form of combination than component-based construction. Meta-analysis and stochastic inference are closely related. As shown in chapter 15, stochastic inference makes use of meta-analysis, and meta-analysis can be expressed in terms of stochastic inference.

16.3 Extending the Semantic Web

We now give a concrete proposal for how the Semantic Web can be augmented to include BNs and stochastic inference. The architecture for the Semantic Web consists of a series of layers, as shown in figure 16.1. This figure was taken from a presentation by Tim Berners-Lee (Berners-Lee 2000a). The layers that are relevant to the BW are the following:

1. The resource description framework (RDF) layer introduces semantics to XML. It makes it possible to link one resource to another resource such that the link and resources may be in different webpages. RDF is a minimalist semantic layer with only the most basic constructs.

2. The Web Ontology Language (OWL) layer expands on the RDF layer by adding more constructs and richer formal semantics.

3. The Logic layer adds inference. At this layer one can have both resources and links that have been inferred. However, the inference is limited by the formal semantics specified by RDF and OWL.

4. The Proof layer adds rules. Rules can take many forms such as logical rules as in the Logic layer, search rules for finding documents that match a query, and domain-specific heuristic rules.

The proposed BW consists of a collection of ontologies that formalize the notion of a BN together with stochastic inference rules. The BW resides primarily on two of the Semantic Web layers: the Web Ontology layer and the Proof layer. The BW ontologies are expressed in OWL on the Web Ontology layer, and the algorithms for the stochastic operations are located on the Proof layer. By splitting the BW into two layers, one ensures that BW information can be processed using generic Semantic Web tools which have no understanding of probability or statistics. The result of processing at the OWL layer is to obtain authenticated and syntactically consistent BNs. The probabilistic and statistical semantics is specified on the Proof layer which requires engines that understand probability and statistics.

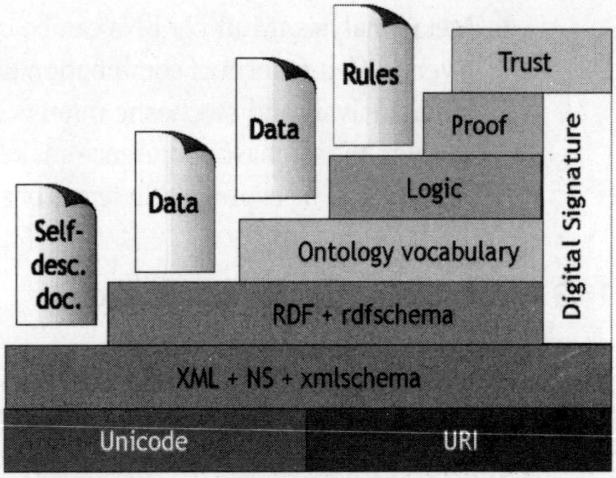

Figure 16.1 The Semantic Web architecture.

16.4 Ontologies for Bayesian Networks

The first requirement for a viable BW is to have a standard interchange format. An example of such a format is the XBN format. The XBN format is an XML DTD. To illustrate this format, the medical diagnosis BN in figure 14.1 is represented as follows:

```
<?XML VERSION="1.0">
<!DOCTYPE ANALYSISNOTEBOOK SYSTEM "xbn.dtd">
<ANALYSISNOTEBOOK
     NAME="Diagnostic Bayesian Network Example"
     ROOT="InfluenzaDiagnosis">
  <BNMODEL NAME="InfluenzaDiagnosis">
    <STATICPROPERTIES>
      <FORMAT VALUE="MSR DTAS XML"/>
      <VERSION VALUE="1.0"/>
      <CREATOR VALUE="Ken Baclawski"/>
    </STATICPROPERTIES>
    <VARIABLES>
      <VAR NAME="Flu" TYPE="discrete">
        <DESCRIPTION>Patient has influenza</DESCRIPTION>
        <STATENAME>Absent</STATENAME>
        <STATENAME>Present</STATENAME>
      </VAR>
```

```
    <VAR NAME="Cold" TYPE="discrete">
      <DESCRIPTION>Patient has mild upper respiratory
                     viral infection</DESCRIPTION>
      <STATENAME>Absent</STATENAME>
      <STATENAME>Present</STATENAME>
    </VAR>
    <VAR NAME="PerceivesFever" TYPE="discrete">
      <DESCRIPTION>
        Patient self-diagnoses a fever
      </DESCRIPTION>
      <STATENAME>Absent</STATENAME>
      <STATENAME>Present</STATENAME>
    </VAR>
    <VAR NAME="Temperature" TYPE="continuous">
      <DESCRIPTION>
        Oral measurement of the body temperature
        of the patient
      </DESCRIPTION>
    </VAR>
  </VARIABLES>
  <STRUCTURE>
    <ARC PARENT="Flu" CHILD="PerceivesFever"/>
    <ARC PARENT="Flu" CHILD="Temperature"/>
    <ARC PARENT="Cold" CHILD="PerceivesFever"/>
    <ARC PARENT="Cold" CHILD="Temperature"/>
  </STRUCTURE>
  <DISTRIBUTIONS>
    <DIST TYPE="discrete">
      <PRIVATE NAME="Flu"/>
      <DPIS>
        <DPI>0.9999 0.0001</DPI>
      </DPIS>
    </DIST>
    <DIST TYPE="discrete">
      <PRIVATE NAME="Cold"/>
      <DPIS>
        <DPI>0.99 0.01</DPI>
      </DPIS>
    </DIST>
```

```
      <DIST TYPE="discrete">
        <CONDSET>
          <CONDELEM NAME="Flu"/>
          <CONDELEM NAME="Cold"/>
        </CONDSET>
        <PRIVATE NAME="PerceivesFever"/>
        <DPIS>
          <DPI INDEXES="0 0">0.99 0.01</DPI>
          <DPI INDEXES="0 1">0.90 0.10</DPI>
          <DPI INDEXES="1 0">0.10 0.90</DPI>
          <DPI INDEXES="1 1">0.05 0.95</DPI>
        </DPIS>
      </DIST>
      <DIST TYPE="gaussian">
        <CONDSET>
          <CONDELEM NAME="Flu"/>
          <CONDELEM NAME="Cold"/>
        </CONDSET>
        <PRIVATE NAME="Temperature"/>
        <DPIS>
         <DPI INDEXES="0 0" MEAN="37" VARIANCE="0.25">
         <DPI INDEXES="0 1" MEAN="37.5" VARIANCE="1.0">
         <DPI INDEXES="1 0" MEAN="39" VARIANCE="2.25">
         <DPI INDEXES="1 1" MEAN="39.2" VARIANCE="2.56">
        </DPIS>
      </DIST>
    </DISTRIBUTIONS>
  </BNMODEL>
</ANALYSISNOTEBOOK>
```

The ANALYSISNOTEBOOK element is the root. It allows one to specify more than one BN in a single XML document. In this case there is just one BN which is specified by the BNMODEL element. The child elements of BNMODEL specify the nodes, edges, and CPDs of the BN. One can also annotate the BN with information such as who created it and which version it is. The annotations are in the STATICPROPERTIES element, the nodes are in the VARIABLES element, the edges are in the STRUCTURE element, and the CPDs are in the DISTRIBUTIONS element. Nodes are specified by VAR elements. Edges are specified by ARC elements. CPDs are specified by DIST elements.

The CPDs are the most complex elements. In general, a CPD is a list of PDs. The list is contained in a `DPIS` element. PDs are specified by `DPI` elements. If a node has no incoming edges, then its CPD is a PD and there is only a single `DPI` element. Nodes with incoming edges must specify several PDs. The published DTD for XBN does not support continuous random variables, so it was necessary to add two attributes to the `DPI` element: the `MEAN` and `VARIANCE`.

The XBN format has a number of limitations as the basis for the BW. In its current published form, it only supports random variables with a finite number of values. It does not support continuous random variables. It should be possible to specify a wide variety of types of PD. Another significant limitation is its lack of a mechanism for referring to external resources or for external documents to refer to the BN. This makes it difficult to use this mechanism to satisfy the requirement for common variables, and there is only limited support for annotation.

These considerations suggest that a better choice of language for the BW is OWL. We now present a series of three OWL ontologies that satisfy the requirements for the BW. We present them in top-down fashion, starting with high-level concepts and successively elaborating them:

1. The ontology of phenomena which can be modeled using BNs

2. The ontology of networks of CPDs

3. The ontology of elementary PDs

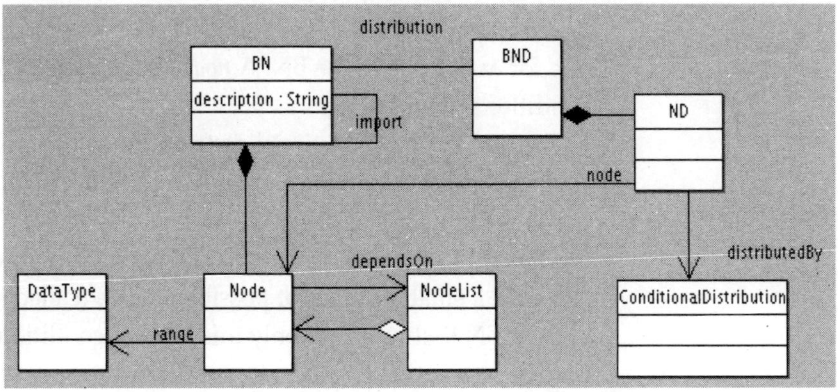

Figure 16.2 Ontology for Bayesian networks.

While one would think that the notion of a random variable is unambiguous, in fact it is a combination of two different concepts. First, there is the phenomenon that is being observed or measured, such as one toss of a coin or the measurement of a person's blood pressure. The second concept is the PD of the phenomenon. It is the combination of these two notions which is the concept of a random variable. The relationship between the phenomenon and its PD is many-to-many. Many phenomena have the same PD, and the same phenomenon can be distributed in many ways. The reason why a phenomenon does not uniquely determine its PD is due to the notion of *conditioning*. As one observes related events, the distribution of a phenomenon changes. The phenomenon is the same; what changes is the knowledge about it.

The top-level concept of the BW is the BN which is used to model networks of more elementary phenomena (see figure 16.2). A BN consists of a collection of *nodes*, each of which represents one elementary phenomenon. Think of a node as a random variable whose PD has not yet been specified. A node has a range of values. For example, the height of a person is a positive real number. A `Node` can *depend on* other `Nodes`. A dependency is called a *dependency arc*. It is convenient to order the dependencies of a single node, so in figure 16.2, a `Node` can depend on a `NodeList`, which consists of a sequence of `Nodes`. The order of the dependencies is used when the conditional probabilities are specified. A BN can *import* another BN. The nodes and dependencies of an imported BN become part of the importing BN.

The most complex part of a BN is its joint probability distribution (JPD) which is specified using a collection of conditional and unconditional PDs. Since a BN can have more than one PD, the notion of a *BN distribution* (BND) is separated from that of the BN. There is a one-to-many relationship between the concepts of BN and BND. A BND consists of a collection of distributions, one for each node in the BN. A *node distribution* (ND) relates one node to its conditional distribution.

The notion of a conditional distribution is the main concept in the conditional probability ontology, as shown in figure 16.3. A conditional distribution has three special cases. It can be a *CPD table* (CPT), a *general stochastic function* (SF), or an unconditional PD. The CPT is used in the case of phenomena with a small number of possible values (called *states* in this case). Most current BN tools support only this kind of conditional probability specification.

A CPT is defined recursively, with one level for each dependency. There is one *conditional probability entry* (CPE) for each value of the first parent node.

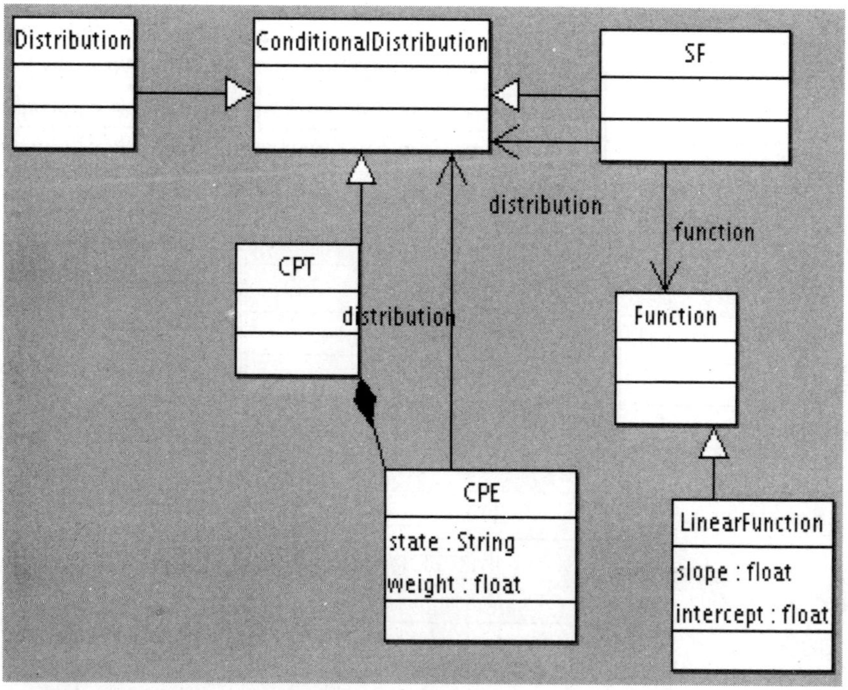

Figure 16.3 Ontology for conditional probability distributions.

Each CPE specifies a weight and a CPT for the remaining parent nodes. Weights are nonnegative real numbers. They need not be normalized. At the last level one uses an unconditional PD.

A SF is also defined recursively, but instead of using an explicit collection of CPEs, it uses one or more functions that specify the parameter(s) of the remaining distributions. The most common function is a linear function, and it is the only one shown in the diagram. Functions are necessary to specify dependencies on continuous phenomena. More general functions can be specified by using the Mathematical Markup Language (MathML) (W3C 2003).

PDs are classified in the PD ontology shown in figure 16.4. This ontology is a hierarchy of the most commonly used PDs. The main classification is between discrete and continuous distributions. Discrete distributions may either be defined by a formula (as in the Poisson and binomial distributions) or explicitly for each value (state). Every continuous distribution can be altered by changing its *scale* or by *translating* it (or both). The most commonly used continuous distributions are the uniform and normal (Gaussian) distributions. The uniform distribution is on the unit interval and the normal has mean 0 and variance 1. Other uniform and normal distributions can be obtained by scaling and translating the standard ones. Other commonly used distributions are the exponential and chi-square distributions as well as Gosset's *t* distribution, and Fisher's *F* distribution.

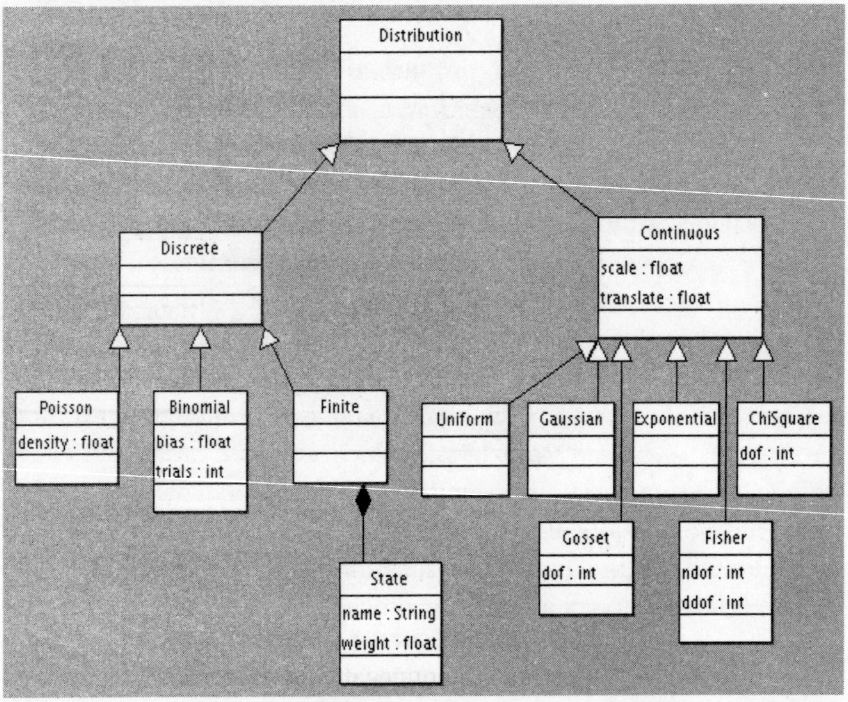

Figure 16.4 Ontology for probability distributions.

17 *Answers to Selected Exercises*

ANSWER TO EXERCISE 1.1

```
<bio_sequence element_id="U83302" sequence_id="MICR83302"
  organism_name="Colaptes rupicola" seq_length="1047" type="DNA"/>
<bio_sequence element_id="U83303" sequence_id="HSU83303"
  organism_name="Homo sapiens" seq_length="3460" type="DNA"/>
<bio_sequence element_id="U83304" sequence_id="MMU83304"
  organism_name="Mus musculus" seq_length="51" type="RNA"/>
<bio_sequence element_id="U83305" sequence_id="MIASSU833"
  organism_name="Accipiter striatus" seq_length="1143" type="DNA"/>
```

ANSWER TO EXERCISE 1.2

```
<!ATTLIST bio_sequence
    element_id    ID        #IMPLIED
    sequence_id   CDATA     #IMPLIED
    organism_name CDATA     #IMPLIED
    seq_length    CDATA     #IMPLIED
    molecule_type (DNA | mRNA | rRNA | tRNA | cDNA | AA)
                            #IMPLIED>
```

This example was taken from the AGAVE DTD (AGAVE 2002). The actual element has some additional attributes, and it differs in a few other ways as well. For example, some of the attributes are restricted to NMTOKEN rather than just CDATA. NMTOKEN specifies text that starts with a letter (and a few other characters, such as an underscore), and is followed by letters and digits. Programming languages such as Perl restrict the names of variables and procedures in this way, and many genomics databases use this same convention for their accession numbers and other identifiers.

ANSWER TO EXERCISE 1.3

```
<physical_unit name="millisecond">
  <factor prefix="milli" unit="second"/>
</physical_unit>
<physical_unit name="per_millisecond">
  <factor prefix="milli" unit="second" exponent="-1"/>
</physical_unit>
<physical_unit name="millivolt">
  <factor prefix="milli" unit="volt"/>
</physical_unit>
<physical_unit name="microA_per_mm2">
  <factor prefix="micro" unit="ampere"/>
  <factor prefix="milli" unit="mitre" exponent="-2"/>
</physical_unit>
<physical_unit name="microF_per_mm2">
  <factor prefix="micro" unit="farad"/>
  <factor prefix="milli" unit="mitre" exponent="-2"/>
</physical_unit>
```

The XML DTD looks like this:

```
<!ELEMENT physical_unit (factor)*>
<!ATTLIST physical_unit name ID #REQUIRED>
<!ELEMENT factor EMPTY>
<!ATTLIST factor
    prefix   CDATA #IMPLIED
    unit     CDATA #REQUIRED
    exponent CDATA "1">
```

ANSWER TO EXERCISE 1.4

```
<component name="membrane">
  <variable name="u" interface="out"/>
  <variable name="Vr" interface="out" initial="-85.0"
            physical_unit="millivolt"/>
  <variable name="Cm" initial="0.01"
            physical_unit="microF_per_mm2"/>
  <variable name="time" interface="in"
            physical_unit="millisecond"/>
```

```
  </component>
  <component name="ionic_current">
    <variable name="I_ion" interface="out"
              physical_unit="microA_per_mm2"/>
    <variable name="v" interface="in"/>
    <variable name="Vth" interface="in"
              physical_unit="millivolt"/>
  </component>
```

The DTD is the following:

```
<!ELEMENT component (variable)*>
<!ATTLIST component name ID #REQUIRED>
<!ELEMENT variable EMPTY>
<!ATTLIST variable
    name           CDATA     #REQUIRED
    initial        CDATA     #IMPLIED
    physical_unit  IDREF     "dimensionless"
    interface      (in|out)  #IMPLIED>
```

IDREF means that the attribute refers to another one elsewhere in the document. In this case it is referring to a physical unit definition in exercise 1.3.

ANSWER TO EXERCISE 2.1

The XML schema can be obtained by translating the molecule DTD in figure 1.6 using dtd2xsd.pl (W3C 2001a). The answer is the following:

```
<schema
  xmlns='http://www.w3.org/2000/10/XMLSchema'
  targetNamespace='http://www.w3.org/namespace/'
  xmlns:t='http://www.w3.org/namespace/'>

 <element name='molecule'>
  <complexType>
   <sequence>
    <element ref='t:atomArray'/>
    <element ref='t:bondArray'/>
   </sequence>
   <attribute name='title' type='string' use='optional'/>
   <attribute name='id' type='string' use='optional'/>
   <attribute name='convention' type='string' use='default' value='CML'/>
   <attribute name='dictRef' type='string' use='optional'/>
   <attribute name='count' type='string' use='default' value='1'/>
  </complexType>
 </element>
```

```
<element name='atomArray'>
 <complexType>
  <sequence>
   <element ref='t:atom' maxOccurs='unbounded'/>
  </sequence>
  <attribute name='title' type='string' use='optional'/>
  <attribute name='id' type='string' use='optional'/>
  <attribute name='convention' type='string' use='default' value='CML'/>
 </complexType>
</element>

<element name='atom'>
 <complexType>
  <attribute name='elementType' type='string' use='optional'/>
  <attribute name='title' type='string' use='optional'/>
  <attribute name='id' type='string' use='optional'/>
  <attribute name='convention' type='string' use='default' value='CML'/>
  <attribute name='dictRef' type='string' use='optional'/>
  <attribute name='count' type='string' use='default' value='1'/>
 </complexType>
</element>

<element name='bondArray'>
 <complexType>
  <sequence>
   <element ref='t:bond' maxOccurs='unbounded'/>
  </sequence>
  <attribute name='title' type='string' use='optional'/>
  <attribute name='id' type='string' use='optional'/>
  <attribute name='convention' type='string' use='default' value='CML'/>
 </complexType>
</element>

<element name='bond'>
 <complexType>
  <attribute name='title' type='string' use='optional'/>
  <attribute name='id' type='string' use='optional'/>
  <attribute name='convention' type='string' use='default' value='CML'/>
  <attribute name='dictRef' type='string' use='optional'/>
  <attribute name='atomRefs' type='string' use='optional'/>
 </complexType>
</element>
</schema>
```

ANSWER TO EXERCISE 2.2

Change the line

```
<attribute name='elementType' type='string' use='optional'/>
```

in the molecule schema to

```
<attribute name='elementType'
  type='elementTypeType' use='optional'/>
```

where `elementTypeType` is defined by

```
  <xsd:simpleType name="elementTypeType">
    <xsd:restriction base="xsd:string">
      <xsd:enumeration value="Ac"/>
      <xsd:enumeration value="Al"/>
      <xsd:enumeration value="Ag"/>
        ...
      <xsd:enumeration value="Zn"/>
      <xsd:enumeration value="Zr"/>
    </xsd:restriction>
  </xsd:simpleType>
```

ANSWER TO EXERCISE 2.3

One possible answer uses an enumeration of cases:

```
  <xsd:simpleType name="DNABase">
    <xsd:restriction base="xsd:string">
      <xsd:enumeration value="A"/>
      <xsd:enumeration value="C"/>
      <xsd:enumeration value="G"/>
      <xsd:enumeration value="T"/>
    </xsd:restriction>
  </xsd:simpleType>
```

Another answer uses a pattern restriction:

```
  <xsd:simpleType name="DNAbase">
    <xsd:restriction base="xsd:string">
      <xsd:pattern value="[ACGT]"/>
    </xsd:restriction>
  </xsd:simpleType>
```

ANSWER TO EXERCISE 2.4

A DNA sequence could be defined using a list of bases as in

```
<simpleType name="DNASequence">
  <list itemType="DNABase"/>
</simpleType>
```

For example, the TATA sequence would be written as `T A T A`. The items in an XML list are separated by spaces. A better answer would be

```
<simpleType name="DNASequence" base="xsd:string">
  <restriction>
    <pattern value="[ACGT]+"/>
  </restriction>
</simpleType>
```

Using this definition, the TATA sequence would be written without spaces as `TATA`, just as one would expect.

ANSWER TO EXERCISE 4.1

In the following answer, it was presumed that the concepts of `atomArray` and `bondArray` were artifacts of the design of the XML DTD and schema and were not fundamental to the meaning of a molecule. Other assumptions would lead to many other designs.

```
<Class rdf:ID="Molecule"/>
<Class rdf:ID="Atom"/>
<Class rdf:ID="Bond"/>
<Property rdf:ID="atom">
  <domain rdf:resource="#Molecule"/>
  <range rdf:resource="#Atom"/>
</Property>
<Property rdf:ID="bond">
  <domain rdf:resource="#Molecule"/>
  <range rdf:resource="#Bond"/>
</Property>
<Property rdf:ID="title"/>
<Property rdf:ID="convention"/>
<Property rdf:ID="dictRef"/>
<Property rdf:ID="count">
  <range rdf:resource=
    "http://www.w3.org/2000/10/XMLSchema#positiveInteger"/>
</Property>
<Property rdf:ID="elementType">
```

```
    <domain rdf:resource="#Atom"/>
    <range rdf:resource=
      "http://ontobio.org/molecule.xsd#elementTypeType"/>
  </Property>
  <Property rdf:ID="atomRef">
    <domain rdf:resource="#Bond"/>
    <range rdf:resource="#Atom"/>
  </Property>
```

ANSWER TO EXERCISE 4.2

Using the ontology in the sample answer above, nitrous oxide would be the following:

```
<Molecule rdf:ID="m1" title="nitrous oxide">
  <atom>
    <Atom rdf:ID="n1" elementType="N"/>
    <Atom rdf:ID="o1" elementType="O"/>
  </atom>
  <bond>
    <Bond>
      <atomRef rdf:resource="n1"/>
      <atomRef rdf:resource="o1"/>
    </Bond>
  </bond>
</Molecule>
```

ANSWER TO EXERCISE 4.3

The bio sequence attributes can be interpreted as properties as follows:

1. `element_id` This is the identifier of the bio sequence, so it should be the URI of the resource. So it corresponds to the `rdf:ID` property.

2. `sequence_id` This is another identifier, but it was not declared to be of type `ID` so it is probably a reference to a URI in some other location. Accordingly, it is interpreted as a functional property.

3. `organism_name` This is a name so it will be interpreted as a string-valued property.

4. `seq_length` This is a length so it will be interpreted as a numerical property.

5. `molecule_type` This has six possible values so it is necessary to introduce an enumerated class for the values of this property.

The OWL ontology looks like the following:

```
<owl:Class rdf:ID="bio_sequence"/>
<owl:ObjectProperty rdf:ID="sequence_id">
  <rdfs:domain rdf:about="#bio_sequence"/>
</owl:ObjectProperty>
<owl:DatatypeProperty rdf:ID="organism_name">
  <rdfs:domain rdf:about="#bio_sequence"/>
  <rdfs:range rdf:about=
    "http://www.w3.org/2000/10/XMLSchema#string"/>
</owl:DatatypeProperty>
<owl:DatatypeProperty rdf:ID="organism_name">
  <rdfs:domain rdf:about="#bio_sequence"/>
  <rdfs:range rdf:about=
    "http://www.w3.org/2000/10/XMLSchema#string"/>
</owl:DatatypeProperty>
<owl:DatatypeProperty rdf:ID="seq_length">
  <rdfs:domain rdf:about="#bio_sequence"/>
  <rdfs:range rdf:about=
    "http://www.w3.org/2000/10/XMLSchema#nonNegativeInteger"/>
</owl:DatatypeProperty>
<owl:ObjectProperty rdf:ID="molecule_type">
  <rdfs:domain rdf:about="#bio_sequence"/>
  <rdfs:range rdf:about="#MoleculeTypes"/>
</owl:ObjectProperty>
<owl:Class rdf:ID="MoleculeTypes">
  <owl:oneOf parseType="Collection">
    <owl:MoleculeTypes rdf:ID="DNA"/>
    <owl:MoleculeTypes rdf:ID="mDNA"/>
    <owl:MoleculeTypes rdf:ID="rDNA"/>
    <owl:MoleculeTypes rdf:ID="tDNA"/>
    <owl:MoleculeTypes rdf:ID="cDNA"/>
    <owl:MoleculeTypes rdf:ID="AA"/>
  </owl:oneOf>
</owl:Class>
```

ANSWER TO EXERCISE 8.1

To solve this exercise extract all interview elements that have attributes with the desired characteristics. Dates always start with the year so the following query gives the desired results:

```
document("healthstudy.xml")
//Interview[starts-with(@Date,"2000") and @BMI>30]
```

Alternatively, one can use the XQuery function that extracts the year from a date as in the following query:

```
document("healthstudy.xml")
//Interview[year-from-dateTime(@Date)=2000 and @BMI>30]
```

ANSWER TO EXERCISE 8.2

First find the insulin gene locus. Then within this locus find all literature entries. An entry is a literature reference if the reference element containing the entry is named "Literature references." Note the use of ".." to obtain the name attribute of the parent element of the entry.

```
for $locus in document("bio.xml")//locus
where $locus/gene/@name = "Insulin gene"
return (for $entry in $locus/reference/db_entry
        where $entry/../@name = "Literature references"
        return $entry)
```

ANSWER TO EXERCISE 8.3

Look for all citations that have a MeSH heading satisfying the criteria. The following query looks at all citations, and then within each citation it looks at every heading. Whenever a heading satisfies the criteria, the citation is returned.

```
for $citation in
       document("pubmed.xml")//MedlineCitation,
    $heading in
       $citation//MeshHeading
where $heading/DescriptorName/@MajorTopicYN="Y"
and $heading/DescriptorName="Glutethimide"
and $heading/QualifierName="therapeutic use"
return $citation
```

In the query above, if a citation has more than one MeSH heading that satisfies the criteria, then the citation will be returned more than once. One can avoid this problem by using a "nested" subquery as in the following query. For each citation, this query runs a separate subsidiary query that finds all headings within the citation that satisfy the criteria. If the nested subquery has one or more results, then the citation is returned.

```
for $citation in document("pubmed.xml")//MedlineCitation
where exists
 (for $heading in $citation//MeshHeading
  where $heading/DescriptorName/@MajorTopicYN="Y"
  and $heading/DescriptorName="Glutethimide"
  and $heading/QualifierName="therapeutic use"
  return $heading)
return $citation
```

ANSWER TO EXERCISE 8.5

To solve this exercise, extract all pairs of interviews for the same subject such that the second interview is more than 2 years later, and the BMI attribute is more than 4.5 larger.

```
for $i in document("healthstudy.xml")//Interview,
    $j in document("healthstudy.xml")//Interview
where $i/@SID = $j/@SID
and $j/@BMI - $i/@BMI > 4.5
and $j/@Date - $i/@Date < "P2Y"
return $i/@SID
```

This query can return the same subject identifier more than once when a subject satisfies the criteria multiple times.

ANSWER TO EXERCISE 8.6

The number of associations is specified by the `n_associations` attribute of the `go:term` element. The term number is specified by the `go:accession` element. The GO namespace must be declared prior to using it in the query.

```
declare namespace
  go="http://www.geneontology.org/dtds/go.dtd#";
document("go.xml")
  //go:term[go:accession="GO:0003673"]/@n_associations
```

Unfortunately, the GO is a very large ontology. When the associations are included, the GO is a 500 megabyte file. A query such as the one above would have to read the entire file, which can take a long time. More efficient techniques for querying such a large ontology were discussed in section 6.3.

ANSWER TO EXERCISE 11.1

Here is one possible solution:

```
<xsl:template match="gene">
  <xsl:copy>
    <xsl:attribute name="locus">
      <xsl:value-of select="../@name"/>
    </xsl:attribute>
    <xsl:apply-templates select="@*|node()"/>
  </xsl:copy>
</xsl:template>
```

This template applies only to gene elements. All other elements are copied exactly. For gene elements, the element itself is copied, then a new attribute is added named locus, having a value equal to the name attribute of its parent element.

ANSWER TO EXERCISE 11.2

```
<xsl:template match="locus">
  <xsl:apply-templates select="gene"/>
</xsl:template>

<xsl:template match="gene">
  <xsl:copy>
    <xsl:apply-templates select="@*|node()"/>
    <xsl:apply-templates select="../reference"/>
  </xsl:copy>
</xsl:template>
```

The first template removes the locus element, along with all of its child elements, except for the gene element. The second template copies all gene elements and adds the reference elements that were removed by the first template.

ANSWER TO EXERCISE 11.3

```
<xsl:template match="organism">
  <organism>
    <xsl:apply-templates select="@*"/>
    <contains>
      <xsl:apply-templates select="node()"/>
    </contains>
  </organism>
</xsl:template>
```

Note that the attributes of `organism` remain in the same element, but the child elements of `organism` are made child elements of the new `contains` element.

ANSWER TO EXERCISE 11.4

```
<xsl:template match="reference">
  <xsl:choose>
    <xsl:when test="@name='Sequence databases'">
      <isStoredIn>
        <xsl:apply-templates select="@*|node()"/>
      </isStoredIn>
    </xsl:when>
    <xsl:when test="@name='Literature references'">
      <isCitedBy>
        <xsl:apply-templates select="@*|node()"/>
      </isCitedBy>
    </xsl:when>
  </xsl:choose>
</xsl:template>
```

ANSWER TO EXERCISE 11.5

```
<xsl:template match="gene">
  <gene>
    <xsl:attribute name="embl">
      <xsl:value-of select=
        "../reference[@name='Sequence databases']
          /db_entry[@name='EMBL sequence']/@entry"/>
    </xsl:attribute>
    <xsl:attribute name="organism">
      <xsl:value-of select="../../../../organism/@name"/>
    </xsl:attribute>
    <xsl:apply-templates select="@*|node()"/>
  </gene>
</xsl:template>
```

ANSWER TO EXERCISE 11.6

```
<xsl:template match="gene">
  <gene>
    <xsl:attribute name="totalExonLength">
      <xsl:value-of
```

```
  select="sum(exon/@end)-sum(exon/@start)+count(exon)"/>
    </xsl:attribute>
    <xsl:apply-templates select="@*|node()"/>
  </gene>
</xsl:template>
```

ANSWER TO EXERCISE 12.1

Why	Assist research
What	High-level view
Who	Researchers
When	A few weeks
How	Help understanding

ANSWER TO EXERCISE 12.2

Consistency checking uses a software tool. This is analogous to the data-mining tool in the diagram. So it should be modeled as an actor. The consistency-checking tool can check the ontology for consistency, and it can also check that the chart database is consistent with the chart ontology. The modified diagram is in figure 17.1

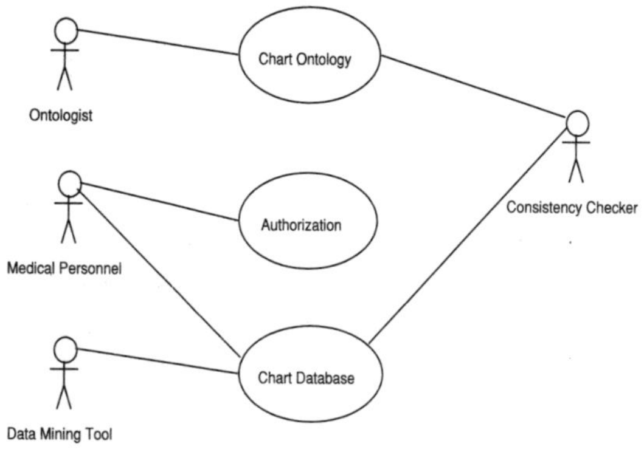

Figure 17.1 Modified use case diagram for the medical chart ontology. The diagram now includes the consistency checking tool.

ANSWER TO EXERCISE 12.3

For this project, one would like to make use of more advanced modeling constructs. This is a good argument in favor of the OWL languages. Incompatibility with the other major web-based ontology language groups is unimportant because the ontology will be used for research purposes by a relatively small group of individuals for a short time. Since the purpose of the project is to improve the understanding of concepts in the domain, it will emphasize concept definitions. Description logic is ideally suited for defining concepts. This suggests that one should use either OWL Lite or OWL-DL. Since one would expect that the concepts will be relatively complex, it is likely that OWL Lite will not be adequate. The most appropriate language for this project is therefore OWL-DL.

ANSWER TO EXERCISE 12.4

It is probably not appropriate to reuse SNPdb. The SNP ontology to be developed is concerned with high-level information about SNPs, not low-level information about the experimental procedures that are needed for finding SNPs. Importing all of the SNPdb would incorporate much more than is required. However, it is certainly useful to look at the SNPdb ontology for design possibilities.

ANSWER TO EXERCISE 12.5

One possible way to classify SNPs is according to whether they affect genes or are between genes. The SNPs that have a genetic effect are further subclassified according to whether they are part of an exon, intron, or the regulatory region of the gene. The class diagram is shown in figure 17.2. Of course, there are many other ways that one could classify SNPs.

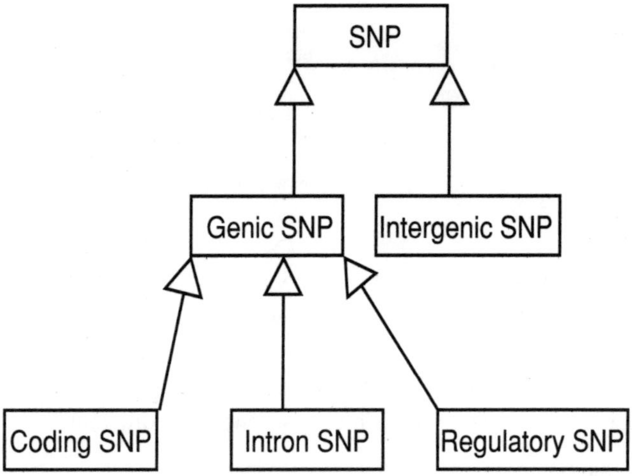

Figure 17.2 Concept hierarchy for the SNP ontology.

References

AGAVE, 2002. The Architecture for Genomic Annotation, Visualization and Exchange. `www.animorphics.net/lifesci.html`.

Al-Shahrour, F., R. Diaz-Uriarte, and J. Dopazo. 2004. FatiGO: a web tool for finding significant associations of Gene Ontology terms with groups of genes. *Bioinformatics* 20:578–580.

Altschul, S.F. 1991. Amino acid substitution matrices from an information theoretic perspective. *J. Mol. Biol.* 219:555–565.

Altschul, S.F., and W. Gish. 1996. Local alignment statistics. *Methods Enzymol.* 266: 460–480.

Altschul, S.F., W. Gish, W. Miller, E.W. Myers, and D.J. Lipman. 1990. Basic local alignment search tool. *J. Mol. Biol.* 215:403–410.

Altschul, S.F., T.L. Madden, A.A. Schaffer, J. Zhang, Z. Zhang, W. Miller, and D.J. Lipman. 1997. Gapped BLAST and PSI-BLAST: a new generation of protein database search programs. *Nucleic Acids Res.* 25:3389–3402.

Andreeva, A., D. Howorth, S.E. Brenner, T.J. Hubbard, C. Chothia, and A.G. Murzin. 2004. SCOP database in 2004: refinements integrate structure and sequence family data. *Nucleic Acids Res.* 32:D226–D229. Database issue.

Aronson, A.R. 2001. Effective mapping of biomedical text to the UMLS Metathesaurus: the MetaMap program. In *Proc. AMIA Symp.*, pp. 17–21.

Asimov, I. 1964. *A Short History of Biology*. London: Thomas Nelson & Sons.

Attwood, T.K., P. Bradley, D.R. Flower, A. Gaulton, N. Maudling, A.L. Mitchell, G. Moulton, A. Nordle, K. Paine, P. Taylor, A. Uddin, and C. Zygouri. 2003. PRINTS and its automatic supplement, prePRINTS. *Nucleic Acids Res.* 31:400–402.

Attwood, T.K., D.R. Flower, A.P. Lewis, J.E. Mabey, S.R. Morgan, P. Scordis, J.N. Selley, and W. Wright. 1999. PRINTS prepares for the new millennium. *Nucleic Acids Res.* 27:220–225.

Baclawski, K., 1997a. Distributed computer database system and method. United States Patent No. 5,694,593. Assigned to Northeastern University, Boston.

Baclawski, K. 1997b. Long time, no see: categorization in information science. In S. Hecker and G.C. Rota (eds.), *Essays on the Future. In Honor of the 80th Birthday of Nick Metropolis*, pp. 11–26. Cambridge, MA: Birkhauser.

Baclawski, K. 2003. Ontology development. Kenote address in *International Workshop on Software Methodologies, Tools and Techniques*, pp. 3–26.

Baclawski, K., J. Cigna, M.M. Kokar, P. Mager, and B. Indurkhya. 2000. Knowledge representation and indexing using the Unified Medical Language System. In *Pacific Symposium on Biocomputing*, vol. 5, pp. 490–501.

Baclawski, K., M. Kokar, P. Kogut, L. Hart, J. Smith, W. Holmes, J. Letkowski, and M. Aronson. 2001. Extending UML to support ontology engineering for the Semantic Web. In M. Gogolla and C. Kobryn (eds.), *Fourth International Conference on the Unified Modeling Language*, vol. 2185, pp. 342–360. Berlin: Springer-Verlag.

Baclawski, K., C. Matheus, M. Kokar, and J. Letkowski. 2004. Toward a symptom ontology for Semantic Web applications. In *ISWC'04*, vol. 3298, pp. 650–667. Springer-Verlag, Berlin.

Bader, G.D., D. Betel, and C.W. Hogue. 2003. BIND: the biomolecular interaction network database. *Nucleic Acids Res.* 31:248–250.

Bairoch, A. 1991. PROSITE: a dictionary of sites and patterns in proteins. *Nucleic Acids Res.* 19:2241–2245.

Baker, P.G., A. Brass, S. Bechhofer, C. Goble, N. Paton, and R. Stevens. 1998. TAMBIS–transparent access to multiple bioinformatics information sources. In *Proc. Int. Conf. Intell. Syst. Mol. Biol.*, vol. 6, pp. 25–34.

Baker, P.G., C.A. Goble, S. Bechhofer, N.W. Paton, R. Stevens, and A. Brass. 1999. An ontology for bioinformatics applications. *Bioinformatics* 15:510–520.

Bateman, A., L. Coin, R. Durbin, R.D. Finn, V. Hollich, S. Griffiths-Jones, A. Khanna, M. Marshall, S. Moxon, E.L. Sonnhammer, D.J. Studholme, C. Yeats, and S.R. Eddy. 2004. The Pfam protein families database. *Nucleic Acids Res.* 32:D138–D141. Database issue.

Benson, D.A., I. Karsch-Mizrachi, D.J. Lipman, J. Ostell, and D.L. Wheeler. 2004. GenBank: update. *Nucleic Acids Res.* 32:D23–D26. Database issue.

Bergamaschi, S., S. Castano, and M. Vincini. 1999. Semantic integration of semistructured and structured data sources. *SIGMOD Rec.* 28:54–59.

Bergman, C.M., B.D. Pfeiffer, D. Rincon-Limas, R.A. Hoskins, A. Gnirke, C.J. Mungall, A.M. Wang, B. Kronmiller, J. Pacleb, S. Park, M. Stapleton, K. Wan, R.A. George, P.J. de Jong, J. Botas, G.M. Rubin, and S.E. Celniker. 2002. Assessing the impact of comparative genomic sequence data on the functional annotation of the *Drosophila* genome. *Genome Biol.* 3:RESEARCH0086.

Berman, H.M., T.N. Bhat, P.E. Bourne, Z. Feng, G. Gilliland, H. Weissig, and J. Westbrook. 2000. The Protein Data Bank and the challenge of structural genomics. *Nat. Struct. Biol.* 7:957–959.

Berman, H.M., W.K. Olson, D.L. Beveridge, J. Westbrook, A. Gelbin, T. Demeny, S.H. Hsieh, A.R. Srinivasan, and B. Schneider. 1992. The nucleic acid database. A comprehensive relational database of three-dimensional structures of nucleic acids. *Biophys. J.* 63:751–759.

Berners-Lee, T., 2000a. Semantic Web - XML2000. `www.w3.org/2000/Talks/1206-xml2k-tbl`.

Berners-Lee, T., 2000b. Why RDF model is different from the XML model. `www.w3.org/DesignIssues/RDF-XML.html`.

BioML, 2003. Biopolymer Markup Language website. `www.rdcormia.com/COIN78/files/XML_Finals/BIOML/Pages/BIOML.htm`.

Birney, E., T.D. Andrews, P. Bevan, M. Caccamo, Y. Chen, L. Clarke, G. Coates, J. Cuff, V. Curwen, T. Cutts, T. Down, E. Eyras, X.M. Fernandez-Suarez, P. Gane, B. Gibbins, J. Gilbert, M. Hammond, H.R. Hotz, V. Iyer, K. Jekosch, A. Kahari, A. Kasprzyk, D. Keefe, S. Keenan, H. Lehvaslaiho, G. McVicker, C. Melsopp, P. Meidl, E. Mongin, R. Pettett, S. Potter, G. Proctor, M. Rae, S. Searle, G. Slater, D. Smedley, J. Smith, W. Spooner, A. Stabenau, J. Stalker, R. Storey, A. Ureta-Vidal, K.C. Woodwark, G. Cameron, R. Durbin, A. Cox, T. Hubbard, and M. Clamp. 2004. An overview of Ensembl. *Genome Res.* 14:925–928.

Bodenreider, O. 2004. The Unified Medical Language System (UMLS): integrating biomedical terminology. *Nucleic Acids Res.* 32:D267–D270. Database issue.

Bodenreider, O., S.J. Nelson, W.T. Hole, and H.F. Chang. 1998. Beyond synonymy: exploiting the UMLS semantics in mapping vocabularies. In *Proc. AMIA Symp.*, pp. 815–819.

Brachman, R., and J. Schmolze. 1985. An overview of the KL-ONE knowledge representation system. *Cognitive Sci* 9:171–216.

Brazma, A., P. Hingamp, J. Quackenbush, G. Sherlock, P. Spellman, C. Stoeckert, J. Aach, W. Ansorge, C.A. Ball, H.C. Causton, T. Gaasterland, P. Glenisson, F.C. Holstege, I.F. Kim, V. Markowitz, J.C. Matese, H. Parkinson, A. Robinson, U. Sarkans, S. Schulze-Kremer, J. Stewart, R. Taylor, J. Vilo, and M. Vingron. 2001. Minimum information about a microarray experiment (MIAME)–toward standards for microarray data. *Nat. Genet.* 29:365–371.

Buck, L. 2000. The molecular architecture of odor and pheromone sensing in mammals. *Cell* 100:611–618.

Buck, L., and R. Axel. 1991. A novel multigene family may encode odorant receptors: a molecular basis for odor recognition. *Cell* 65:175–187.

Bunge, M. 1977. *Treatise on Basic Philosophy. III: Ontology: The Furniture of the World.* Dordrecht, Netherlands: Reidel.

Bunge, M. 1979. *Treatise on Basic Philosophy. IV: Ontology: A World of Systems.* Dordrecht, Netherlands: Reidel.

Camon, E., M. Magrane, D. Barrell, D. Binns, W. Fleischmann, P. Kersey, N. Mulder, T. Oinn, J. Maslen, A. Cox, and R. Apweiler. 2003. The Gene Ontology Annotation (GOA) project: implementation of GO in SWISS-PROT, TrEMBL, and InterPro. *Genome Res.* 13:662–672.

Celis, J.E., M. Ostergaard, N.A. Jensen, I. Gromova, H.H. Rasmussen, and P. Gromov. 1998. Human and mouse proteomic databases: novel resources in the protein universe. *FEBS Lett.* 430:64–72.

CellML, 2003. CellML website. `www.cellml.org`.

Chakrabarti, S., B. Dom, D. Gibson, J. Kleinberg, P. Raghavan, and S. Rajagopalan. 1998. Automatic resource list compilation by analyzing hyperlink structure and associated text. In *Proc.* 7th *Int. World Wide Web Conf.*

Chen, R.O., R. Felciano, and R.B. Altman. 1997. RIBOWEB: linking structural computations to a knowledge base of published experimental data. In *Proc. Int. Conf. Intell. Syst. Mol. Biol.*, vol. 5, pp. 84–87.

Cheng, J., S. Sun, A. Tracy, E. Hubbell, J. Morris, V. Valmeekam, A. Kimbrough, M.S. Cline, G. Liu, R. Shigeta, D. Kulp, and M.A. Siani-Rose. 2004. NetAffx Gene Ontology Mining Tool: a visual approach for microarray data analysis. *Bioinformatics* 20:1462–1463.

Cleverdon, C., and E. Keen. 1966. Factors determining the performance of indexing systems. Vol. 1: Design, Vol. 2: Results. Technical report, Aslib Cranfield Research Project, Cranfield, UK.

Clocksin, W., C. Mellish, and W. Clocksin. 2003. *Programming in PROLOG*. New York: Springer-Verlag.

CML, 2003. Chemical Markup Language website. `www.xml-cml.org`.

Conde, L., J.M. Vaquerizas, J. Santoyo, F. Al-Shahrour, S. Ruiz-Llorente, M. Robledo, and J. Dopazo. 2004. PupaSNP Finder: a web tool for finding SNPs with putative effect at transcriptional level. *Nucleic Acids Res.* 32:W242–W248. Web server issue.

Cooper, D.N. 1999. *Human Gene Evolution*. San Diego: Academic Press.

Crasto, C., L. Marenco, P. Miller, and G. Shepherd. 2002. Olfactory Receptor Database: a metadata-driven automated population from sources of gene and protein sequences. *Nucleic Acids Res.* 30:354–360.

Dayhoff, M.O., R.M. Schwartz, and B.C. Orcutt. 1978. A model of evolutionary change in proteins. In M.O. Dayhoff (ed.), *Atlas of Protein Sequence and Structure*, vol. 5, pp. 345–352. Washington, DC: National Biomedical Research Foundation.

De Finetti, B. 1937. La prévision: ses lois logiques, ses sources subjectives. *Ann. Inst. Henri Poincaré* 7:1–68.

Decker, S., D. Brickley, J. Saarela, and J. Angele. 1998. A query and inference service for RDF. In *QL'98 - The Query Language Workshop*.

Dennis, G., Jr., B.T. Sherman, D.A. Hosack, J. Yang, W. Gao, H.C. Lane, and R.A. Lempicki. 2003. DAVID: Database for annotation, visualization, and integrated discovery. *Genome Biol.* 4:P3.

Denny, J.C., J.D. Smithers, and R.A. Miller. 2003. "Understanding" medical school curriculum content using KnowledgeMap. *J. Am. Med. Inf. Assoc.* 10:351–362.

Denny, M., 2002a. Ontology building: a survey of editing tools. `www.xml.com/pub/a/2002/11/06/ontologies.html`.

Denny, M., 2002b. Ontology editor survey results. `www.xml.com/2002/11/06/Ontology_Editor_Survey.html`.

Ding, Z., and Y. Peng. 2004. A probabilistic extension to ontology language OWL. In *Proc. 37th Hawaii Int. Conf. on Systems Science*.

Do, H., S. Melnik, and E. Rahm. 2002. Comparison of schema matching evaluations. In *Proc. GI-Workshop "Web and Databases,"* vol. 2593, Erfurt, Germany. Springer-Verlag.

Do, H., and E. Rahm. 2002. COMA - a system for flexible combination of schema matching approaches. In *Proc. VLDB*.

Dodd, I.B., and J.B. Egan. 1990. Improved detection of helix-turn-helix DNA-binding motifs in protein sequences. *Nucleic Acids Res.* 18:5019–5026.

Draghici, S., P. Khatri, P. Bhavsar, A. Shah, S.A. Krawetz, and M.A. Tainsky. 2003. Onto-Tools, the toolkit of the modern biologist: Onto-Express, Onto-Compare, Onto-Design and Onto-Translate. *Nucleic Acids Res.* 31:3775–3781.

DUET, 2002. DAML UML enhanced tool (DUET). `grcinet.grci.com/maria/www/CodipSite/Tools/Tools.html`.

Dwight, S.S., R. Balakrishnan, K.R. Christie, M.C. Costanzo, K. Dolinski, S.R. Engel, B. Feierbach, D.G. Fisk, J. Hirschman, E.L. Hong, L. Issel-Tarver, R.S. Nash, A. Sethuraman, B. Starr, C.L. Theesfeld, R. Andrada, G. Binkley, Q. Dong, C. Lane, M. Schroeder, S. Weng, D. Botstein, and J.M. Cherry. 2004. *Saccharomyces* genome database: underlying principles and organisation. *Brief Bioinform.* 5:9–22.

EcoCyc, 2003. Encyclopedia of *Escherichia coli* Genes and Metabolism. `www.ecocyc.org`.

Eddy, S.R. 1998. Profile hidden Markov models. *Bioinformatics* 14:755–763.

Embley, D., et al. 2001. Multifaceted exploitation of metadata for attribute match discovery in information integration. In *International Workshop on Information Integration on the Web*.

Euler, 2003. Euler proof mechanism. `www.agfa.com/w3c/euler/`.

ezOWL, 2004. ezOWL website. `iweb.etri.re.kr/ezowl/`.

Fauconnier, G., and M. Turner. 1998. Conceptual integration networks. *Cognitive Sci.* 22:133–187.

Fauconnier, G., and M. Turner. 2002. *The Way We Think: Conceptual Blending and The Mind's Hidden Complexities.* New York: Basic Books.

Fenyo, D. 1999. The Biopolymer Markup Language. *Bioinformatics* 15:339–340.

Fitzhugh, R.A. 1961. Impulses and physiological states in theoretical models of nerve membrane. *Biophys. J.* 1:445–466.

FlyBase. 2002. The FlyBase database of the *Drosophila* genome projects and community literature. *Nucleic Acids Res.* 30:106–108.

Forgy, C. 1982. Rete: a fast algorithm for the many pattern/many object pattern match problem. *Artif. Intell.* 19:17–37.

FOWL, 2003. F-OWL: an OWL inference engine in Flora-2. `fowl.sourceforge.net`.

Fraser, A.G., and E.M. Marcotte. 2004. A probabilistic view of gene function. *Nat. Genet.* 36:559–564.

Friedman, N. 2004. Inferring cellular networks using probabilistic graphical models. *Science* 303:799.

GAME, 2002. Genome Annotation Markup Elements. `www.fruitfly.org/comparative`.

Genesereth, M., 1998. Knowledge Interchange Format draft proposed American National Standard (dpANS) NCITS.T2/98-004. `logic.stanford.edu/kif/dpans.html`.

Ghosh, D. 1993. Status of the transcription factors database (tfd). *Nucleic Acids Res.* 21:3117–3118.

Ghosh, D. 2000. Object-oriented transcription factors database (ooTFD). *Nucleic Acids Res.* 28:308–310.

Gibson, D., J. Kleinberg, and P. Raghavan. 1998. Inferring Web communities from link topology. In *Proc. 9th ACM Conf. on Hypertext and Hypermedia.*

Gilmour, R. 2000. Taxonomic Markup Language: applying XML to systematic data. *Bioinformatics* 16:406–407.

Gish, W., and D.J. States. 1993. Identification of protein coding regions by database similarity search. *Nat. Genet.* 3:266–272.

Glymour, C., and G. Cooper (eds.). 1999. *Computation, Causation and Discovery.* Cambridge, MA: MIT Press.

GO, 2003. Gene ontology website. www.geneontology.org.

GO. 2004. The Gene Ontology (GO) database and informatics resource. *Nucleic Acids Res.* 32:D258–D261.

GOA, 2003. Gene Ontology Annotation website. www.ebi.ac.uk/GOA.

Goddard, N.H., M. Hucka, F. Howell, H. Cornelis, K. Shankar, and D. Beeman. 2001. NeuroML: model description methods for collaborative modelling in neuroscience. *Philos. Trans. R. Soc. Lond. B. Biol. Sci.* 356:1209–1228.

Goguen, J. 1999. Semiotic morphism, representations, and blending for interface design. *Formal Aspects of Computing* 11:272–301.

Goguen, J., and D. Harrell. 2004. Foundations for active multimedia narrative: semiotic spaces and structural blending. Technical report, University of California, San Diego. www.cs.ucsd.edu/users/goguen/pps/narr.pdf.

Gough, J., K. Karplus, R. Hughey, and C. Chothia. 2001. Assignment of homology to genome sequences using a library of hidden Markov models that represent all proteins of known structure. *J. Mol. Biol.* 313:903–919.

Grant, J.D., R.L. Dunbrack, F.J. Manion, and M.F. Ochs. 2002. BeoBLAST: distributed BLAST and PSI-BLAST on a beowulf cluster. *Bioinformatics* 18:765–766.

Guarino, N., and P. Giaretta. 1995. Ontologies and knowledge bases: towards a terminological clarification. In N. Mars (ed.), *Towards Very Large Knowledge Bases*. Amsterdam: IOS Press.

Guo, J., K. Araki, K. Tanaka, J. Sato, M. Suzuki, A. Takada, T. Suzuki, Y. Nakashima, and H. Yoshihara. 2003. The latest MML (Medical Markup Language) version 2.3–XML-based standard for medical data exchange/storage. *J. Med. Syst.* 27:357–366.

Hadley, C., and D.T. Jones. 1999. A systematic comparison of protein structure classifications: SCOP, CATH and FSSP. *Struct. Fold. Des.* 7:1099–1112.

Haft, D.H., J.D. Selengut, and O. White. 2003. The TIGRFAMs database of protein families. *Nucleic Acids Res.* 31:371–373.

Han, K., B. Park, H. Kim, J. Hong, and J. Park. 2004. HPID: the Human Protein Interaction Database. *Bioinformatics* 20:2466–2470.

Hanisch, D., R. Zimmer, and T. Lengauer. 2002. ProML–the Protein Markup Language for specification of protein sequences, structures and families. *In Silico Biol.* 2:313–324.

Hayes, P., 2004. RDF semantics. www.w3.org/TR/rdf-mt/.

Heflin, J., J. Hendler, and S. Luke. 1999. Coping with changing ontologies in a distributed environment. In *AAAI-99 Workshop on Ontology Management*. MIT Press, Cambridge, MA.

Heflin, J., J. Hendler, and S. Luke. 2000. SHOE: a knowledge representation language for Internet applications. Technical Report www.cs.umd.edu/projects/-plus/SHOE, Institute for Advanced Studies, University of Maryland, College Park.

Helsper, E., and L. van der Gaag. 2001. Ontologies for probabilistic networks: A case study in oesophageal cancer. In B. Kröse, M. de Rijke, G. Schreiber, and M. van Someren (eds.), *Proc. 13th Belgium-Netherlands Conference on Artificial Intelligence*, Amsterdam, pp. 125–132.

Helsper, E., and L. van der Gaag. 2002. A case study in ontologies for probabilistic networks. In M. Bramer, F. Coenen, and A. Preece (eds.), *Research and Development in Intelligent Systems XVIII*, pp. 229–242. London: Springer-Verlag.

Henikoff, J.G., E.A. Greene, S. Pietrokovski, and S. Henikoff. 2000. Increased coverage of protein families with the blocks database servers. *Nucleic Acids Res.* 28:228–230.

Henikoff, S., and J.G. Henikoff. 1991. Automated assembly of protein blocks for database searching. *Nucleic Acids Res.* 19:6565–6572.

Henikoff, S., and J.G. Henikoff. 1992. Amino acid substitution matrices from protein blocks. *Proc. Natl. Acad. Sci. U.S.A.* 89:10915–10919.

Henikoff, S., and J.G. Henikoff. 1994. Protein family classification based on searching a database of blocks. *Genomics* 19:97–107.

Henikoff, S., S. Pietrokovski, and J.G. Henikoff. 1998. Superior performance in protein homology detection with the Blocks database servers. *Nucleic Acids Res.* 26:309–312.

Henrion, M., M. Pradhan, B. del Favero, K. Huang, G. Provan, and P. O'Rorke. 1996. Why is diagnosis using belief networks insensitive to imprecision in probabilities? In *Proc. 12th Conf. Uncertainty in Artificial Intelligence*, pp. 307–314.

Hertz, G.Z., G.W. Hartzell III, and G.D. Stormo. 1990. Identification of consensus patterns in unaligned DNA sequences known to be functionally related. *Comput. Appl. Biosci.* 6:81–92.

Hertz, G.Z., and G.D. Stormo. 1999. Identifying DNA and protein patterns with statistically significant alignments of multiple sequences. *Bioinformatics* 15:563–577.

Hoebeke, M., H. Chiapello, P. Noirot, and P. Bessieres. 2001. SPiD: a subtilis protein interaction database. *Bioinformatics* 17:1209–1212.

Holm, L., C. Ouzounis, C. Sander, G. Tuparev, and G. Vriend. 1992. A database of protein structure families with common folding motifs. *Protein Sci.* 1:1691–1698.

Holm, L., and C. Sander. 1998. Touring protein fold space with Dali/FSSP. *Nucleic Acids Res.* 26:316–319.

Howard, R., and J. Matheson. 1981. Influence diagrams. In R. Howard and J. Matheson (eds.), *Readings on the Principles and Applications of Decision Analysis*, vol. 2, pp. 721–762. Menlo Park, CA: Strategic Decisions Group.

Hubbard, T., D. Barker, E. Birney, G. Cameron, Y. Chen, L. Clark, T. Cox, J. Cuff, V. Curwen, T. Down, R. Durbin, E. Eyras, J. Gilbert, M. Hammond, L. Huminiecki, A. Kasprzyk, H. Lehvaslaiho, P. Lijnzaad, C. Melsopp, E. Mongin, R. Pettett, M. Pocock, S. Potter, A. Rust, E. Schmidt, S. Searle, G. Slater, J. Smith, W. Spooner, A. Stabenau, J. Stalker, E. Stupka, A. Ureta-Vidal, I. Vastrik, and M. Clamp. 2002. The Ensembl genome database project. *Nucleic Acids Res.* 30:38–41.

Hucka, M., A. Finney, H.M. Sauro, H. Bolouri, J.C. Doyle, H. Kitano, A.P. Arkin, B.J. Bornstein, D. Bray, A. Cornish-Bowden, A.A. Cuellar, S. Dronov, E.D. Gilles, M. Ginkel, V. Gor, I.I. Goryanin, W.J. Hedley, T.C. Hodgman, J.H. Hofmeyr, P.J. Hunter, N.S. Juty, J.L. Kasberger, A. Kremling, U. Kummer, N. Novere Le, L.M. Loew, D. Lucio, P. Mendes, E. Minch, E.D. Mjolsness, Y. Nakayama, M.R. Nelson, P.F. Nielsen, T. Sakurada, J.C. Schaff, B.E. Shapiro, T.S. Shimizu, H.D. Spence, J. Stelling, K. Takahashi, M. Tomita, J. Wagner, and J. Wang. 2003. The Systems Biology Markup Language (SBML): a medium for representation and exchange of biochemical network models. *Bioinformatics* 19:524–531. SBML forum.

Hughes, J.D., P.W. Estep, S. Tavazoie, and G.M. Church. 2000. Computational identification of cis-regulatory elements associated with groups of functionally related genes in *Saccharomyces cerevisiae*. *J. Mol. Biol.* 296:1205–1214.

Hulo, N., C.J. Sigrist, V. Le Saux, P.S. Langendijk-Genevaux, L. Bordoli, A. Gattiker, E. De Castro, P. Bucher, and A. Bairoch. 2004. Recent improvements to the PROSITE database. *Nucleic Acids Res.* 32:D134–D137. Database issue.

IHMC. 2003. The International HapMap Project. *Nature* 426:789–796.

Indurkhya, B. 1992. *Metaphor and Cognition*. Dordrecht, Netherlands: Kluwer Academic.

Indurkhya, B. 2002. On the philosophical foundation of Lyee: interaction theories and Lyee. In H. Fujita and P. Johannesson (eds.), *New Trends in Software Methodologies, Tools and Techniques*, pp. 45–51. Amsterdam: IOS Press.

ISMWG. 2001. A map of human genome sequence variation containing 1.42 million single nucleotide polymorphisms. *Nature* 409:928–933.

Jaakkola, T., and M. Jordan. 1999. Variational probabilistic inference and the QMR-DT network. *J. of Artif. Intell. Res.* 10:291–322.

Jain, A., and R. Dubes. 1988. *Algorithms for Clustering Data*. Englewood Cliffs, NJ: Prentice Hall.

Jarg, 2005. SemanTx Life Sciences, a division of Jarg Corporation. `www.semantxls.com`.

Jung, E., A.L. Veuthey, E. Gasteiger, and A. Bairoch. 2001. Annotation of glycoproteins in the SWISS-PROT database. *Proteomics* 1:262–268.

Jurowski, P., R. Gos, A. Kapica, and M. Zdzieszynska. 2004. Secondary glaucoma due to progressive iris atrophy–a century after the first description. Case report. *Klin. Oczna* 106:80–82.

Kanehisa, M., and S. Goto. 2000. KEGG: Kyoto encyclopedia of genes and genomes. *Nucleic Acids Res.* 28:27–30.

Kanehisa, M., S. Goto, S. Kawashima, and A. Nakaya. 2002. The KEGG databases at GenomeNet. *Nucleic Acids Res.* 30:42–46.

Karlin, S., and S.F. Altschul. 1990. Methods for assessing the statistical significance of molecular sequence features by using general scoring schemes. *Proc. Natl. Acad. Sci. U.S.A.* 87:2264–2268.

Karlin, S., and S.F. Altschul. 1993. Applications and statistics for multiple high-scoring segments in molecular sequences. *Proc. Natl. Acad. Sci. U.S.A.* 90:5873–5877.

Karp, P.D., S. Paley, and P. Romero. 2002a. The Pathway Tools software. *Bioinformatics* 18:S225–S232.

Karp, P.D., M. Riley, S.M. Paley, and A. Pellegrini-Toole. 2002b. The MetaCyc database. *Nucleic Acids Res.* 30:59–61.

Karp, P.D., M. Riley, M. Saier, I.T. Paulsen, J. Collado-Vides, S.M. Paley, A. Pellegrini-Toole, C. Bonavides, and S. Gama-Castro. 2002c. The EcoCyc database. *Nucleic Acids Res.* 30:56–58.

Kelley, B.P., B. Yuan, F. Lewitter, R. Sharan, B.R. Stockwell, and T. Ideker. 2004. PathBLAST: a tool for alignment of protein interaction networks. *Nucleic Acids Res.* 32: W83–W88. Web server issue.

Kent, W.J. 2002. BLAT–the BLAST-like alignment tool. *Genome Res.* 12:656–664.

King, O.D., R.E. Foulger, S.S. Dwight, J.V. White, and F.P. Roth. 2003. Predicting gene function from patterns of annotation. *Genome Res.* 13:896–904.

Kleinberg, J. 1998. Authoritative sources in a hyperlinked environment. In *Proc. ACM-SIAM Symp. on Discrete Algorithms*.

Know-Me, 2004. Know-Me website. `www.nbirn.net/Resources/Users/Applications/KnowMe/Know-ME.htm`.

Kogut, P., S. Cranefield, L. Hart, M. Dutra, K. Baclawski, M. Kokar, and J. Smith. 2002. UML for ontology development. *Knowledge Eng. Rev.* 17:61–64.

Kohane, I.S., A.T. Kho, and A.J. Butte. 2003. *Microarrays for an Integrative Genomics*. Cambridge, MA: MIT Press.

Kohonen, T. 1997. *Self-organizating maps*. New York: Springer-Verlag.

Kokar, M., J. Letkowski, K. Baclawski, and J. Smith, 2001. The ConsVISor consistency checking tool. `www.vistology.com/consvisor/`.

Kolchanov, N.A., E.V. Ignatieva, E.A. Ananko, O.A. Podkolodnaya, I.L. Stepanenko, T.I. Merkulova, M.A. Pozdnyakov, N.L. Podkolodny, A.N. Naumochkin, and A.G. Romashchenko. 2002. Transcription Regulatory Regions Database (TRRD): its status in 2002. *Nucleic Acids Res.* 30:312–317.

Koller, D., A. Levy, and A. Pfeffer. 1997. P-Classic: a tractable probabilistic description logic. In *Proc. 14th National Conf. on Artificial Intelligence*, Providence, RI, pp. 390–397.

Koller, D., and A. Pfeffer. 1997. Object-oriented Bayesian networks. In *Proc. 13th Ann. Conf. on Uncertainty in Artificial Intelligence*, Providence, RI, pp. 302–313.

Korf, I., and W. Gish. 2000. MPBLAST : improved BLAST performance with multiplexed queries. *Bioinformatics* 16:1052–1053.

Krishnan, V.G., and D.R. Westhead. 2003. A comparative study of machine-learning methods to predict the effects of single nucleotide polymorphisms on protein function. *Bioinformatics* 19:2199–2209.

Kulikova, T., P. Aldebert, N. Althorpe, W. Baker, K. Bates, P. Browne, A. van den Broek, G. Cochrane, K. Duggan, R. Eberhardt, N. Faruque, M. Garcia-Pastor, N. Harte, C. Kanz, R. Leinonen, Q. Lin, V. Lombard, R. Lopez, R. Mancuso, M. McHale, F. Nardone, V. Silventoinen, P. Stoehr, G. Stoesser, M.A. Tuli, K. Tzouvara, R. Vaughan, D. Wu, W. Zhu, and R. Apweiler. 2004. The EMBL nucleotide sequence database. *Nucleic Acids Res.* 32:D27–D30. Database issue.

Kuter, I. 1999. Breast cancer highlights. *Oncologist* 4:299–308.

Lakoff, G. 1987. *Women, Fire, and Dangerous Things: What Categories Reveal about the Mind.* Chicago: University of Chicago Press.

Lassila, O., and R. Swick, 1999. Resource description framework (RDF) model and syntax specification. `www.w3.org/TR/REC-rdf-syntax`.

Lawrence, C., S. Altschul, M. Boguski, J. Liu, A. Neuwald, and J. Wootton. 1993. Detecting subtle sequence signals: a Gibbs sampling strategy for multiple alignment. *Science* 262:208–214.

Leibniz, G. 1998. Monadology. In *G.W. Leibniz Philosophical Texts (1714)*, pp. 267–281. Translated and edited by R. Woolhouse and R. Francks. New York: Oxford University Press.

Leif, R.C., S.B. Leif, and S.H. Leif. 2003. CytometryML, an XML format based on DICOM and FCS for analytical cytology data. *Cytometry* 54A:56–65.

Letunic, I., R.R. Copley, S. Schmidt, F.D. Ciccarelli, T. Doerks, J. Schultz, C.P. Ponting, and P. Bork. 2004. SMART 4.0: towards genomic data integration. *Nucleic Acids Res.* 32:D142–D144. Database issue.

Leung, Y.F., and C.P. Pang. 2002. EYE on bioinformatics: dissecting complex disease traits in silico. *Appl. Bioinformatics* 1:69–80.

Li, W., and C. Clifton. 2000. Semint: a tool for identifying attribute correspondences in heterogeneous databases using neural network. *Data and Knowledge Engineering* 33:49–84.

Lindberg, D.A., B.L. Humphreys, and A.T. McCray. 1993. The Unified Medical Language System. *Methods Inf. Med.* 32:281–291.

Liu, J.S., A.F. Neuwald, , and C.E. Lawrence. 1995. Bayesian models for multiple local sequence alignment and Gibbs sampling strategies. *J. Am. Statis. Assoc.* 90: 1156–1170.

Liu, X., D.L. Brutlag, and J.S. Liu. 2001. BioProspector: discovering conserved DNA motifs in upstream regulatory regions of co-expressed genes. In *Pac. Symp. Biocomput.*, pp. 127–138.

Lutteke, T., M. Frank, and C.W. von der Lieth. 2004. Data mining the protein data bank: automatic detection and assignment of carbohydrate structures. *Carbohydr. Res.* 339:1015–1020.

Lynch, M., and J.S. Conery. 2000. The evolutionary fate and consequences of duplicate genes. *Science* 290:1151–1155.

MacKay, D., 2004. Bayesian methods for neural networks - FAQ. `www.inference.phy.cam.ac.uk/mackay/Bayes_FAQ.html`.

MacQueen, J. 1967. Some methods for classification and analysis of multivariate observations. In L. Le Cam and J. Neyman (eds.), *Proc. Fifth Berkeley Symp. Math. Statis. and Prob.*, vol. 1, pp. 281–297, Berkeley, CA. University of California Press.

Madera, M., C. Vogel, S.K. Kummerfeld, C. Chothia, and J. Gough. 2004. The SUPERFAMILY database in 2004: additions and improvements. *Nucleic Acids Res.* 32: D235–D239. Database issue.

Madhavan, J., P. Bernstein, and E. Rahm. 2001. Generic schema matching with Cupid. In *Proc. VLDB*.

MAGE-ML, 2003. MicroArray Gene Expression Markup Language website. `www.mged.org`.

Marchler-Bauer, A., A.R. Panchenko, B.A. Shoemaker, P.A. Thiessen, L.Y. Geer, and S.H. Bryant. 2002. CDD: a database of conserved domain alignments with links to domain three-dimensional structure. *Nucleic Acids Res.* 30:281–283.

Maybeck, P. 1979. *Stochastic models, estimation and control*, vol. 1. New York: Academic Press.

McCray, A.T., O. Bodenreider, J.D. Malley, and A.C. Browne. 2001. Evaluating UMLS strings for natural language processing. In *Proc. AMIA Symp.*, pp. 448–452.

McGinnis, S., and T.L. Madden. 2004. BLAST: at the core of a powerful and diverse set of sequence analysis tools. *Nucleic Acids Res.* 32:W20–W25. Web server issue.

McGuinness, D., R. Fikes, J. Rice, and S. Wilder. 2000. An environment for merging and testing large ontologies. In *Proceedings of the 7th International Conference on Principles of Knowledge Representation and Reasoning (KR2000)*, Breckenridge, CO.

Mellquist, J.L., L. Kasturi, S.L. Spitalnik, and S.H. Shakin-Eshleman. 1998. The amino acid following an asn-X-Ser/Thr sequon is an important determinant of N-linked core glycosylation efficiency. *Biochemistry* 37:6833–6837.

Mewes, H.W., C. Amid, R. Arnold, D. Frishman, U. Guldener, G. Mannhaupt, M. Munsterkotter, P. Pagel, N. Strack, V. Stumpflen, J. Warfsmann, and A. Ruepp. 2004. MIPS: analysis and annotation of proteins from whole genomes. *Nucleic Acids Res.* 32:D41–D44. Database issue.

Miller, E., R. Swick, D. Brickley, and B. McBride, 2001. Semantic Web activity page. `www.w3.org/2001/sw/`.

Miller, R., L. Haas, and M. Hernandez. 2000. Schema mapping as query discovery. In *Proc. VLDB*, pp. 77–88.

Mitra, P., G. Wiederhold, and J. Jannink. 1999. Semi-automatic integration of knowledge sources. In *Proc. 2nd International Conf. on Information Fusion*.

Miyazaki, S., H. Sugawara, K. Ikeo, T. Gojobori, and Y. Tateno. 2004. DDBJ in the stream of various biological data. *Nucleic Acids Res.* 32:D31–D34. Database issue.

Mulder, N.J., R. Apweiler, T.K. Attwood, A. Bairoch, D. Barrell, A. Bateman, D. Binns, M. Biswas, P. Bradley, P. Bork, P. Bucher, R.R. Copley, E. Courcelle, U. Das, R. Durbin, L. Falquet, W. Fleischmann, S. Griffiths-Jones, D. Haft, N. Harte, N. Hulo, D. Kahn, A. Kanapin, M. Krestyaninova, R. Lopez, I. Letunic, D. Lonsdale, V. Silventoinen, S.E. Orchard, M. Pagni, D. Peyruc, C.P. Ponting, J.D. Selengut, F. Servant, C.J. Sigrist, R. Vaughan, and E.M. Zdobnov. 2003. The InterPro database, 2003 brings increased coverage and new features. *Nucleic Acids Res* 31:315–318.

Muller, A., R.M. MacCallum, and M.J. Sternberg. 1999. Benchmarking PSI-BLAST in genome annotation. *J. Mol. Biol.* 293:1257–1271.

Muller, A., R.M. MacCallum, and M.J. Sternberg. 2002. Structural characterization of the human proteome. *Genome Res.* 12:1625–1641.

Murphy, K., 1998. A brief introduction to graphical models and Bayesian networks. `www.ai.mit.edu/~murphyk/Bayes/bnintro.html`.

Murray-Rust, P., and H.S. Rzepa. 2003. Chemical Markup, XML, and the World Wide Web. 4. CML Schema. *J. Chem. Inf. Comput. Sci.* 43:757–772.

Murzin, A.G., S.E. Brenner, T. Hubbard, and C. Chothia. 1995. SCOP: a structural classification of proteins database for the investigation of sequences and structures. *J. Mol. Biol.* 247:536–540.

Nagumo, J. 1962. An active pulse transmission line simulating nerve axon. *Proc. Inst. Radio Eng.* 50:2061–2070.

Nam, Y., J. Goguen, and G. Wang. 2002. A metadata integration assistant generator for heterogeneous distributed databases. In *Proc. Int. Conf. Ontologies, Databases, and Applications of Semantics for Large Scale Information Systems*, vol. 2519, pp. 1332–1344. Springer-Verlag, New York.

NCHS, 2003. National hospital discharge survey, 1988-2002.

Needleman, S.B., and C.D. Wunsch. 1970. A general method applicable to the search for similarities in the amino acid sequence of two proteins. *J. Mol. Biol.* 48:443–453.

Neil, M., N. Fenton, and L. Nielsen. 2000. Building large-scale Bayesian networks. *Knowledge Eng. Rev.* 15:257–284.

Ng, P.C., and S. Henikoff. 2002. Accounting for human polymorphisms predicted to affect protein function. *Genome Res.* 12:436–446.

Ng, P.C., and S. Henikoff. 2003. SIFT: predicting amino acid changes that affect protein function. *Nucleic Acids Res.* 31:3812–3814.

NHS, 2004. The Nurses' Health Study. `www.channing.harvard.edu/nhs`.

NIH, 2004a. NCBI reference sequences. `www.ncbi.nlm.nih.gov/RefSeq/`.

NIH, 2004b. PubMed Central (PMC). `www.pubmedcentral.nih.gov`.

Niu, T. 2004. Algorithms for inferring haplotypes. *Genet. Epidemiol.* 27:334–347.

Niu, T., K. Baclawski, Y. Feng, and H. Wang. 2003. Database schema for management and storage of single nucleotide polymorphism. *Am. J. Human Genetics* 73 (Suppl): A467.

Noirot-Gros, M.F., E. Dervyn, L.J. Wu, P. Mervelet, J. Errington, S.D. Ehrlich, and P. Noirot. 2002. An expanded view of bacterial DNA replication. *Proc. Natl. Acad. Sci. U.S.A.* 99:8342–8347.

Notredame, C., D.G. Higgins, and J. Heringa. 2000. T-Coffee: a novel method for fast and accurate multiple sequence alignment. *J. Mol. Biol.* 302:205–217.

Noy, N.F., M. Crubezy, R.W. Fergerson, H. Knublauch, S.W. Tu, J. Vendetti, and M.A. Musen. 2003. Protégé-2000: an open-source ontology-development and knowledge-acquisition environment. In *Proc. AMIA Annual Symp.*, p. 953.

Noy, N., and D. McGuinness, 2001. Ontology 101. `protege.stanford.edu/publications/ontology_development/ontology101.html`.

Noy, N., and M. Musen. 2000. PROMPT: algorithm and tool for automated intelligence. In *AAAI-2000*, Austin, TX.

NRC. 1992. *Combining Information: Statistical Issues and Opportunities for Research.* Washington, DC: National Academy Press.

Ogren, P.V., K.B. Cohen, G.K. Acquaah-Mensah, J. Eberlein, and L. Hunter. 2004. The compositional structure of Gene Ontology terms. In *Pac. Symp. Biocomput.*, pp. 214–225.

Opdahl, A. Henderson-Sellers, B., and F. Barbier. 2000. An ontological evaluation of the OML metamodel. In E. Falkenberg, K. Lyytinen, and A. Verrijn-Stuart (eds.), *Information System Concepts: An Integrated Discipline Emerging*, vol. 164, pp. 217–232. Dordrect, Netherlands: Kluwer.

Orengo, C.A., A.D. Michie, S. Jones, D.T. Jones, M.B. Swindells, and J.M. Thornton. 1997. CATH–a hierarchic classification of protein domain structures. *Structure* 5: 1093–1108.

Orengo, C.A., F.M. Pearl, and J.M. Thornton. 2003. The CATH domain structure database. *Methods Biochem. Anal.* 44:249–271.

Packer, B.R., M. Yeager, B. Staats, R. Welch, A. Crenshaw, M. Kiley, A. Eckert, M. Beerman, E. Miller, A. Bergen, N. Rothman, R. Strausberg, and S.J. Chanock. 2004. SNP500Cancer: a public resource for sequence validation and assay development for genetic variation in candidate genes. *Nucleic Acids Res.* 32:D528–D532. Database issue.

Page, L., and S. Brin, 2004. Google page rank algorithm. `www.google.com/technology`.

Pandey, A., and F. Lewitter. 1999. Nucleotide sequence databases: a gold mine for biologists. *Trends Biochem. Sci.* 24:276–280.

Patel-Schneider, P., P. Hayes, and I. Horrocks, 2004. OWL web ontology language semantics and abstract syntax. `www.w3.org/TR/owl-semantics/`.

Pearl, J. 1988. *Probabilistic Reasoning in Intelligent Systems: Networks of Plausible Inference.* San Francisco: Morgan Kaufmann.

Pearl, J. 1998. Graphical models for probabilistic and causal reasoning. In D. Gabbay and P. Smets (eds.), *Handbook of Defeasible Reasoning and Uncertainty Management Systems, Volume 1: Quantified Representation of Uncertainty and Imprecision*, pp. 367–389. Dordrecht, Netherlands: Kluwer Academic.

Pearl, J. 2000. *Causality: Models, Reasoning and Inference.* Cambridge, UK: Cambridge University Press.

Pearson, W.R., and D.J. Lipman. 1988. Improved tools for biological sequence comparison. *Proc. Natl. Acad. Sci. U.S.A.* 85:2444–2448.

Pellet, 2003. Pellet OWL reasoner. `www.mindswap.org/2003/pellet/`.

Perez, A., and R. Jirousek. 1985. Constructing an intensional expert system (INES). In *Medical Decision Making*. Amserdam: Elsevier.

Peri, S., J.D. Navarro, T.Z. Kristiansen, R. Amanchy, V. Surendranath, B. Muthusamy, T.K. Gandhi, K.N. Chandrika, N. Deshpande, S. Suresh, B.P. Rashmi, K. Shanker, N. Padma, V. Niranjan, H.C. Harsha, N. Talreja, B.M. Vrushabendra, M.A. Ramya, A.J. Yatish, M. Joy, H.N. Shivashankar, M.P. Kavitha, M. Menezes, D.R. Choudhury, N. Ghosh, R. Saravana, S. Chandran, S. Mohan, C.K. Jonnalagadda, C.K. Prasad, C. Kumar-Sinha, K.S. Deshpande, and A. Pandey. 2004. Human protein reference database as a discovery resource for proteomics. *Nucleic Acids Res.* 32:D497–D501. Database issue.

Piaget, J. 1971. *The Construction of Reality in the Child.* New York: Ballantine Books.

Piaget, J., and B. Inhelder. 1967. *The Child's Conception of Space*. New York: Norton.

Piaget, J., B. Inhelder, and A. Szeminska. 1981. *The Child's Conception of Geometry*. New York, NY: Norton.

Pingoud, A., and A. Jeltsch. 2001. Structure and function of type II restriction endonucleases. *Nucleic Acids Res.* 29:3705–3727.

Pradhan, M., M. Henrion, G. Provan, B. del Favero, and K. Huang. 1996. The sensitivity of belief networks to imprecise probabilities: an experimental investigation. *Artif. Intell.* 85:363–397.

Pradhan, M., G. Provan, B. Middleton, and M. Henrion. 1994. Knowledge engineering for large belief networks. In *Proc. Tenth Annual Conf. on Uncertainty in Artificial Intelligence (UAI–94)*, pp. 484–490, San Mateo, CA. Morgan Kaufmann.

Rahm, E., and P. Bernstein. 2001. On matching schemas automatically. Technical report, Dept. of Computer Science, University of Leipzig. `dol.uni-leipzig.de/pub/2001-5/en`.

Ramensky, V., P. Bork, and S. Sunyaev. 2002. Human non-synonymous SNPs: server and survey. *Nucleic Acids Res.* 30:3894–3900.

Roberts, R.J., T. Vincze, J. Posfai, and D. Macelis. 2003. REBASE: restriction enzymes and methyltransferases. *Nucleic Acids Res.* 31:418–420.

Rosch, E., and B. Lloyd (eds.). 1978. *Cognition and Categorization*. Hillsdale, NJ: Lawrence Erlbaum.

Roth, F.R., J.D. Hughes, P.E. Estep, and G.M. Church. 1998. Finding DNA regulatory motifs within unaligned non-coding sequences clustered by whole-genome mRNA quantitation. *Nat. Biotechnol.* 16:939–945.

Salton, G. 1989. *Automatic Text Processing*. Reading, MA: Addison-Wesley.

Salton, G., E. Fox, and H. Wu. 1983. Extended boolean information retrieval. *Comm. ACM* 26:1022–1036.

Salton, G., and M. McGill. 1986. *Introduction to Modern Information Retrieval*. New York: McGraw-Hill.

Salwinski, L., C.S. Miller, A.J. Smith, F.K. Pettit, J.U. Bowie, and D. Eisenberg. 2004. The database of interacting proteins: 2004 update. *Nucleic Acids Res.* 32:D449–D451. Database issue.

Saracevic, T. 1975. Relevance: a review of and a framework for the thinking on the notion in information science. *J. Am. Soc. Info. Sci.* 26:321–343.

Sarle, W., 2002. Neural network FAQ. `www.faqs.org/faqs/ai-faq/neural-nets`.

SBML, 2003. The Systems Biology Markup Language website. `www.sbw-sbml.org`.

Schaffer, A.A., L. Aravind, T.L. Madden, S. Shavirin, J.L. Spouge, Y.I. Wolf, E.V. Koonin, and S.F. Altschul. 2001. Improving the accuracy of PSI-BLAST protein database searches with composition-based statistics and other refinements. *Nucleic Acids Res.* 29:2994–3005.

Schofield, P.N., J.B. Bard, C. Booth, J. Boniver, V. Covelli, P. Delvenne, M. Ellender, W. Engstrom, W. Goessner, M. Gruenberger, H. Hoefler, J. Hopewell, M. Mancuso, C. Mothersill, C.S. Potten, L. Quintanilla-Fend, B. Rozell, H. Sariola, J.P. Sundberg, and A. Ward. 2004. Pathbase: a database of mutant mouse pathology. *Nucleic Acids Res.* 32:D512–D515. Database issue.

Servant, F., C. Bru, S. Carrere, E. Courcelle, J. Gouzy, D. Peyruc, and D. Kahn. 2002. ProDom: automated clustering of homologous domains. *Brief Bioinform.* 3:246–251.

Shafer, G. 1976. *A Mathematical Theory of Evidence*. Princeton, NJ: Princeton University Press.

Sherry, S.T., M.H. Ward, M. Kholodov, J. Baker, L. Phan, E.M. Smigielski, and K. Sirotkin. 2001. dbSNP: the NCBI database of genetic variation. *Nucleic Acids Res.* 29:308–311.

Shipley, B. 2000. *Cause and Correlation in Biology*. Cambridge, UK: Cambridge University Press.

Shortliffe, E. 1976. *Computer-Based Medical Consultation: MYCIN*. New York: Elsevier.

Sigrist, C.J., L. Cerutti, N. Hulo, A. Gattiker, L. Falquet, M. Pagni, A. Bairoch, and P. Bucher. 2002. PROSITE: a documented database using patterns and profiles as motif descriptors. *Brief Bioinform.* 3:265–274.

Smith, T.F., and M.S. Waterman. 1981. Identification of common molecular subsequences. *J. Mol. Biol.* 147:195–197.

Software, Hit, 2004. Hit Software XML utilities. `www.hitsw.com/xml_utilites/`.

Spellman, P.T., M. Miller, J. Stewart, C. Troup, U. Sarkans, S. Chervitz, D. Bernhart, G. Sherlock, C. Ball, M. Lepage, M. Swiatek, W.L. Marks, J. Goncalves, S. Markel, D. Iordan, M. Shojatalab, A. Pizarro, J. White, R. Hubley, E. Deutsch, M. Senger, B.J. Aronow, A. Robinson, D. Bassett, C.J. Stoeckert, Jr., and A. Brazma. 2002. Design and implementation of microarray gene expression markup language (MAGE-ML). *Genome Biol.* 3:RESEARCH0046.

Spinoza, B. 1998. *The Ethics (1677)*. Translated by R. Elwes. McLean, VA: IndyPublish.com.

Spirtes, P., C. Glymour, and R. Scheines. 2001. *Causation, Prediction and Search*. Cambridge, MA: MIT Press.

States, D.J., and W. Gish. 1994. Combined use of sequence similarity and codon bias for coding region identification. *J. Comput. Biol.* 1:39–50.

Steinberg, A., C. Bowman, and F. White. 1999. Revisions to the JDL data fusion model. In *SPIE Conf. Sensor Fusion: Architectures, Algorithms and Applications III*, vol. 3719, pp. 430–441.

Stock, A., and J. Stock. 1987. Purification and characterization of the CheZ protein of bacterial chemotaxis. *J. Bacteriol.* 169:3301–3311.

Stoeckert, C.J., Jr., H.C. Causton, and C.A. Ball. 2002. Microarray databases: standards and ontologies. *Nat. Genet.* 32 (Suppl):469–473.

Stormo, G.D., and G.W. Hartzell III. 1989. Identifying protein-binding sites from unaligned DNA fragments. *Proc. Natl. Acad. Sci. U.S.A.* 86:1183–1187.

Strausberg, R.L. 2001. The Cancer Genome Anatomy Project: new resources for reading the molecular signatures of cancer. *J. Pathol.* 195:31–40.

Strausberg, R.L., S.F. Greenhut, L.H. Grouse, C.F. Schaefer, and K.H. Buetow. 2001. In silico analysis of cancer through the Cancer Genome Anatomy Project. *Trends Cell Biol.* 11:S66–S71.

Tatusov, R.L., M.Y. Galperin, D.A. Natale, and E.V. Koonin. 2000. The COG database: a tool for genome-scale analysis of protein functions and evolution. *Nucleic Acids Res.* 28:33–36.

Tatusova, T.A., and T.L. Madden. 1999. BLAST 2 sequences, a new tool for comparing protein and nucleotide sequences. *FEMS Microbiol. Lett.* 174:247–250.

Taylor, W.R. 1986. Identification of protein sequence homology by consensus template alignment. *J. Mol. Biol.* 188:233–258.

Thompson, J.D., D.G. Higgins, and T.J. Gibson. 1994. CLUSTAL W: improving the sensitivity of progressive multiple sequence alignment through sequence weighting, position-specific gap penalties and weight matrix choice. *Nucleic Acids Res.* 22: 4673–4680.

Thorisson, G.A., and L.D. Stein. 2003. The SNP Consortium website: past, present and future. *Nucleic Acids Res.* 31:124–127.

Tigris, 2004. ArgoUML website. `argouml.tigris.org/`.

Tuttle, M.S., D. Sheretz, M. Erlbaum, N. Olson, and S.J. Nelson. 1989. Implementing Meta-1: the first version of the UMLS Metathesaurus. In L.C. Kingsland (ed.), *Proc. 13th Annual Symp. Comput. App. Med. Care*, Washington, DC, pp. 483–487. New York: IEEE Computer Society Press,

UML, 2004. Introduction to OMG's Unified Modeling Language. `www.omg.org/gettingstarted/what_is_uml.htm`.

Uschold, M., and M. Gruninger. 1996. Ontologies: principles, methods and applications. *Knowledge Eng. Rev.* 11:93–155.

van Harmelen, F., J. Hendler, I. Horrocks, D. McGuinness, P. Patel-Schneider, and L. Stein, 2003. OWL web ontology language reference. `www.w3.org/TR/owl-ref/`.

Villanueva, J., J. Philip, D. Entenberg, C.A. Chaparro, M.K. Tanwar, E.C. Holland, and P. Tempst. 2004. Serum peptide profiling by magnetic particle-assisted, automated sample processing and MALDI-TOF mass spectrometry. *Anal. Chem.* 76:1560–1570.

Volinia, S., R. Evangelisti, F. Francioso, D. Arcelli, M. Carella, and P. Gasparini. 2004. GOAL: automated Gene Ontology analysis of expression profiles. *Nucleic Acids Res.* 32:W492–W499. Web server issue.

vOWLidator, 2003. BBN OWL validator. `owl.bbn.com/validator/`.

W3C, 1999. XML Path language. `www.w3.org/TR/xpath`.

W3C, 2001a. A conversion tool from DTD to XML Schema. `www.w3.org/2000/04/schema_hack/`.

W3C, 2001b. eXtensible Markup Language website. `www.w3.org/XML/`.

W3C, 2001c. XML Schema website. `www.w3.org/XML/Schema`.

W3C, 2001d. XML Stylesheet Language website. `www.w3.org/Style/XSL`.

W3C, 2003. W3C Math Home. `w3c.org/Math`.

W3C, 2004a. Resource description framework (RDF): concepts and abstract syntax. `www.w3.org/TR/rdf-concepts/`.

W3C, 2004b. XML information set (second edition). `www.w3.org/TR/2004/REC-xml-infoset-20040204`.

W3C, 2004c. XML Query (XQuery) website. `www.w3.org/XML/Query`.

Wain, H.M., E.A. Bruford, R.C. Lovering, M.J. Lush, M.W. Wright, and S. Povey. 2002. Guidelines for human gene nomenclature. *Genomics* 79:464–470.

Wall, L., T. Christiansen, and R. Schwartz. 1996. *Programming Perl*. Sebastopol, CA: O'Reilly & Associates.

Wand, Y. 1989. A proposal for a formal model of objects. In W. Kim and F. Lochovsky (eds.), *Object-Oriented Concepts, Databases and Applications*, pp. 537–559. Reading, MA: Addison-Wesley.

Wang, G., J. Goguen, Y. Nam, and K. Lin. 2004. Data, schema and ontology integration. In *CombLog'04 Workshop*, Lisbon.

Wang, L., J.J. Riethoven, and A. Robinson. 2002. XEMBL: distributing EMBL data in XML format. *Bioinformatics* 18:1147–1148.

Waugh, A., P. Gendron, R. Altman, J.W. Brown, D. Case, D. Gautheret, S.C. Harvey, N. Leontis, J. Westbrook, E. Westhof, M. Zuker, and F. Major. 2002. RNAML: a standard syntax for exchanging RNA information. *RNA* 8:707–717.

Westbrook, J.D., and P.E. Bourne. 2000. STAR/mmCIF: an ontology for macromolecular structure. *Bioinformatics* 16:159–168.

Whewell, W. 1847. *The Philosophy of the Inductive Sciences*. London: Parker. 2nd ed.

Wingender, E., P. Dietz, H. Karas, and R. Knuppel. 1996. TRANSFAC: A database on transcription factors and their DNA binding sites. *Nucleic Acids Res.* 24:238–241.

Wittgenstein, L. 1922. *Tractatus Logico-Philosophicus.* London: Routledge and Kegan Paul. Translated by C. Ogden.

Wittgenstein, L. 1953. *Philosophical Investigations.* New York: Macmillan.

WonderWeb, 2004. WonderWeb OWL Ontology Validator. `phoebus.cs.man.ac.uk:9999/OWL/Validator`.

Wroe, C.J., R. Stevens, C.A. Goble, and M. Ashburner. 2003. A methodology to migrate the gene ontology to a description logic environment using DAML+OIL. *Pac. Symp. Biocomput.* pp. 624–635.

XBN, 1999. XML Belief Network file format. `research.microsoft.com/dtas/bnformat/xbn_dtd.html`.

Xerlin, 2003. Xerlin XML Modeling Application website. `www.xerlin.org`.

XML, 2004. Survey of XML editors. `www.xml.com/pub/rg/XML_Editors`.

XTM, 2000. The XTM website. `topicmaps.org`.

Yandell, M.D., and W.H. Majoros. 2002. Genomics and natural language processing. *Nat. Rev. Genet.* 3:601–610.

Zadeh, L. 1965. Fuzzy sets. *Information and Control* 8:338–353.

Zadeh, L. 1981. Possibility theory and soft data analysis. In L. Cobb and R. Thrall (eds.), *Mathematical Frontier of the Social and Policy Sciences*, pp. 69–129. Boulder, CO: Westview.

Zadeh, L. 1984. A mathematical theory of evidence [book review]. *Artif. Intell.* 5: 81–83.

Zanzoni, A., L. Montecchi-Palazzi, M. Quondam, G. Ausiello, M. Helmer-Citterich, and G. Cesareni. 2002. MINT: a Molecular INTeraction database. *FEBS Lett.* 513: 135–140.

Zeeberg, B.R., W. Feng, G. Wang, M.D. Wang, A.T. Fojo, M. Sunshine, S. Narasimhan, D.W. Kane, W.C. Reinhold, S. Lababidi, K.J. Bussey, J. Riss, J.C. Barrett, and J.N. Weinstein. 2003. GoMiner: a resource for biological interpretation of genomic and proteomic data. *Genome Biol.* 4:R28.

Zhang, B., D. Schmoyer, S. Kirov, and J. Snoddy. 2004. GOTree Machine (GOTM): a web-based platform for interpreting sets of interesting genes using Gene Ontology hierarchies. *BMC Bioinformatics* 5:16.

Zhang, Z., A.A. Schaffer, W. Miller, T.L. Madden, D.J. Lipman, E.V. Koonin, and S.F. Altschul. 1998. Protein sequence similarity searches using patterns as seeds. *Nucleic Acids Res.* 26:3986–3990.

Zhang, Z., S. Schwartz, L. Wagner, and W. Miller. 2000. A greedy algorithm for aligning DNA sequences. *J. Comput. Biol.* 7:203–214.

Index